D1261961

Organic Coatings: Science and Technology

SPE MONOGRAPHS

Injection Molding
Irvin I. Rubin

Nylon Plastics
Melvin I. Kohan

Introduction to Polymer Science and Technology: An SPE Textbook
Edited by Herman S. Kaufman and Joseph J. Falcetta

Principles of Polymer Processing
Zehev Tadmor and Costas G. Gogos

Coloring of Plastics
Edited by Thomas G. Webber

The Technology of Plasticizers
J. Kern Sears and Joseph R. Darby

Fundamental Principles of Polymeric Materials
Stephen L. Rosen

Plastics Polymer Science and Technology
Edited by M. D. Baijal

Plastics vs. Corrosives
Raymond B. Seymour

Handbook of Plastics Testing Technology
Vishu Shah

Aromatic High-Strength Fibers
H. H. Yang

Giant Molecules: An SPE Textbook
Raymond B. Seymour and Charles E. Carraher

Analysis and Performance of Fiber Composites, 2nd ed.
Bhagwan D. Agarwal and Lawrence J. Broutman

Impact Modifiers for PVC: The History and Practice
John T. Lutz, Jr. and David L. Dunkelberger

The Development of Plastics Processing Machinery and Methods
Joseph Fred Chabot, Jr.

Organic Coatings: Science and Technology. Volume I: Film Formation, Components, and Appearance
Zeno W. Wicks, Jr., Frank N. Jones, and S. Peter Pappas

Organic Coatings: Science and Technology. Volume II: Applications, Properties, and Performance
Zeno W. Wicks, Jr., Frank N. Jones, and S. Peter Pappas

Organic Coatings: Science and Technology

Volume II: Applications, Properties, and Performance

ZENO W. WICKS, JR.
North Dakota State University

FRANK N. JONES
Eastern Michigan University

S. PETER PAPPAS
Polychrome Corporation

A WILEY-INTERSCIENCE PUBLICATION

JOHN WILEY & SONS, INC.

New York • Chichester • Brisbane • Toronto • Singapore

This text is printed on acid-free paper.

Library of Congress Cataloging in Publication Data:

Wicks, Zeno W.
Organic coatings.

(SPE monographs)
"A Wiley-Interscience publication."
Includes bibliographical references and index.
Contents: v. 1. Film formation, components, and appearance—v. 2. Applications, properties, and performance.
1. Plastic coatings. I. Jones, Frank N., 1936– . II. Pappas, S. Peter (Socrates Peter), 1936– . III. Title.
TP1175.S6W53 1992 667′.9 92-214
ISBN 0-471-59893-3 (v. 2)

Printed in the United States of America

10 9 8 7 6 5 4 3 2 1

Foreword

The Society of Plastics Engineers (SPE) is pleased to sponsor *Organic Coatings: Science and Technology*. This volume combines a concise review of theory with numerous practical applications to produce a basic understanding of coatings technology.

While the book can be utilized as a basic text, it provides in addition a ready industrial reference tool for the engineer and manager. The outstanding organization and presentation reflect upon the extensive background of the three authors, all of whom are recognized experts in the field.

SPE, through its Technical Volumes Committee, has long sponsored books on various aspects of plastics and polymers. Its involvement has ranged from identification of needed volumes to recruitment of authors. An ever-present ingredient, however, is review of the final manuscript to insure accuracy of the technical content.

This technical competence pervades all SPE activities, not only in publication of books but also in other activities such as technical conferences and educational programs. In addition, the Society publishes periodicals—*Plastics Engineering, Polymer Engineering and Science, Polymer Processing and Rheology, Journal of Vinyl Technology*, and *Polymer Composites*—as well as conference proceedings and other selected publications, all of which are subject to the same rigorous technical review procedure.

The resource of some 38,000 practicing plastics engineers has made SPE the largest organization of its type worldwide. Further information is available from the Society at 14 Fairfield Drive, Brookfield, Connecticut 06804.

EUGENE D. DE MICHEL

Executive Director
Society of Plastics Engineers

Technical Volumes Committee
Claire Bluestein
Captan Associates, Inc.

Preface

Volume I covered film formation, components, and some aspects of appearance of coatings. Volume II covers primarily application, properties, and performance of coatings. The contents of Volumes I and II are interrelated; the index to Volume II is an inclusive index with listings for both volumes.

Chapters 19–27 cover basic principles involved in the application and performance of coatings. We have attempted to discuss the various subjects from a basic scientific point of view relating available scientific understanding with practical coatings experience. As in Volume I, in some cases, where there is a lack of scientific studies of the aspects of certain problems, we have speculated as to possible explanations for behaviors. Such speculations are based on our understanding of related phenomena and our cumulative experience acquired over several decades in the field. We recognize the risk that speculation tends to increase in scientific stature with passing time and may even be cited without qualification as evidence or adopted as an experimentally supported hypothesis. It is our intent rather that such speculations promote the advancement of coatings science and technology by stimulating discussion which leads to experimentation designed to disprove or support the speculative proposal. Obviously, we believe that the latter purpose outweighs the former risk, and we will endeavor to identify the speculative proposals as such.

Chapters 28–32 deal with solvent-borne, water-borne, electrodeposition, powder, and radiation cure coatings. We have attempted to bring out the advantages and disadvantages of each class and to look into our crystal balls as to possible future needs and developments. Chapters 33–36 deal with various end uses for coatings. Formulations actually used are proprietary and frequently even broad descriptions of compositions are not available to the public. We have attempted to emphasize the principles involved in selecting various coatings for particular end uses. While in some cases we have provided examples of formulations, these are not included as recommendations for actual use but rather to illustrate the types of components in some formulations. This organization by classes of coatings and then by classes of end uses leads to duplication; we have attempted to minimize repetition by cross-references.

Finally Chapter 37 deals with the important topic of productivity and creativity—How can a formulator work more effectively and efficiently?

We want to acknowledge the many people who have helped in the development of this text. We particularly appreciate the efforts of Werner Funke, Archie Garner, Peter Gribble, Harold Haag, Loren Hill, Wayne Knaus, Josef Jilek, Alex Milne, Clifford Schoff, Eric Urruti, Joseph Webster, and Douglas Wicks in reviewing individual chapters. We acknowledge the assistance of Rebecca Doll for preparing figures and Steve Kloos for preparing the organic structural formulas in both volumes. The hundreds of students who have taken our classes in both universities and industry have made enormous contributions through their questions and specific knowledge.

ZENO W. WICKS, JR.
FRANK N. JONES
S. PETER PAPPAS

Las Cruces, New Mexico
Ypsilanti, Michigan
Carlstadt, New Jersey
September 1993

Contents—Volume II

Contents—Volume I

Symbols

A	Arrhenius preexponential term
C	Concentration—weight per unit volume of solution
c	Concentration—moles per liter
CPVC	Critical pigment volume concentration (content)
CRH	Critical relative humidity
°C	degrees Celsius
E	Modulus
E'	Storage modulus (elastic modulus)
E''	Loss modulus
E	Relative evaporation rate
E_a	Thermal coefficient of reaction rate (Arrhenius activation energy)
F	Functionality of a monomer
$\overline{F}$	Average functionality of a monomer mixture
f	Functionality of a polymer (resin)
$\overline{f}_n$	Number average functionality of a polymer (resin)
G	Free energy
G	Small's molar association constant
g	gram
g	Gravitational constant
H	Enthalpy
i	Angle of incidence
K	Kelvin temperature
K	Absorption coefficient
K_E	Einstein constant
k	Rate constant for a specific reaction
kg	Kilogram
KU	Krebs units
M	Molecular weight
$\overline{M}_c$	Molecular weight between cross-links
$\overline{M}_n$	Number average molecular weight
$\overline{M}_W$	Weight average molecular weight
MFT	Minimum film formation temperature
N	Newton
N	Number of moles
n	Refractive index
NVV	Nonvolatile volume (volume percent solids)
NVW	Nonvolatile weight (weight percent solids)
OA	Oil absorption
P	Vapor pressure
P	Degree of polymerization

$\overline{P}_n$	Number average degree of polymerization
$\overline{P}_W$	Weight average degree of polymerization
p	Extent of reaction
p_g	Extent of reaction at gelation onset
Pa	Pascal
Pa·s	Pascal second = 10 poise
mPa·s	Milipascal second = 1 centipoise
PDI	Polydispersity index = $\overline{M}_W/\overline{M}_n$
PV	Pigment volume in a dry film
PVC	Pigment volume concentration (content)
R	Gas constant
RH	Relative humidity
r	Angle of reflection or angle of refraction
S	Entropy
S	Scattering coefficient
s	second
T	Temperature (K if not otherwise specified)
T_b	Brittle–ductile transition temperature
T_g	Glass transition temperature
T_m	Melting point
t	Time
tan δ	tan delta, loss tangent, E''/E'
V	Molar volume
V_i	Volume fraction of internal phase
VOC	Volatile organic compound
w	Weight fraction
X	Film thickness
x	Mole fraction
x	Optical path length
XLD	Cross-link density
γ	Surface tension
$\dot{\gamma}$	Shear rate
δ	Solubility parameter
δ	Phase shift in viscoelastic deformation
ε	Molar absorbance
η	Absolute shear viscosity
η_e	External phase viscosity
η_r	Relative viscosity = η/η_s
η_s	Viscosity of solvent
η^*	Extensional viscosity
$[\eta]$	Intrinsic viscosity
$[\eta]_w$	Weight intrinsic viscosity
$[\eta]_\theta$	Intrinsic viscosity under theta conditions
λ	Wave length
ϕ	Packing factor
ρ	Density
τ	Shear stress
τ_0	Yield value

θ	Contact angle
υ	Kinematic viscosity
ν_e	Mole of elastically effective network chains per cm^3
χ	Activity coefficient

All equations using logarithms are given in terms of natural logarithms (ln). Equations from the literature using base 10 logarithms have been restated in terms of ln.

SI Units are used throughout with conversion factors given at the site of first usage.

Organic Coatings: Science and Technology

CHAPTER XIX

Flow

Rheology is the science of flow and deformation. In this chapter we deal only with flow; deformation aspects of rheology are discussed in Chapter 24. Flow properties of coatings are critical to obtaining proper application and appearance characteristics of films. For example, in brush application of a flat wall paint, the flow properties govern how much paint is picked up on the brush, the film thickness of the coating applied, the leveling of the applied film, and the control of sagging of the film.

Depending on how stress is applied to a fluid, there are several types of flow. Of major importance in coatings is flow under a shear stress. We consider shear flow first and then, more briefly, other types of flow.

19.1. SHEAR FLOW

To understand and define *shear flow*, consider the model shown in Figure 19.1. The lower plate is stationary and the upper parallel plate is movable. The plates are separated by a layer of liquid with thickness x. Force F is applied to the top movable plate having area A, so that the plate slides sidewise with velocity v. When the plate moves, the layers of liquid also move. The top layer moves with a velocity approaching that of the movable plate and the bottom layer moves with a velocity approaching zero. If the liquid is ideal, the velocity gradient dv/dx for any section of the liquid is constant and, therefore, also equal to v/x. This ratio is defined as *shear rate* ($\dot{\gamma}$). The units of shear rate are reciprocal seconds (s^{-1}).

$$\dot{\gamma} = \frac{dv}{dx} = \frac{v}{x}; \qquad \frac{\text{cm s}^{-1}}{\text{cm}} = \text{s}^{-1}$$

The result of force F acting on the top plate of area A is stress (F/A); the ratio is *shear stress* (τ). The units of shear stress are pascals (Pa).

$$\tau = \frac{F}{A} = \text{m kg s}^{-2}/\text{m}^2; \qquad \frac{\text{N}}{\text{m}^2} = \text{Pa}$$

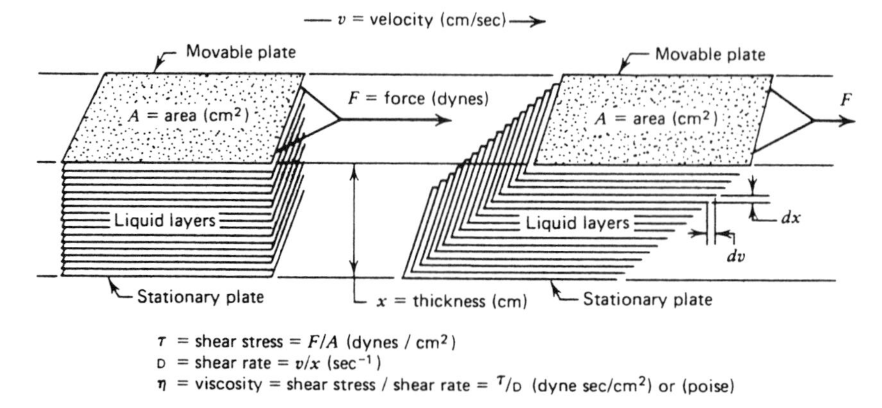

Figure 19.1. Schematic model of shear flow on an ideal liquid. (From Ref. [4], with permission.)

The liquid exerts a resistance to flow called *viscosity* (η) defined as the ratio shear stress/shear rate. This type of viscosity is more correctly called *absolute shear viscosity* but, since it is the most widely encountered type of viscosity, it is almost always just called viscosity. The units are pascal seconds (Pa·s). The older, and still commonly used, units are poises (P). One Pa·s equals 10 P and 1 mPa·s equals 1 cP.

$$\eta = \frac{\tau}{\dot{\gamma}}; \qquad \frac{\text{Pa}}{\text{s}^{-1}} = \text{Pa·s} \tag{19.1}$$

When the applied stress is due to gravity, the resulting resistance to shear flow is called *kinematic viscosity*. The symbol for kinematic viscosity is ν with units of $m^2\ s^{-1}$, formerly stokes, where 1 $m^2\ s^{-1}$ equals 10^4 stokes. Kinematic viscosity equals absolute shear viscosity divided by density (ρ) of the liquid.

$$\nu = \frac{\eta}{\rho} \tag{19.2}$$

19.2. TYPES OF SHEAR FLOW

In the case of ideal liquids, the ratio of shear stress to shear rate is constant, that is, viscosity is independent of shear rate (or shear stress). Liquids that show this type of shear flow are called *Newtonian* liquids. If one determines shear stress as a function of shear rate, one will obtain a linear data plot as shown in Figure 19.2*a*. When plotted as shown in Figure 19.2, the slope of the line equals the reciprocal of the viscosity. Sometimes such plots appear in the literature with axes opposite to those shown here. In those cases the slope is, of course, the viscosity. Always check the axes labels when looking at curves in the literature.

Newtonian flow is exhibited by all liquids composed of miscible small molecules. Solutions of resins in good solvents, that is, solvents that interact more strongly with resin molecules than resin molecules interact with each other, also exhibit

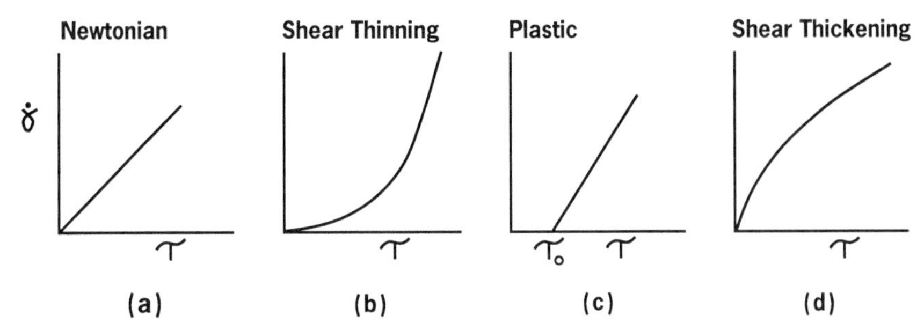

Figure 19.2. Schematic plots of flow of different types of liquids.

Newtonian flow. Dispersions of rigid particles in a Newtonian liquid, when there is no particle–particle interaction, also exhibit Newtonian flow.

Some liquids are non-Newtonian, that is, the ratio of shear stress to shear rate is not constant. One class of non-Newtonian liquids exhibits decreasing viscosity as shear rate (or shear stress) increases. A schematic plot of such behavior is given in Figure 19.2*b*. It should be noted that the curve becomes linear at higher shear rates. Liquids that show this decrease in viscosity as shear increases are called *shear thinning*. In the older literature they are called *pseudoplastic*. Dispersed systems where the dispersed phase is not rigid (such as emulsions), or where there are particle–particle interactions, can exhibit shear thinning.

In some cases no detectable flow occurs within the measurement time span until some minimum shear stress is exceeded. Such materials are sometimes called *Bingham bodies*; they are also said to exhibit *plastic flow*. The minimum shear stress required is called the *yield value* or *yield stress* and is designated by the symbol τ_0. A schematic flow diagram is shown in Figure 19.2*c*. It can be argued that plastic flow is merely found with a shear thinning flow system where the time interval required for flow is long at shear stresses below the yield value, or where instrument limitations do not permit measurement of a shear rate at low shear stresses. Hence it is fairly common to extrapolate the linear part of the curve of a shear thinning coating to the intercept with the shear stress axis and call the intercept a yield value.

Another class of liquids exhibits increasing viscosity as shear rate (or shear stress) increases. A schematic plot of such behavior is given in Figure 19.2*d*. Liquids that show this increase in viscosity as shear increases are called *shear thickening*. Such liquids are also called *dilatant*. Strictly speaking, dilatant systems show the further characteristic of increasing volume when shear is exerted on them. Most but not all shear thickening systems are dilatant. Dilatant systems have dispersed phases that become less ordered and hence occupy more volume when exposed to increasing shear. Quicksand is an example. If you stand on quicksand, you slowly sink in; but if you try to pull your foot out rapidly (exert higher shear), the viscosity increases, and it becomes more difficult to move your foot. Some pigment and resin dispersions exhibit shear thickening.

The Casson equation (Eq. 19.3) is widely used to express the extent of shear thinning or thickening exhibited by a fluid. In the equation η_∞ is the viscosity at infinite shear rate. As noted earlier, the plot of the flow curve of a shear thinning (or thickening) fluid becomes linear as the shear rate is increased, permitting the calculation of η_∞. Also, as noted above, most systems do not have a true yield

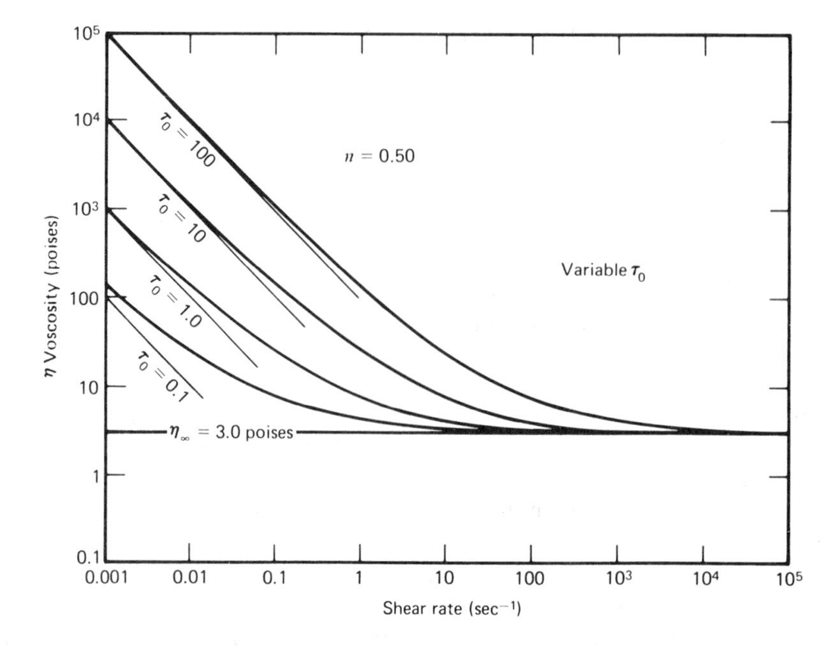

Figure 19.3. Casson plot of viscosity as a function of shear rate showing the dependence on τ_0 with constant η and η_∞. (From Ref. [1], with permission.)

value. The τ_0 in the Casson equation is the equivalent of extrapolation of the linear segment of a flow diagram to the intercept with the shear stress axis.

$$\eta^n = \eta_\infty^n + \left(\frac{\tau_0}{\dot{\gamma}}\right)^n \tag{19.3}$$

In most cases the value of n is 0.5 and commonly the Casson equation is shown with just the half-power relationship. It is common to plot log viscosity against shear rate. In such plots the degree of curvature is related to the value of τ_0. In the case of a Newtonian fluid, τ_0 equals zero and the plot will be a straight line parallel to the shear rate axis. A schematic plot is given in Figure 19.3 where the values of η and η_∞ are held constant to show the effect of changes in τ_0 on the flow response.

A further complication in considering flow is the possibility of time dependence or shear history dependence of the viscosity. The effect is illustrated in Figure 19.4*a*. The curves in the figure result from shear stress readings taken at successively higher shear rates to some upper limit (right-hand curve), followed immediately by shear stress readings taken at successively lower shear rates (left-hand curve). At any shear rate on the initial curve, the stress would decrease with time to an equilibrium value between the two curves; that is, the viscosity would decrease. On the other hand, if such a system had been exposed to a high rate of shear and then the shear rate decreased, the shear stress would increase to an equilibrium

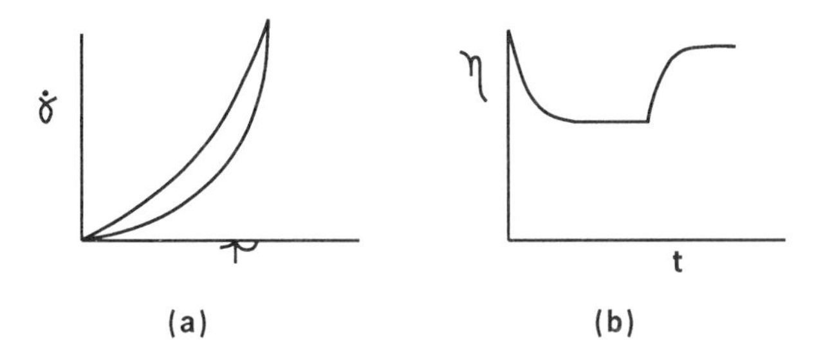

Figure 19.4. Schematic plots of systems exhibiting thixotropic flow. (*a*) The curve to the right is based on readings taken as shear rate was being increased and the curve to the left is based on readings taken as shear rate was then being decreased. (*b*) The viscosity drops as shear continues then increases as the shear rate is decreased.

value as the measurement was continued; that is, the viscosity would increase with time. Systems that show such flow behavior are said to be *thixotropic*. Thixotropic systems are shear thinning; their viscosity is also dependent on time and prior shear history. All thixotropic systems are shear thinning, but not all shear thinning systems are thixotropic. Unfortunately, the term thixotropy is often improperly used as a synonym for shear thinning.

It is commonly said that thixotropic structure is broken down by applying shear for sufficient time and that the structure reforms over time when shear is stopped. It is difficult to quantify a degree of thixotropy. Some thixotropic systems undergo rapid viscosity reduction to equilibrium values in short time periods and recover their viscosity rapidly when shearing is stopped; other systems change more slowly with time. In early work, comparisons of the size of the hysteresis loop, such as shown in Figure 19.4*a*, were proposed. However, the size of such loops is dependent on the time intervals between successive measurements. Perhaps the most useful way to represent the effect is to plot viscosity at a series of shear rates as a function of time, as illustrated in Figure 19.4*b*.

Another way that the effect of shear on a thixotropic system can be shown is by a different type of Casson plot, as shown in Figure 19.5. The square root of the viscosity is plotted against the square root of the reciprocal of the shear rate; the steeper the slope, the greater the degree of shear thinning. If the sample had been sheared until all the thixotropic structure was broken down and if the measurements could be made before any structure buildup occurred, the plot would be linear and parallel to the X axis. Although comparisons of the differences in the slopes of such lines give a qualitative expression of the extent of thixotropy, the slopes of the curves are dependent on prior shear history, rate of acceleration of shear, and length of time that the sample was exposed to the highest shear rate.

It is critical to understand that in Newtonian systems there is *a* viscosity but in all the other systems just discussed there is not *a* viscosity. Viscosity in these systems depends on shear rate (shear stress) and often on shear history. Sometimes the term apparent viscosity, η_a, is used where it is the ratio of shear stress to shear rate under some specified conditions. Usually it is best to describe the viscous behavior of a non-Newtonian liquid by some form of flow diagram.

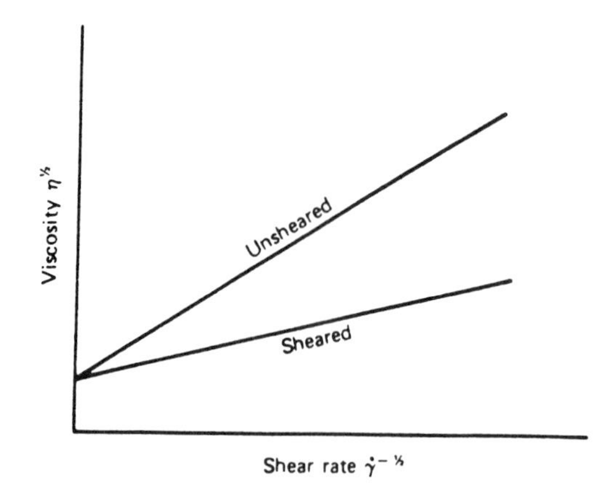

Figure 19.5. Schematic Casson plots of a sheared and unsheared thixotropic coating. The degree of divergence gives an estimate of the degree of thixotropy. (From Ref. [2], with permission.)

19.3. DETERMINATION OF SHEAR VISCOSITY

A large variety of instruments is available to determine viscosity. They vary widely in cost, time required for measurements, operator skill required, susceptibility to abuse, precision, accuracy, and ability to measure shear rate variability or time dependence effects. Data obtained with different instruments and by different operators with the same instruments can vary substantially, especially with shear thinning systems at low rates of shear [3]. Some of the variation can result from lack of attention to details, especially temperature control and possible solvent loss, but major sources of error result from comparing samples that have had different shear histories. Caution is urged in reaching conclusions based on relatively small differences in viscosities.

Viscometers can be divided into three broad classes: (1) those that permit quite accurate viscosity determinations, (2) those that permit determination of reasonable approximations of viscosity, and (3) those that provide flow data casually related to viscosity. Space limitations restrict our discussion to the major examples of measurement methods and not the details of the instruments or the mathematical relationships involved in their use. The four general references listed at the end of the chapter are good sources of further information.

Since viscosity depends on temperature, it is critical to be sure that the sample has reached a constant, known temperature. When high viscosity fluids are sheared at high shear rates, heat is evolved and the temperature of the sample increases unless the heat exchange efficiency of the viscometer is adequate. If viscosities are determined as both shear rate and temperature are increasing, it is not possible to tell whether or not a system is shear thinning.

19.3.1. Capillary Viscometers

In *capillary viscometers*, the time required for a known amount of liquid to flow through a capillary tube is measured. Most commonly the driving force is gravity.

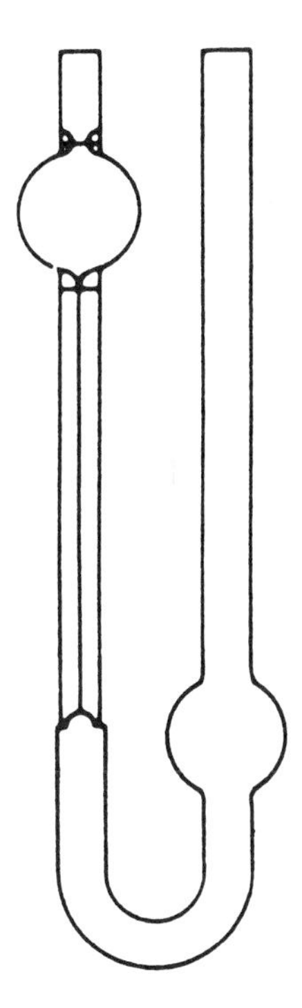

Figure 19.6. Ostwald capillary viscometer. (From Ref. [4], with permission.)

An example of a capillary viscometer is shown in Figure 19.6. While viscosities can be calculated based on the diameter of the capillary, more usually the instruments are standardized with liquids of known viscosity. Then calculation is simply based on instrument constants and time. It must be emphasized that one obtains kinematic viscosity that must be corrected for density if the absolute viscosity is desired.

Capillary viscometers with a wide range of diameters are available that permit the determination of viscosities ranging from about 10^{-7} to 10^{-1} $m^2 s^{-1}$. For liquids with a density of 1, these values correspond to a range of 1 mPa·s to 1000 Pa·s. For higher viscosity liquids, instruments have been designed that force the liquid through a capillary with pressure rather than just relying on the force of gravity.

Capillary viscometers are applicable only to Newtonian liquids. Furthermore, the liquids must be sufficiently transparent to permit observation of the time when the fluid meniscus passes the two volume marking lines. In such situations capillary viscometers are the instruments of choice for research work as the accuracy that can be attained is high. They are also appropriate for routine work; however, determinations are relatively time consuming, especially when the temperature dependence of viscosity data is desired. Relatively long times are required for temperature equilibration because of the large sample sizes and the low rate of heat transfer by glass. Capillary viscometers are particularly appropriate for use in determining the viscosity of volatile liquids or solutions in volatile solvents since they are essentially closed systems.

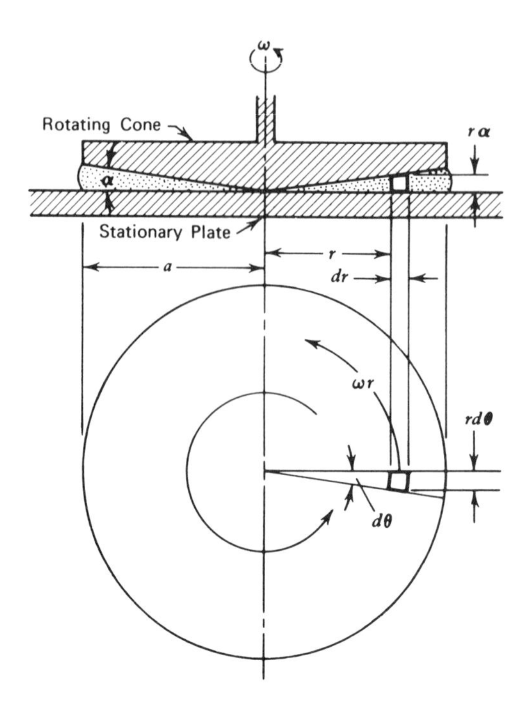

Figure 19.7. Schematic representation of cone and plate viscometer geometry. (From Ref. [1], with permission.)

19.3.2. Cone and Plate Viscometers

For non-Newtonian liquids, including pigmented liquids, the highest accuracy over a wide range of shear rates is obtained with *cone and plate viscometers*. The sample is placed on the plate that is then raised to a level with small clearance from the cone. The cone can be rotated at any desired speed (rpm) and the torque is measured. The angle of the cone is designed to make the shear stress independent of the distance from the center of the cone. The shear rate is proportional to the revolutions per minute and the shear stress is related to the torque. A schematic representation is shown in Figure 19.7. Temperature is controlled by passing temperature controlled water through the plate; temperature control problems are minimized by the small sample size.

A wide variety of cone and plate instruments is available. They vary in the range of shear rates that can be used and in the speed with which shear rate can be increased and decreased. The least expensive instruments are sufficiently rugged, simple to use, and fast enough for quality control applications. The most versatile ones are sensitive scientific instruments requiring skill in use and are most appropriate for research applications. In some instruments, the edge of the liquid sample is exposed to the atmosphere and solvent can evaporate. When such an instrument is used with solutions containing volatile solvents, the cone and plate unit should be shrouded in an atmosphere saturated with solvent vapor. Cone and plate viscometers are the instruments of choice for research studies of flow of coatings. Two types are available: *controlled strain* and *controlled stress* instruments. The latter type offers advantages for coatings as it is generally superior for measurements

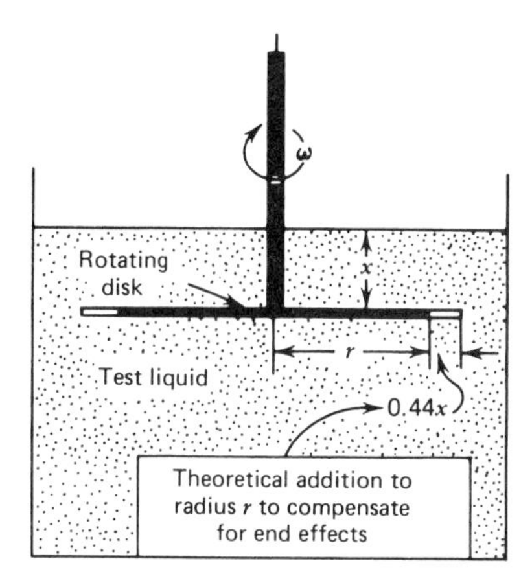

Figure 19.8. Schematic drawing of a disk viscometer. (From Ref. [1], with permission.)

at very low shear rates. Many variations in the design of cone and plate viscometers are available, as well as a variety of other rotational viscometers (see Refs. [2] and [4] for further discussion).

19.3.3. Rotating Disk Viscometers

Rotating disk viscometers, such as a Brookfield Viscometer, have a motor that can rotate a disk in a liquid at a range of revolutions per minute. The resulting torque is measured. A schematic representation is shown in Figure 19.8. The instruments must be calibrated with standards. The measurements should be made in a container of the same dimensions as that in which the standardization was carried out since the distance of the disk below the surface of the liquid, above the bottom of the container, and from the side walls can affect the response. When properly used, the instruments provide relatively accurate viscosity measurements for Newtonian liquids. With non-Newtonian liquids, however, the viscosity reading represents an average response corresponding to the viscosities resulting from a wide span of shear stresses.

Rotating disk viscometers can detect whether a system is shear thinning or shear thickening by measurements carried out at different rpm settings. They can detect thixotropy by a change in response over time at the same rpm setting. They cannot provide absolute viscosity data as a function of shear rate. Thus, they are appropriate for quality control work but caution should be applied in their use for research and development purposes. Results from comparisons of systems with different degrees of shear thinning or thixotropy could be misleading.

19.3.4. Mixing Rheometers

The viscosity of highly viscous materials can be determined at high rates of shear by the use of *mixing rheometers* that are small, heavy duty mixers. The test sample is confined to a relatively small space and subjected to intense mixing by dual rotors in the form of sigma-shaped blades. A dynamometer measures the work

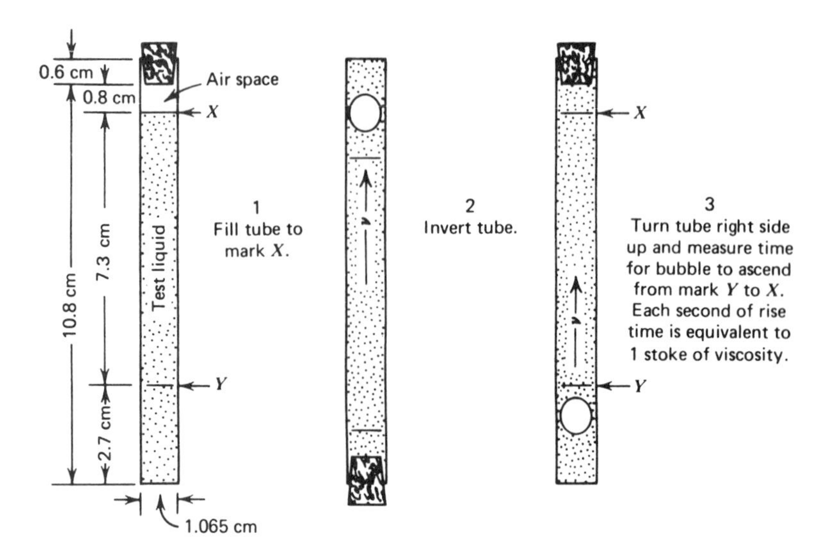

Figure 19.9. Determination of viscosity with a bubble tube. (From Ref. [1], with permission.)

input through a reaction torque that is converted to a strip chart readout. Speed is set by a tachometer. These instruments were originally designed for studying the molding of plastics but are also used in studying the effect of various pigments on viscosity. The heat buildup can be substantial with high viscosity fluids.

19.3.5. Bubble Viscometers

The *bubble viscometer* is based on the rate of rise of an air bubble in a tube of liquid. The higher the viscosity, of course, the slower the bubble rises. A glass tube is filled with a liquid to a graduation mark and stoppered so that a definite amount of air is enclosed at the top, as shown in Figure 19.9. Provided the length of the bubble is greater than its diameter, the rate of rise is found to be substantially independent of the bubble size. Sets of standards of known viscosities from about 10^{-5} to 0.1 $m^2\ s^{-1}$ are available. Note that density of the liquid affects the rate of rise of the bubble so kinematic viscosity is measured.

Sets of bubble tubes are widely used as control instruments in measuring the viscosities of resin solutions. They are only appropriate for Newtonian, transparent fluids. They are low cost, with no moving parts (except air bubbles) to get out of commission, and are very simple to use. Since the sample size is relatively large and heat transfer through glass is relatively slow, a significant time is required to reach temperature equilibrium and hence to make meaningful measurements.

19.3.6. Efflux Cups

By far the most widely used control device for measuring the flow of industrial coatings, especially for spray application, is the *efflux cup*. A wide variety of efflux cups is used (Refs. [2] and [4] provide a comparison of about two dozen that are

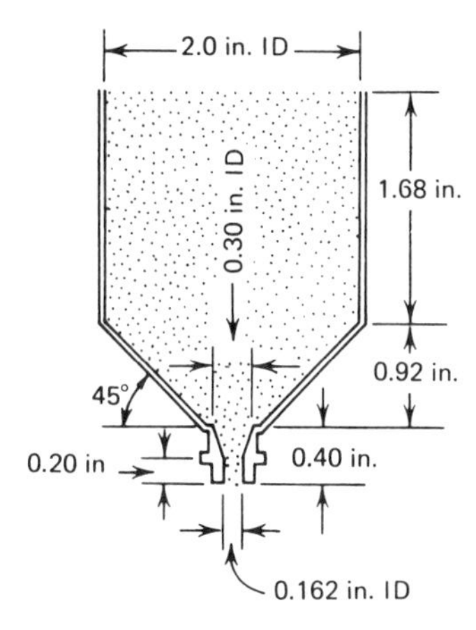

Figure 19.10. Schematic diagram of a Ford No. 4 efflux cup. (From Ref. [1], with permission.)

used commercially). A schematic diagram of one of the most common, the Ford No. 4 cup, is shown in Figure 19.10.

One holds a thumb over the hole in the bottom, fills the cup with coating, removes one's thumb, and presses a stop watch simultaneously. The stop watch is pressed again when the stream of coating flowing through the hole breaks. Flow rate is expressed in seconds. The data should not be converted into kinematic viscosity numbers since a significant amount of the force is converted into kinetic energy, especially with low viscosity coatings. The method is not appropriate for non-Newtonian systems, although efflux cups are frequently used for coatings that exhibit a small degree of shear thinning. Despite their limitations, efflux cups are useful quality control devices. They are low in cost and rugged in construction, results are simply and quickly obtained, and they are easily cleaned. Reproducibility of measurements has been reported to be in the range of 18–20% [4].

The proper way to use an efflux cup in control of viscosity for spraying, for example, is to adjust the viscosity of the coating by solvent addition until the coating sprays properly, then measure the time to flow through the efflux cup. This time can then be used as the standard for spraying that coating through that spray gun at that distance from the object being sprayed. Reversing this order, that is, adding solvent until the viscosity is such that the coating will take some fixed number of seconds to flow through the cup, is not appropriate. Proper viscosity for spraying depends on the coating and upon the application system; different efflux flow times are appropriate for different situations.

19.3.7. Paddle Viscometers

The most widely used "viscometer" in the United States in architectural paint formulation is the *Stormer Viscometer*. The instrument paddle is immersed in the paint and rotated at 200 rpm. The force required to maintain this rotation rate is measured by adding weights to a platform at the end of a cord over a pulley connected by a simple gear train to the paddle. A schematic diagram is shown in

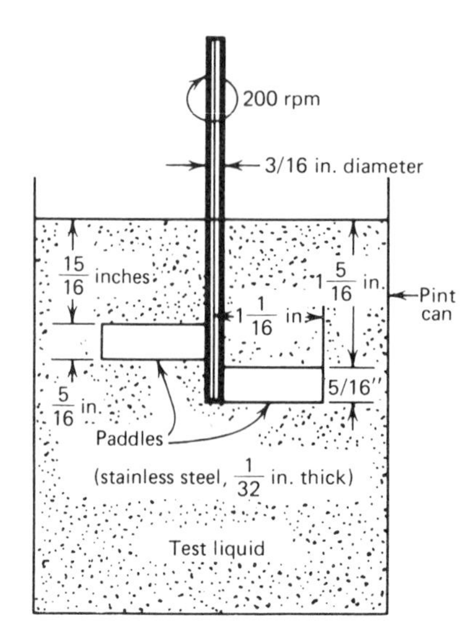

Figure 19.11. Schematic diagram of a paddle viscometer. (From Ref. [1], with permission.)

Figure 19.11. The weight loading is converted into *Krebs Units* (KU) by an arbitrary conversion scale.

Originally a KU value of 100 was supposed to correspond to good brushing consistency. In current practice, paints are usually formulated with somewhat lower KUs. The instrument is of little utility with Newtonian fluids; and the readings have no real meaning for non-Newtonian systems typical of most architectural paints. Even when used only for quality control, the paddle viscometer is not very satisfactory. The proper way to establish the flow properties for a trade sales paint is to apply it with a brush or roller and adjust until the best compromise of such properties as ease of brushing, leveling, sagging, settling, and so forth has been achieved. Having done this, a quality control test should be set up. Appropriate instruments are cone and plate viscometers (some of which are low cost) or, lacking that, a rotating disk viscometer. The fact that Stormer Viscometers are very widely used in paint laboratories does not make them any good. The director of research of one of the largest manufacturers of trade sales paints in the United States said some years ago that the Stormer Viscometer was responsible for setting back the formulation of one-coat hiding latex paints by 20 years.

19.4. SHEAR VISCOSITY OF RESIN SOLUTIONS

The viscosity of liquids depends on the amount of free volume available. Somewhat simplistically, there are free volume holes opening and closing in the liquid; molecules move randomly through these free volume holes. When a stress is applied, movements in the direction to relieve the stress are favored and the liquid flows. Therefore, factors that control viscosity of resin solutions are those that control the availability of free volume.

Many coatings are concentrated solutions of polymers and oligomers. The variables that affect flow behavior of these concentrated solutions are not fully understood. The variables that govern the flow of very dilute polymer solutions have been extensively studied and are better understood. This understanding can be useful in considering the concentrated solutions although it is not always directly applicable. Following a discussion of temperature effects in Section 19.4.1, factors affecting flow of dilute solutions are discussed in Section 19.4.2, and more concentrated solutions in Section 19.4.3. The effects of solvent are then discussed in Section 19.4.4.

19.4.1. Temperature Dependence of Viscosity

The temperature dependence of viscosity for a wide range of resins and resin solutions has been shown to fit a WLF relationship [5,6] (see Section 2.3). In Eq. 19.4, T_r, the reference temperature, is simply the lowest temperature for which experimental data are available. Except for very dilute solutions, data also fit Eq. 19.4 when the reference temperature is T_g and the viscosity at T_g is assumed to be a constant, 10^{12} Pa·s for all systems, as shown [5].

$$\ln \eta = \ln \eta_r - \frac{c_1(T - T_r)}{c_2 + (T - T_r)} \cong 27.6 - \frac{A(T - T_g)}{B + (T - T_g)} \tag{19.4}$$

It is common to see the statement in the literature that the temperature dependence of viscosity fits the form of an Arrhenius equation (Eq. 19.5), frequently with profound discussions of the "activation energy" for viscous flow, E_v.

$$\ln \eta = K + \frac{B}{T} = \ln A + \frac{E_v}{RT} \tag{19.5}$$

[Note that constants A and B are not the same constants as in the WLF equation (Eq. 19.4).]

When careful plots are made, Arrhenius plots of ln η as a function of $1/T$ have been found in all cases investigated to be curved and not linear [5]. On the other hand, the data do fit a WLF equation. From a practical view point, the differences in the models are small if the temperature range is small. However, over a wide range of temperatures, the differences are relatively large. Figure 19.12 shows plots of the temperature dependence of viscosity of commercial standard liquid bisphenol A epoxy resin (see Section 11.1.1) calculated from both Arrhenius and WLF equations, together with experimental data points, that are shown to fit the WLF equation.

It can be seen that there are two other factors in the WLF equation in addition to $(T - T_g)$ that affect the free volume and viscosity of the systems. Constant A is known to depend on the difference in thermal expansion coefficients above and below T_g, but no studies have been reported on the structural factors that control these coefficients. Constant B is the value of $(T_g - T)$ at which viscosity goes to infinity. The so-called universal value of this constant is 51.6°, but the "constant" varies with composition. No studies have been reported on the relationship between structure and the value of constant B.

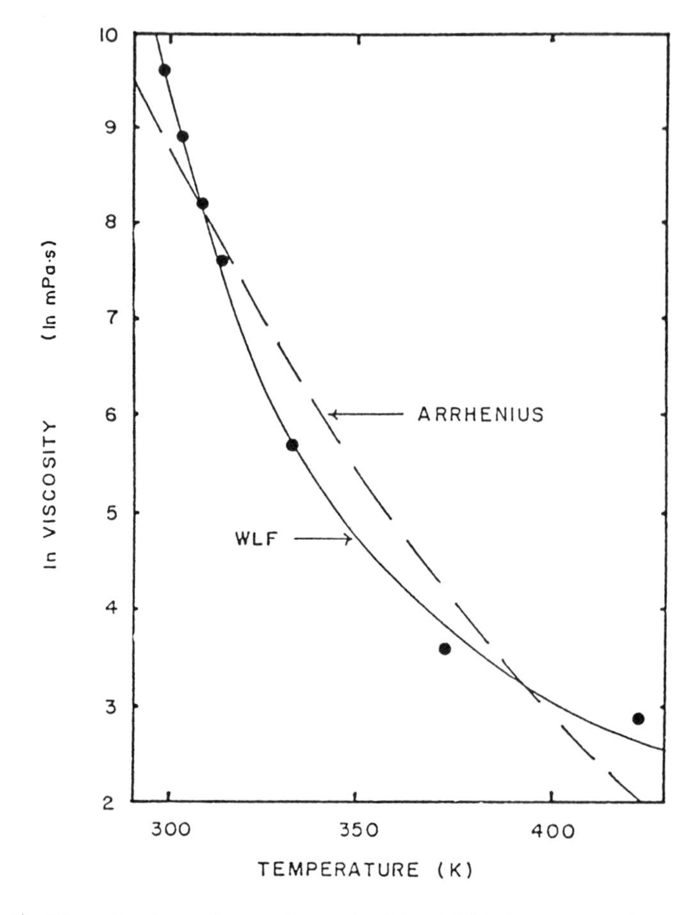

Figure 19.12. Viscosity dependence of standard liquid BPA epoxy resin on temperature. (From Ref. [5], with permission.)

Thus, a major factor controlling the viscosity of resin solutions is $(T - T_g)$, but it is not the only factor. In general, however, in designing resins it is reasonable to predict that lower T_g will lead to lower viscosity of the resins and of their resin solutions. (See Section 2.1.4 for a discussion of the factors controlling the T_g of polymers.) When the differences in T_g are small, differences in constants A and B may overshadow the small difference in $(T - T_g)$. Linear poly(dimethylsiloxanes) have low T_g values and low viscosities. Linear polyethylene glycols have almost as low T_g values and viscosities. Poly(methyl methacrylate) resin solutions have higher T_g values and viscosities than comparable poly(methyl acrylate) resin solutions. The BPA epoxy resins have higher T_g values and viscosities than corresponding hydrogenated derivatives. An apparent exception to this generalization about the effect of T_g has been reported in high solids acrylic resins [7]. Acrylic resins made with a comonomer with a branched, bulky group, such as 3,3,5-trimethylcyclohexyl methacrylate, are reported to have low viscosity at high solids although they have high T_g values; no explanation of this effect has been advanced.

19.4.2. Dilute Solution Viscosity

If the concentration is low enough that individual polymer molecules and their associated solvent molecules are isolated from each other, the *relative viscosity* (η_r) of the solution can be expressed by the following Huggins' equation (Eq. 19.6), where $[\eta]$ is *intrinsic viscosity*, sometimes called the *limiting viscosity number*, and C is the concentration of the polymer solution expressed in units of weight of polymer per unit volume of polymer solution. Relative viscosity is the ratio (unitless) of solution viscosity to the viscosity of the solvent. In recent literature, the units of C most commonly used are grams of polymer per milliliter of solution, but in older literature grams per deciliter is common, so one must be careful to check units. When C is expressed in g mL^{-1}, the units of intrinsic viscosity are mL g^{-1}. Intrinsic viscosity is obtained by extrapolating a plot of ln η_r/C as a function of C to zero concentration. It is related to the *hydrodynamic volume* of the sphere swept out by the isolated polymer molecule and its associated solvent as it moves through the dilute solution.

$$\ln \eta_r = [\eta]C + [\eta]^2C^2 \qquad (19.6)$$

Intrinsic viscosity depends on temperature. As temperature increases, the coil size of the polymer molecule usually increases, more solvent is entrapped, and the intrinsic viscosity increases. The intrinsic viscosity is also affected by solvent/polymer interactions. The greater the extent of solvent association with a polymer molecule, the more the coil expands, and therefore the higher the intrinsic viscosity. Intrinsic viscosity depends upon molecular weight: the higher the molecular weight, the larger the intrinsic viscosity. Another factor is the rigidity of the chain. Everything else being equal, polymers with completely flexible, randomly kinked chains have lower intrinsic viscosities than those with rigid, rod-like structures. The relationship between intrinsic viscosity and molecular weight (M) is expressed by the *Mark–Houwink equation* (Eq. 19.7) where K and a are constants.

$$[\eta] = KM^a \qquad (19.7)$$

If the solvent is too poor or the temperature is too low, the polymer molecules precipitate rather than staying in solution. The combination of minimum temperature and poorest solvent that just maintains solubility is called a *theta condition*. Under theta conditions, the intrinsic viscosity $[\eta]_\theta$ for that polymer is at a minimum. If the chains are completely flexible, as for example with acrylic polymers, $[\eta]_\theta$ is proportional to the half power of molecular weight as shown in Eq. 19.8. Note that Eqs. 19.7 and 19.8 are based on narrow molecular weight distribution samples of polymers.

$$[\eta]_\theta = K_\theta M^{1/2} \qquad (19.8)$$

19.4.3. Concentrated Solution Viscosity

There have been relatively few fundamental studies of the factors controlling viscosity of more concentrated solutions of polymers and resins such as are used in the coatings field. Several empirical equations have been proposed to express the relationships.

One such relationship for the concentration dependence of relative viscosity is shown in Eq. 19.9, where w_r is weight fraction resin and the ks are constants.

$$\ln \eta_r = \frac{w_r}{k_1 - k_2 w_r + k_3 w_r^2} \qquad (19.9)$$

Nonlinear regression analysis of all the limited number of sets of data available in the literature fit Eq. 19.9 over a very wide range of concentrations [5]. Even with this many terms, there is some systematic deviation from the model at very low concentrations. Constant k_1 is the reciprocal of *weight intrinsic viscosity*, $[\eta]_w$, which, although formally unitless, is based on grams of solution containing a gram of resin. Weight intrinsic viscosity can be converted to the more familiar volume intrinsic viscosity $[\eta]$ by dividing by the density of the solution at concentration $w_r = k_1$. No physical significance of the other two constants, k_2 and k_3, has been elucidated; they are presumably related to further solvent/resin interactions and to free volume availability.

At low concentrations, the k_2 and k_3 terms become negligible and Eq. 19.9 reduces to Eq. 19.10.

$$\ln \eta_r = \frac{w_r}{k_1} = [\eta]_w w_r \qquad (19.10)$$

The simpler, one-term equation (Eq. 19.10) has been extensively used to calculate approximate relative viscosities over a narrow range of concentrations having viscosities from 0.01 to 10 Pa·s. It is convenient to remember that through such a range, the logarithm of relative viscosity of resin solutions increases approximately linearly with concentration.

A word of caution—it is common to "eliminate" the effect of solvent viscosity by normalization to relative viscosity, η_r. This is very useful in studies of the effect of solvent/resin interactions, but it must be remembered that actual flow is related to viscosity, not to relative viscosity.

Concentration can be expressed in a variety of ways. In Eqs. 19.9 and 19.10 the concentrations are expressed in weight fractions. In relationships such as Eq. 19.6, used in studies of dilute solutions of high polymers, concentrations are expressed in terms of weight of polymer per unit volume of polymer solution. In air pollution regulations, concentrations are expressed in terms of weight of solvent per volume unit of coating. In the coatings field, it is common to refer to *volume solids*, that is, the volume of nonvolatile material per volume of coating (NVV), and *weight solids*, that is, weight of nonvolatile material per unit weight (NVW). In some cases people speak of volume fractions. While such concentration relationships may very well be important, there is no easy way to determine volume fractions in most resin solutions. If two liquids give an ideal mixture, the total volume of a mixture will equal the sum of the volumes of the components. However, few solutions are ideal; in nonideal solutions there really is no way to define the volume fraction of solvent and the volume fraction of resin in a solution.

One must be very careful in making comparisons to be sure that one is always comparing results based on the same concentration units. Although much of the data used in the coatings field are based on NVW or NVV, in the light of air pollution regulation requirements, it is often more useful to think in terms of weight of solvent per volume of solution.

Solvent density must be considered in making decisions on solvent selection. Solvents are commonly purchased by the price per unit weight, but coatings are generally sold by the price per unit volume. Obviously, the critical solvent cost is the cost per unit volume. Since VOC regulations are based on weight of VOC per unit volume of coating, the density becomes a critical variable. It is common to see discussions of solvent efficiency based on weight concentration data; this is seldom the important basis for comparison.

Viscosity generally increases with increasing molecular weight, however, the relationship can be complex. Equations 19.7 and 19.8 are based on narrow molecular weight distribution samples of polymers. As shown in Eq. 19.8 under theta conditions, that is, in poor solvent, log viscosity and intrinsic viscosity of dilute solutions of polymers with relatively flexible backbones depend on the square root of molecular weight. In better solvents, the isolated polymer coils expand, intrinsic viscosity increases, and the dependence on molecular weight increases to a power of 0.8 or greater.

In more concentrated solutions, in contrast to dilute solutions, relative viscosity is higher in poor solvents than in good solvents. In the few cases reported in the literature, log of relative viscosity increases with the square root of molecular weight of resins dissolved in good solvents at these higher concentrations [5,8]. It is said that intrinsic viscosities of oligomers exhibit theta condition response, that is, relative viscosity of oligomer solutions is proportional to the square root of molecular weight [9]. In solutions with concentrations that give viscosities between about 0.01 Pa·s and 10 Pa·s in good solvents, this appears to be true, but much further research is needed.

However, viscosity of resin solutions also depends on the dispersity of the molecular weight. A relatively few very high molecular weight molecules disproportionately increase the viscosity of a solution; hence, the broader the molecular weight distribution, the higher is the viscosity.

The T_g of resin solutions is another important factor that affects viscosity. Viscosity increases as $(T - T_g)$ decreases and free volume decreases. As resin solution concentration increases, at some point the T_g of the solution becomes the factor that dominates resin solution viscosity; above that concentration, viscosity increases rapidly. The T_g of resin solutions is controlled by resin structure, molecular weight, and concentration, solvent structure, and interactions between solvent and resin, as discussed further in Section 19.4.4.

19.4.4. Solvent Effects on Viscosity of Resin Solutions

As can be seen in the equations relating viscosity to concentration, a factor affecting the viscosity of resin solutions is the viscosity of the solvent. At first glance, it might appear that a small difference in the low viscosity of the solvent would have a trivial affect on the higher viscosity of the solution. However, there are examples where the difference in solvent viscosity is as little as 0.2 mPa·s, whereas the corresponding difference in viscosity of 50 wt% resin solutions in the same solvents is as great as 2 Pa·s, a difference of four orders of magnitude.

More important are the effects of resin/solvent interactions. In good solvents, there are stronger interactions between solvent molecules and resin molecules than in poor solvents. In very dilute solutions, this means that the chains will become more extended and sweep out larger hydrodynamic volumes in good solvents than in poor solvents. However, in more concentrated solutions, resin molecules are forced to be within the hydrodynamic volumes swept out by the other resin molecules. In such a case, if the interaction between solvent and resin is strong, there will be weaker resin/resin interactions and the molecules can flow easily through the hydrodynamic volumes swept out by other molecules (provided free volume is adequate). On the other hand, in the case of a poor solvent/resin combination, there will be stronger resin/resin interactions when resin molecules try to flow through the hydrodynamic volume swept out by other resin molecules. More or less transient clusters of resin molecules form and viscosity is higher. In solutions in good solvents, flow is Newtonian. In many cases, flow of more concentrated resin "solutions" in poor solvents behave like dispersion systems and are non-Newtonian since shear can break up or distort resin clusters.

Although the difference in viscosity of resin solutions in good and poor solvents is reasonably well understood, there is little definitive work in the literature on comparisons between solutions in various good solvents where some of the solvents are "more good" than others. Erickson [10] studied relative viscosities of solutions of several low molecular weight resins in a fairly wide range of solvents. He concluded that relative viscosities decrease as one changes from a very good solvent to a good solvent, pass through a minimum, and then increase rapidly in very poor solvents. As can be seen in Eqs. 19.9 and 19.10, which relate relative viscosity to concentration, hydrodynamic volume of the isolated resin molecule and its associated solvent molecules is a factor in determining the viscosity not just of very dilute solutions but more concentrated ones too. As we change from a very good to a good solvent, intrinsic viscosity and hence relative viscosity should decrease; this fits in with Erickson's hypothesis. The range of error in Erickson's work is not small enough to establish his conclusions beyond doubt. He may well be right; but there is need for much further research.

Solvent effects on hydrogen bonding between resin molecules can be substantial. Figure 19.13 shows the viscosity of solutions of an acrylated epoxidized oil in three solvents chosen because of their similar viscosities but very different potential effects on hydrogen bonding. The resin molecules have multiple hydroxyl groups. Note that the viscosities of the solutions in xylene are highest. Xylene is a poor hydrogen-bond acceptor and hence promotes intermolecular hydrogen bonding between the resin molecules. Methyl ethyl ketone (MEK) is a good hydrogen-bond acceptor molecule and reduces the viscosity more effectively than xylene by reducing intermolecular hydrogen bonding. Although methyl alcohol is a much stronger hydrogen-bonding solvent than MEK, it is only marginally better at reducing viscosity. Methyl alcohol is both a hydrogen-bond donor and acceptor. Possibly, methyl alcohol can bridge resin molecules by functioning as a hydrogen-bond donor with one resin molecule and a hydrogen-bond acceptor with the other; such bridging would reduce the effectiveness of viscosity reduction.

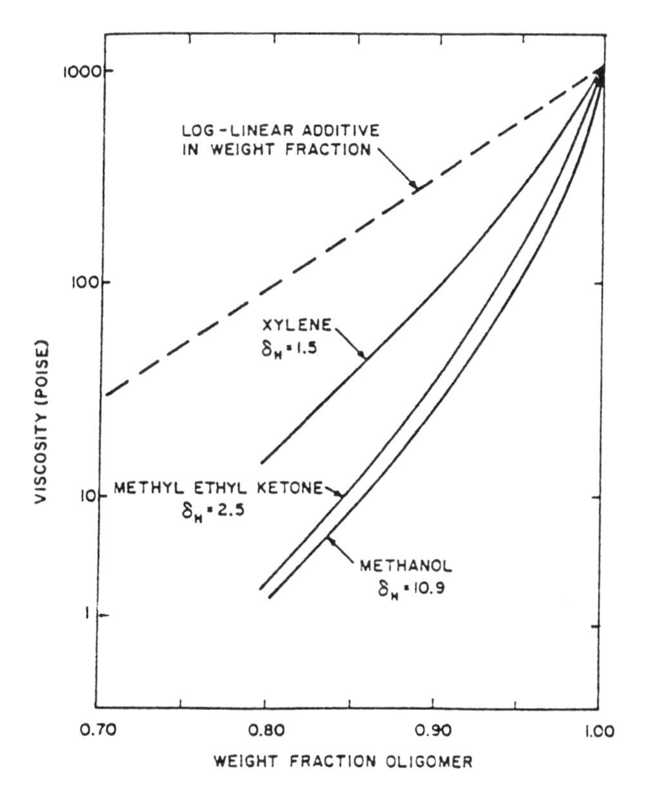

Figure 19.13. Viscosity reduction of an hydroxy-functional UV curable oligomer with xylene, MEK, and methyl alcohol compared to predicted viscosity if the viscosity reduction were a log-linear additive relationship by weight. (From Ref. [11], with permission.)

Intermolecular hydrogen bonding between carboxylic acid-functional resin molecules is particularly strong. Correspondingly, solvent effects on viscosity with such acid-substituted resins are particularly large [12]. This should be no surprise since it has been known for many years that simple carboxylic acids such as acetic acid exist as dimers in poor hydrogen-bond acceptor solvents like benzene, whereas the dimers are dissociated in good hydrogen-bond acceptor solvents like acetone. The effect in resin solutions was demonstrated in a study of a primarily monocarboxylic acid-substituted acrylic oligomer [13]. Relative viscosity in xylene was much higher than in methyl isobutyl ketone (MIBK). Molecular weight, as determined by vapor pressure depression, was much lower in acetone than in benzene. In the hydrocarbon solvents, hydrogen bonding between carboxylic acid molecules was promoted; in the ketone solvents, the predominant hydrogen bonding is between ketones and carboxylic acid groups rather than between carboxylic acids.

Another factor in the effect of solvent choice on viscosity is the solvent T_g. The T_g of resin solutions depends on concentration and the T_g of both the resin and the solvent. This effect has been recognized in the addition of plasticizers to polymers but has not been widely studied in resin solutions with concentrations and viscosities in the range of interest in coatings. In one study [5] it was found that

Table 19.1. Effects of Molecular Weight and Functional Group Content on Viscosity

Characteristics	SAA-I	SAA-II
$\overline{M}_n$	1,600	1,150
$\overline{M}_w/\overline{M}_n$	1.5	1.5
OH content (wt%)	5.7	7.7
	Viscosities (mPa·s)	
80 wt% in MEK	10,000	6,500
70 wt% in MEK	300	230
60 wt% in MEK	80	65
50 wt% in MEK	34	30
50 wt% in toluene	760	3,840

the data fit Eq. 19.11, where T_{gs} is the T_g of the solvent and T_{gr} is the T_g of the solvent-free resin:

$$\frac{1}{T_g} = \frac{w_s}{T_{gs}} + \frac{w_r}{T_{gr}} + kw_r w_s \tag{19.11}$$

In this study, Eq. 19.11 fit the data over the whole range of concentrations from pure solvent to pure resin. Equation 19.11 needs to be tested with other systems.

Since it is common to use mixed solvents in coating formulations, further studies of the effects of mixed solvents on resin solution viscosity would be particularly useful. It has been reported [14] that viscosity of mixtures of hydrogen-bond acceptor solvents like ketones and esters are nearly ideal in their effects on viscosity; that is, the viscosity of a mixture can be estimated quite well by calculating the weighted average viscosity from those of the components. However, in the case of alcohols with other solvents, the viscosities of mixtures varied substantially from ideal behavior. The deviation can be attributed to reduction of intermolecular hydrogen bonding of alcohols by the other solvents, and was particularly pronounced when water was one of the solvents in a mixture.

While these results with solvent mixtures add to our understanding of the effects of intermolecular interactions on viscosity, they are not directly applicable to the problem of mixed solvent effects on viscosity of resin solutions. Since most resins have multiple hydrogen-bond donor and acceptor sites, the interactions with solvent are greater and more complex than in solvent blends. Only one paper [15] has been published on this important question. The authors suggest that the "best" solvent in the mixture dominates in determining the effect on the intrinsic viscosity of the resin solution. The rationale is that the "best" solvent interacts most strongly with the resin molecules and hence controls the degree to which the resin molecules are extended. It is to be hoped that much further research will be reported in the future.

The relationships are further complicated because viscosity can be affected by interactions between molecular weight, number of polar groups per molecule, and solvent/resin interactions. For example, consider the data in Table 19.1 on the viscosity of solutions of a pair of styrene/allyl alcohol (SAA) copolymers in methyl

ethyl ketone (MEK) and toluene [11]. As shown SAA-I has a higher molecular weight but a lower functional group content than SAA-II. In MEK, which accepts hydrogen bonds effectively, the effect on viscosity of the OH content is diminished so that the higher molecular weight of SAA-I results in a somewhat higher viscosity compared with SAA-2. In toluene, which does not hydrogen bond effectively, the difference in OH content dominates over the difference in molecular weight so that the SAA-II solutions have the higher viscosity. Comparison of the 50 wt% solutions in MEK and toluene shows that the hydrogen-bonding solvent is much more effective for viscosity reduction of both SAAs. See Table 15.2 in Section 15.3 for examples of the effects of solvent on viscosity of solutions of a high solids acrylic resin.

19.5. VISCOSITY OF LIQUIDS WITH DISPERSED PHASES

Since most coatings contain dispersed pigment and/or resin particles, it is important to consider the effect of dispersed phases on the viscosity of liquids. When a small amount of a dispersed phase is present, there is only a small effect on the viscosity (unless the dispersed phase is flocculated); however, as the volume of dispersed phase increases there is a sharply increasing effect. More and more energy is diverted into making the particles rotate and the presence of the particles interferes increasingly with the ability of other particles to move. When the system becomes closely packed with particles, the viscosity approaches infinity.

Extending earlier work by Einstein, Mooney developed an equation that models the effect of a dispersed phase on viscosity [16]. Equation 19.12 is a useful form of the Mooney equation for understanding the effects of variables on viscosity where η_e is viscosity of the continuous or external phase, K_E is a shape constant, V_i is the volume fraction of internal phase, and ϕ is the packing factor.

$$\ln \eta = \ln \eta_e + \frac{K_E V_i}{1 - (V_i/\phi)} \tag{19.12}$$

The packing factor (ϕ) is the maximum volume fraction of internal phase that can be fit into the system when the particles are randomly close packed and the external phase just fills all the interstices between the particles. When the volume fraction (V_i) equals ϕ, the viscosity of the system approaches infinity. Two major assumptions are involved in the Mooney equation: (1) the particles are rigid and (2) there are no particle–particle interactions. Figure 19.14 shows an example of a plot of the relationship between log viscosity and V_i for a dispersed-phase system.

The shape constant (K_E) or the Einstein constant has a value of 2.5 for spheres. Many of the particles in coatings are spheres, or are reasonably close to being spheres, but there are exceptions. In the case of uniform diameter spheres, that is, monodisperse systems, the value of ϕ is 0.637. This is the packing factor that has been calculated for a random mixture of cubical and hexagonal close-packed spheres and has been confirmed experimentally.

To the surprise of many first considering the question, the packing factor of monodisperse spheres is independent of particle size. The packing factor for basketballs is 0.637; the packing factor for tennis balls is 0.637; the packing factor for

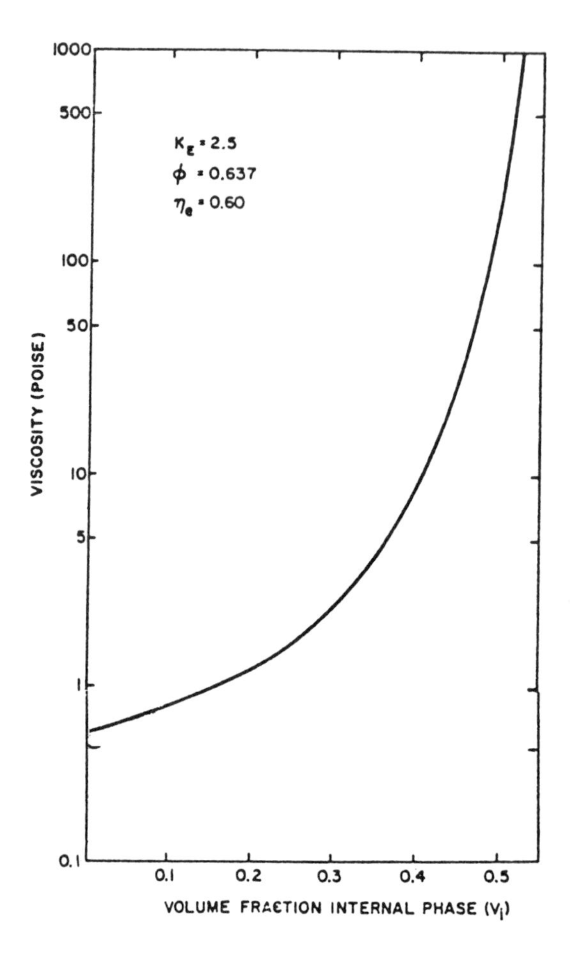

Figure 19.14. The effect of increasing volume fraction of noninteracting spherical particles on the viscosity of a dispersion. (From Ref. [11], with permission.)

monodisperse latex particles is 0.637. However, tennis balls will fit into the interstices between close-packed basketballs, and latex particles would fit in between close-packed tennis balls. In other words, the packing factor depends strongly on particle size distribution—the broader the particle size distribution, the higher the packing factor.

The viscosity of dispersions of nonrigid particles does not follow the Mooney equation. When a shear stress is applied to such a system, for example, an emulsion, the particles can distort. When the particles are distorted, the shape constant changes to a lower value and viscosity decreases. It seems probable that the packing factor increases with distortion of the spheres, also leading to a decrease in viscosity. Commonly, such systems are thixotropic. This is logical since, depending on the difference in viscosity of the internal and external phases, there would be time dependency of the distortion of the particles and hence a decrease in viscosity as a function of time at a given shear rate. There are modifications of the Mooney equation that consider the viscosities of the two phases [17], but not the time dependency.

Examples of dispersed-phase systems that can be relatively readily distorted are emulsions, nonaqueous dispersions (NAD), water-reducible acrylic, polyester, and urethane systems, some latexes where the outer layer of the latex particles and layers adsorbed on them are highly swollen by water, and some pigment dispersions with comparatively thick adsorbed layers of polymer swollen with solvent. While there may be other factors involved, many so-called thixotropic agents act by creating a swollen dispersed phase that can be distorted. For example, very small particle size SiO_2 adsorbs a layer of polymer swollen by solvent that is thick compared to the pigment and is distortable in a shear field. The degree of distortion, up to some point, increases as the shear stress increases and/or as the time of shearing increases. When shearing is stopped or decreased, the polymer layers recover their equilibrium shape and the viscosity increases. In other cases, lightly cross-linked polymer gel particles are used; these swell with solvent in the coating, giving a distortable dispersed phase.

The viscosity of dispersed phase systems is also affected by particle–particle interactions. If clusters of particles form when stirring of a system is stopped, the viscosity of the system increases; if these clusters separate again when shear is exerted, the viscosity drops. Examples of such shear thinning systems are flocculated pigment dispersions and flocculated latexes. Another is the "gelation," really flocculation, induced by water in coatings containing treated clay dispersions [18]. When clusters of particles form, continuous phase is trapped in the clusters; as a result, at low shear rates V_i is high. At high shear rates, when the clusters break up, the value of V_i is reduced to just that of individual particles of the dispersion without trapped continuous phase. As V_i increases, viscosity increases and vice versa.

One can also consider the system from just the point of view of the value of V_i of the primary particles, adjusting K_E and ϕ in Eq. 19.12 to account for the aggregation rather than adjusting V_i. An example is shown in Figure 19.15 [19]. The vertical axis is the ratio of viscosity/viscosity of the external phase (η/η_e). It can be seen that the viscosity increases very rapidly as the number of particles in aggregates increases.

Polymer solutions containing dispersed phases are complex physical systems whose flow properties are still the subject of continuing research. Interactions among dispersed particles, such as discussed in Section 19.4.3, can have substantial effects. In this brief discussion, we have used the Mooney equation; alternative treatments, such as that of Krieger and Dougherty [20], are also useful.

19.6. OTHER MODES OF FLOW

Although flow from application of shear stress is the most common type encountered in making and using coatings, other modes of flow are sometimes involved. This section briefly discusses these types of flow.

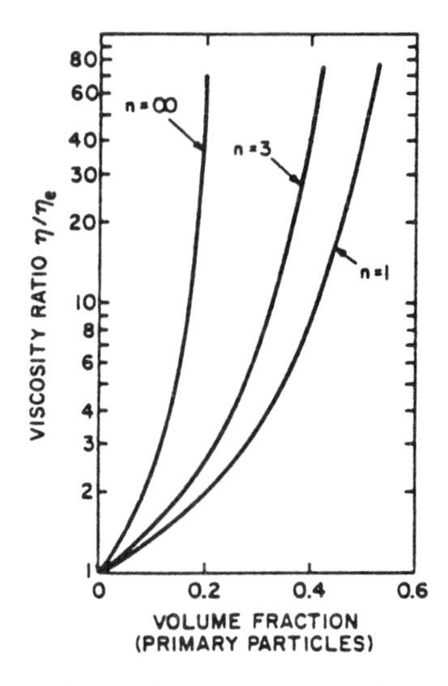

Figure 19.15. Effect of cluster formation on viscosity. (From Ref. [11], with permission.)

19.6.1. Turbulent Flow

Turbulent flow occurs at very high rates of shear or in irregularly shaped containers and pipes. At low shear rates, flow behaves in a laminar fashion, as illustrated schematically in Figure 19.1. However, as shear rate increases, a critical point is reached where flow suddenly becomes chaotic. Laminar flow is disrupted; swirling eddies and vortices occur and flow changes to turbulent flow. Even with Newtonian fluids, viscosity increases more than proportionally with shear rates above this critical value.

19.6.2. Normal Force Flow

When Newtonian fluids are stirred with a rotary stirrer, the liquid level becomes low in the center near the shaft of the stirrer and high on the walls of the vessel as a result of centrifugal force. This is shown schematically in Figure 19.16*a*. However, some liquids climb the shaft of the stirrer, shown in Figure 19.16*b*, rather than the sides of the vessel. Such flow is normal (perpendicular) to the plane of force.

This *normal force flow* behavior is typical of systems that are starting to gel. In the early stages of cooking resins, the flow pattern is as shown in Figure 19.16*a*; but if cross-linking starts and gelation begins, the flow pattern can change abruptly to that in Figure 19.16*b*. If this occurs, it is time to drop the heating mantle and dump the reaction mixture out as quickly as possible before the gel becomes intractable. One can envision the phenomenon as resulting from an end of a long molecule with many branches getting wrapped around the stirrer shaft and then winding up the shaft carrying other material with it.

Normal force flow effects have been detected in some coatings. In effect, the coating flow also has an elastic component to its distortion. It seems logical to assume that a significant extent of elasticity could affect atomization in spraying,

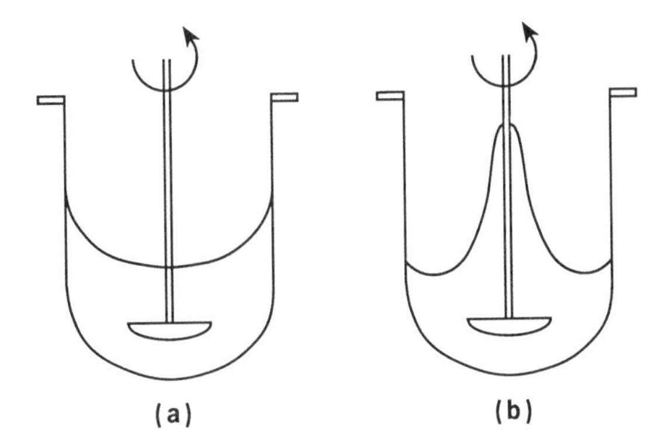

Figure 19.16. (*a*) Conventional compared to (*b*) normal direction flow of liquids on stirring.

film splitting in roller coating, and leveling; however, there have been few studies of correlation between normal force flows and coatings performance [4]. At least part of the reason for the lack of studies is that specialized and expensive instrumentation, such as oscillatory motion viscometers, is required for measurements (see Ref. [4]).

19.6.3. Extensional Flow

Another mode of flow encountered in some methods of coating application is *extensional flow*. Extensional flow occurs when fluid deformation is the result of a stretching motion. Various types of stretching motion are possible. In spin coating, extension occurs in two dimensions. The extensional flow of greatest importance in most other coating is uniaxial, that is, in one direction. In uniaxial flow, the viscosity is properly called *dynamic uniaxial extensional viscosity* (DUEV); we simply use the term extensional viscosity, but it should be remembered that there are other types of extensional viscosity.

The difference between extensional flow and shear flow was first observed in fiber drawing. When the fiber material passes through the spinneret, the mode is shear flow. However, as the fiber is pulled after leaving the spinneret, there is no further shearing action; rather the fiber is extended. The flow is extensional flow and the resistance to flow is *extensional viscosity*. The symbol used for extensional viscosity is η^*. In the case of Newtonian fluids, $\eta^* = 3\,\eta$.

An example of extensional flow in coating application is encountered in roller coating. The material to be coated is passed through the nip between two rollers, one of which is covered with a layer of coating. In the nip, the coating is under pressure; as the coating comes out of the nip, the pressure is suddenly released and cavitation occurs. Cavitation is the formation of small holes inside the liquid coating. As the sheet continues beyond the nip, the distance between the roller and the coated sheet increases and the coating film splits at the weak points (holes) that result from the cavitation. One can think of the holes elongating and the walls of the holes becoming equivalent to fibers as the extension continues. If the extensional viscosity is relatively low, these fibers break quickly; however, with higher extensional viscosity, the fibers grow longer. Figure 19.17 shows an extreme case

Figure 19.17. Fiber development in roll coating a high extensional viscosity paint. (From Ref. [21], with permission.)

of fiber development in roll coating of paint with an impracticably high extensional viscosity [21]. References [22] and [23] discuss the relationship of variables and extensional viscosity effects in roll coating.

Extensional flow can also be encountered in spray application. If, for example, a solution of a thermoplastic acrylic resin with $\overline{M}_w$ above about 100,000 is sprayed, instead of droplets coming out of the orifice of the spray gun, fibers emerge. The

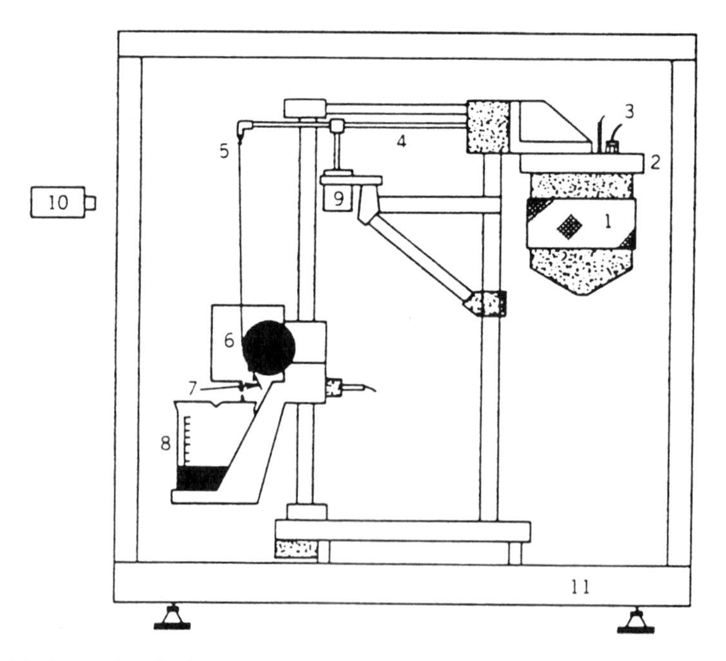

Figure 19.18. Extensional viscometer: 1, reservoir; 2, mount; 3, compressed airline (for forcing sample through system); 4, thin-walled delivery tube; 5, nozzle (spinneret); 6, take-up drum; 7, scraper; 8, beaker; 9, transducer (to measure the deflection of the tube as a result of the forces generated during spinning); 10, video camera (permits determination of filament width); 11, environmental chamber (for temperature control). (From Ref. [2], with permission.)

"strength" of the solution is high enough that the stream of coating stays as a fiber rather than forming droplets. As the fiber extends, the mode of flow is extension. The behavior is called cobwebbing. While cobwebbing is undesirable if you are painting a car, it can give a very desirable effect for applying decoration to a Christmas tree. Reference [24] discusses other possible extensional viscosity phenomena in spray application.

Extensional viscosity is measured by forming a fiber, wrapping it around a drum, and measuring the rate of extension and the force required for further extension. A schematic diagram of such a device that can be used with liquid paint is shown in Figure 19.18.

GENERAL REFERENCES

H. A. Barnes, J. F. Hutton, and K. Walters, *An Introduction to Rheology*, Elsevier, Amsterdam, 1989.

T. C. Patton, *Paint Flow and Pigment Dispersion*, 2nd ed., Wiley-Interscience, New York, 1979.

C. K. Schoff, "Rheological Measurements," in *Encyclopedia of Polymer Science and Engineering*, 2nd ed., Vol. 14, Wiley, New York, 1988, pp. 454–540.

C. K. Schoff, *Rheology*, Federation of Societies for Coatings Technology, Blue Bell, PA, 1991.

REFERENCES

1. T. C. Patton, *Paint Flow and Pigment Dispersion*, 2nd ed., Wiley-Interscience, New York, 1979.
2. C. K. Schoff, "Rheological Measurements," in *Encyclopedia of Polymer Science and Engineering*, 2nd ed., Vol. 14, Wiley, New York, 1988, pp. 454–540.
3. F. Anwari, B. J. Carlozzo, C. J. Knauss, R. B. Krafcik, P. Rozick, P. M. Slifko, and J. C. Weaver, *J. Coat. Technol.*, **61** (774), 41 (1989).
4. C. K. Schoff, *Rheology*, Federation of Societies for Coatings Technology, Blue Bell, PA, 1991.
5. Z. W. Wicks, Jr., G. F. Jacobs, I. C. Lin, E. H. Urruti, and L. G. Fitzgerald, *J. Coat. Technol.*, **57** (725), 51 (1985).
6. A. Toussaint and I. Szigetvari, *J. Coat. Technol.*, **59** (750), 49 (1987).
7. K. J. H. Kruithof and H. J. W. van den Haak, *J. Coat. Technol.*, **62** (790), 47 (1990).
8. P. R. Sperry and A. Mercurio, *ACS Coat. Plast. Chem. Prepr.*, **43**, 427 (1978).
9. W. A. Lee and R. A. Rutherford, "The Glass Transition Temperatures of Polymers," in J. Branderup and E. H. Immergut, Eds., *Polymer Handbook*, 2nd ed. Wiley, New York, 1975, p. III–141.
10. J. R. Erickson, *J. Coat. Technol.*, **48** (620), 58 (1976).
11. L. W. Hill and Z. W. Wicks, Jr., *Prog. Org. Coat.*, **10**, 55 (1982).
12. M. A. Sherwin, J. V. Koleske, and R. A. Taller, *J. Coat. Technol.*, **53** (683), 35 (1981).
13. Z. W. Wicks, Jr. and L. G. Fitzgerald, *J. Coat. Technol.*, **57**, (730), 45 (1985).
14. A. L. Rocklin and G. D. Edwards, *J. Coat. Technol.*, **48** (620), 68 (1976).
15. J. R. Erickson and A. W. Garner, *ACS Org. Coat. Plast. Chem. Prepr.*, **37** (1), 447 (1977).
16. M. Mooney, *J. Colloid Sci.*, **6**, 162 (1951).
17. L. E. Nielsen, *Polymer Rheology*, Marcel Dekker, New York, 1977, pp. 56–61.
18. S. J. Kemnetz, A. L. Still, C. A. Cody, and R. Schwindt, *J. Coat. Technol.*, **61** (776), 47 (1989).
19. T. B. Lewis and L. E. Nielsen, *Trans. Soc. Rheol.*, **12**, 421 (1968).

20. I. M. Krieger and T. J. Dougherty, *Trans. Soc. Rheology*, **III**, 137 (1959); G. M. Choi and I. M. Krieger, *J. Colloid Interface Sci.*, **113**, 94, 101 (1986).
21. J. E. Glass, *J. Coat. Technol.*, **50** (641), 56 (1978).
22. D. A. Soules, R. H. Fernando, and J. E. Glass, *J. Rheology*, **32**, 181 (1988).
23. R. H. Fernando and J. E. Glass, *J. Rheology*, **32**, 199 (1988).
24. D. A. Soules, G. P. Dinga, and J. E. Glass, in *Polymers as Rheology Modifiers*, J. E. Glass, Ed., American Chemical Society, Washington, DC, 1991, pp. 322–332.

CHAPTER XX

Pigment Dispersion

A majority of coatings contain one or more pigments. In most cases, it is necessary for the coatings manufacturer to disperse the dry pigment or pigments into part of the vehicle of the coating. Dry pigments consist of aggregates of crystals of pigment. The pigment manufacturer designed the particle size of the crystals to give the best compromise of properties. The last step in most pigment manufacturing processes is filtering from a reaction mixture and then drying the product. In this drying process, pigment particles are cemented together into aggregates. The coatings manufacturer then has to make a stable dispersion of the pigment particles. This involves three aspects: (1) *wetting*, (2) *separation*, and (3) *stabilization*. Most authors agree there are three aspects to dispersion but many different terms are used, sometimes with conflicting meanings. Be careful in reading papers in the field to know how an author is using the terms.

20.1. DISPERSIONS IN ORGANIC MEDIA

Because of the effects of the high surface tension and polarity of water, it is appropriate to separate our consideration of pigment dispersion into two sections: dispersion in organic media and dispersion in aqueous media.

20.1.1. Wetting

As we are using the term, wetting is the displacement of air (and sometimes of water or of other surface contaminants) from the surface of the pigment particles and aggregates by the vehicle in which dispersion is to be effected. Some authors use the term wetting for some undefined combination of what we are calling wetting and what we are calling stabilization.

Wetting requires that the surface tension of the vehicle be lower than that of the pigment. In the dispersion of pigments in organic media, this is usually the case; any inorganic pigment and most organic pigments have higher surface tensions than any organic vehicle. If the pigment has an especially low surface tension, it is necessary to use a medium with still lower surface tension. When the word wetting is used to include not only wetting but some aspect of stabilization, authors say that wetting is the key requirement for pigment dispersion. It is essential but, in the case of organic media, it is seldom a problem (if the term wetting is used as we use it).

There can be important differences in the rate of wetting. When a dry pigment is added to a vehicle to make a mill base (i.e., the mixture put into a piece of premix or dispersion equipment), it tends to clump up in clusters of pigment aggregates. For wetting to occur, the vehicle must penetrate completely through these clusters and, in so far as possible, into the pigment aggregates. It is desirable for this to happen quickly. The rate of wetting is dominantly controlled by the viscosity of the vehicle; lower viscosity leads to more rapid wetting.

20.1.2. Separation

The objective is to separate the pigment aggregates into the individual crystals but not to grind the individual crystals to smaller particle size crystals. It is generally undesirable to decrease the crystal size. Many different types of machinery are used to carry out this separation stage. Some of the most important examples are described in Section 20.5. Dispersion machines work by applying a shear stress to the aggregates that are suspended in the vehicle. If the aggregates are easily separated, machinery that is only able to exert a comparatively small shear stress is adequate. If on the other hand the aggregates require a relatively large force for separation, then machinery that can apply a higher shear stress is required. Pigment manufacturers have been increasingly successful in processing pigments so that their aggregates are relatively easily separated.

Recall from Eq. 19.1 that shear stress is equal to shear rate times the viscosity of the mill base. The applicable shear rate for any particular dispersion equipment is set by the machine design. For the fastest rate of separation of aggregates, the mill base should have as high a viscosity as that type of equipment can handle efficiently. Then, the highest shear stress will be exerted on the pigment aggregates and, therefore, the separation stage will be accomplished in the minimum time. Any pigment aggregates can be separated if the proper machinery is used. The formulator must select appropriate dispersion machinery that can transfer sufficient shear stress to the aggregates and formulate a mill base for the most efficient use of the selected machinery. The engineering theory and equations modeling the forces for separation are discussed in an excellent series of three papers by Winkler and co-workers [1].

20.1.3. Stabilization

Wetting and separation are important steps in making a pigment dispersion, but it is seldom a problem to carry out these two stages. On the other hand, stabilization can frequently be a serious problem and is usually the key to making good pigment dispersions. If the dispersion is not stabilized, the pigment particles will be attracted to each other and will undergo *flocculation*. Flocculation is a type of aggregation but the aggregates formed are not cemented together like the aggregates in the dry pigment powder. Although substantial shear stress is required to separate the original aggregates, flocculation can usually be readily reversed by applying relatively low levels of shear stress. Nevertheless, flocculation is almost always undesirable. With light scattering pigments, the larger particle size resulting from flocculation reduces scattering and, therefore, reduces hiding. With color pigments, the larger particle size reduces both light absorption and color strength. The larger size floccules in the final film tend to reduce gloss. Flocculation of pigments, including inert pigments, can change critical pigment volume concentration (CPVC) and thus affect properties of the coating films (see Section 21.2). Flocculated dispersions are shear thinning and have higher viscosities at low shear rates than well-stabilized dispersions. Flocculated dispersions do have the advantage that they settle to form soft pigment-bearing sediments that are easily stirred back to uniformity. However, settling problems can usually be minimized in other ways without the adverse results of flocculation.

Two broad mechanisms have been proposed for stabilization: charge repulsion and entropic repulsion. (See Section 2.5.1 for a general discussion of the stabilization of colloidal dispersions, including a discussion of the alternative terminology for entropic repulsion, namely steric repulsion and osmotic repulsion.) In charge repulsion, particles with like electrostatic charges repel each other. While charge repulsion certainly can occur in organic media, it is probable that entropic repulsion is more important. Furthermore, the ways to increase charge repulsion tend to leave ionic substances in the final film which would tend to reduce water resistance. Charge repulsion is more important in aqueous dispersions, as discussed in Section 20.4.

Entropic repulsion is a term used to describe the repelling effect of layers of adsorbed material on the surface of the particles of the dispersion that prevent the particles from getting close enough together for flocculation to occur. In many dispersions of pigments in organic media, the adsorbed layer consists of resin molecules swollen with solvent. The particles are in rapid, random (Brownian) motion. As they approach each other, their adsorbed layers become crowded; there is a reduction in the number of possible conformations of molecules of resin and associated solvent in the adsorbed layers. The resulting decrease in disorder constitutes a reduction of entropy. Reduction of entropy corresponds to an increase in energy and requires force; hence resistance to the reduction of entropy leads to repulsion. Similarly, compression of the layers could lead to a more ordered system by squeezing out solvent. The accompanying reduction in entropy would again lead to repulsion.

Much of our understanding of entropic stabilization of pigment dispersions flows from the seminal work of Rehacek [2]. He devised an experimental technique that permits determination of the thickness and composition of the adsorbed layer on the surface of a pigment dispersed in a resin solution. First, he made a series of solutions of the resin in a solvent with different concentrations (c_1). He then dispersed a known amount of pigment in each of these solutions with different resin concentrations. Samples of the pigment dispersions were then centrifuged until a pigment-free layer formed. The concentration of resin, c_2, in this supernatant layer was determined. He plotted $(c_1 - c_2)/P)$, P = grams of pigment, for each sample against c_2. [Unfortunately, he labeled $(c_1 - c_2)/P$ as "adsorption." It would have made it easier to understand his papers if he had labeled this axis of his graphs "apparent adsorption."]

A schematic representation of typical data is presented in Figure 20.1; in all cases, the shape of the curve is the same. Rehacek extrapolated the straight line portion of the curve to the intercepts with both axes. The intercept with the c_2 axis

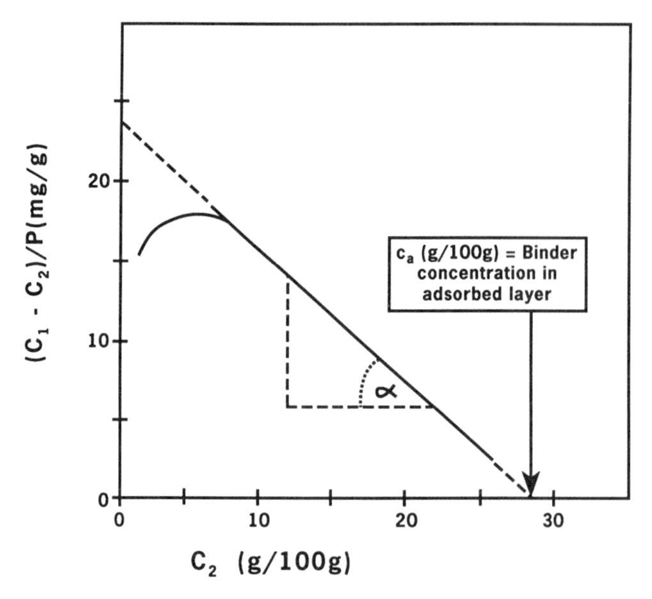

Figure 20.1. Schematic plot of $(c_1 - c_2)/P$ versus c_2 in studies of adsorption of resin and solvent on pigment surfaces. (Adapted from Ref. [2], with permission.)

represents the concentration of resin in the adsorbed layer on the pigment surface. The intercept with the $(c_1 - c_2)/P$ axis represents the number of milligrams of resin solids adsorbed per gram of pigment. It can be proven mathematically that these two values remain constant at all points on the linear portion of the curve. At the intercept with the c_2 axis, apparent adsorption is zero; this does not mean that no resin is adsorbed. It means that the concentration of resin in the adsorbed layer is the same as the concentration in the initial resin solution, c_1, and in the resin solution after pigment was dispersed in the resin solution, c_2. Thus, the amount of resin and solvent in the adsorbed layer is established. The intercept with the $(c_1 - c_2)/P$ axis, corresponds to the $(c_1 - c_2)/P$ value when c_2 equals zero; in other words, c_1 is the amount of resin solids adsorbed per gram of pigment. This gives the milligrams of resin in the adsorbed layer per gram of pigment through the linear part of the curve. The surface area of a gram of pigment can be determined by nitrogen adsorption. Using the densities of the pigment and the resin solution, the thickness of the adsorbed layer of resin and solvent in the linear range of the plot can be calculated.

In Rehacek's work, and also that of others [3,4], it has been found that, if the adsorbed layer thickness of resin plus solvent is less than 9–10 nm, the dispersion is not stable and flocculation occurs. This adsorbed layer thickness is an average layer thickness, presumably with some portions thinner and others thicker. Similar studies in mixed solvent systems with no resin present show that the adsorbed layer thickness of a combination of solvents is of the order of 0.6–0.8 nm, clearly not sufficient to stabilize against flocculation [4].

When monofunctional surfactants are used, the adsorbed layer can be thinner and still protect against flocculation. In one case, McKay [5] showed that an adsorbed layer thickness of 4.5 nm of surfactant and associated solvent was adequate to give stabilization. In contrast to the adsorbed layer of resin that is nonuniform in thickness, the surfactant layer is comparatively uniform so that it does not have to be as thick to provide the same degree of stabilization as an irregular adsorbed resin/solvent layer. The advantages and disadvantages of resin/solvent versus surfactant/solvent are discussed later.

Referring back to Figure 20.1, it is of interest to understand why the plot deviates from linearity at lower values of c_2. As can be seen, the values of $(c_1 - c_2)/P$ are lower at low values of c_2. In other words, there is less resin adsorbed per unit area of pigment surface. This can be explained by a competition between resin and solvent adsorption that, it can be speculated, depends on both the relative affinity of resin and solvent molecules for the pigment surface and the concentration of resin. If the concentration of resin is high enough, the resin "wins" and there is a full adsorbed resin layer swollen with solvent. However, at lower concentrations, both solvent and resin are adsorbed on the particle surface so that the average layer thickness is insufficient to prevent flocculation.

Rehacek observed two other differences in behavior of the dispersions with c_2 values above and below the lower end of the linear section of the curve. The low shear viscosities of the dispersions below the critical concentration were higher than of those above it. Also, the separation of pigment during centrifugation was much more rapid and the bulk of the centrifugate formed was much greater. This behavior indicates that the system is flocculated below the critical concentration. Viscosity at low shear rates increases and the system becomes shear thinning when flocculation occurs (see Section 19.5). Also, under low shear conditions, in a flocculated system the particle sizes are much larger than in a nonflocculated system; this causes more rapid settling or centrifugation. Furthermore, the floccules occupy more volume than the stable dispersion of the same amount of pigment since continuous phase is trapped inside the floccules; this leads to the bulkier layer of sediment.

What controls the thickness of the adsorbed layer? In the case of a surfactant with the polar end adsorbed on the surface of a polar pigment, the length of the nonpolar aliphatic chain is the primary factor. In the case of resins with several adsorption sites, the largest single factor is probably molecular weight. For example, Saarnak [3] showed that the adsorbed layer thickness on TiO_2, dispersed in a series of BPA epoxy resins in MEK, increased from 7 to 25 nm as the molecular weight of the epoxy resin increased. With the lowest molecular weight resin the layer thickness, 7 nm, was insufficient to prevent flocculation. Dispersions in solutions of the higher molecular weight epoxy resins were stable.

The adsorbed layer thickness is also affected by the pigment surface. The relatively less polar organic pigments are generally more likely to give significant differences in adsorbed layer thickness with different resin/solvent combinations than are the more polar inorganic pigments. However, even with inorganic pigments, significant differences can be encountered. For example, it has been shown that a TiO_2 surface treated with alumina forms a more stable dispersion than a TiO_2 with a silica surface treatment in the same long oil alkyd solution [7]. The authors propose that the adsorbed layer is more compact on the silica-treated TiO_2. Commonly the interactions between pigment surfaces and adsorbed molecules are considered to be hydrogen-bond interactions, but some authors prefer to interpret the interactions as acid/base interactions [6,8].

The spacing and number of functional groups along the resin chain will affect layer thickness. Taking an extreme example, a linear aliphatic chain resin with a polar group on every other carbon atom, would be expected to adsorb strongly on the surface of a polar pigment like TiO_2. At equilibrium, adsorption of single molecules with interaction of successive polar groups with the pigment surface would be favored, resulting in a thin adsorbed layer. However, if the resin had only occasional polar groups along the chains, at equilibrium the longer segments between polar groups would give loops and tails of resin swollen with solvent projecting out from the pigment surface, hence a thicker adsorbed layer.

The layer thickness can also be affected by solvent/resin interaction. If the loops and tails interact strongly with the solvent, there will be more solvent molecules in the layer and the average conformation of the resin will be more extended. Thus the layer thickness will be greater than using the same resin with a solvent having a weaker solvent/resin interaction.

The solvent/resin combination can also affect resin adsorption. Resin molecules having multiple adsorbing groups have an advantage in the competition, but if the solvent has a functional group that interacts strongly with the pigment surface and the resin has only functional groups that interact weakly with the surface, the more numerous solvent molecules will "win" the contest. For example, it has been found that toluene favors adsorption of macromolecules such as nitrocellulose, polyurethanes, and phenoxy resins on magnetic iron oxide pigment particles as compared with tetrahydrofuran [8].

Addition of solvent to a stable pigment dispersion can, in some circumstances, lead to flocculation. If the added solvent is more strongly adsorbed than the resin molecules, it will displace the resin from the surface, resulting in an unstable dispersion. If the ratio of resin to solvent is just sufficient to allow adequate adsorption of resin to stabilize the dispersion, addition of more of the same solvent can shift the equilibrium, displacing part of the resin, reducing the average adsorbed layer thickness below the critical level for stabilization, resulting in flocculation. In such cases, the dispersion is said to have been subjected to *solvent shock*.

For most conventional solvent-borne coatings, the resin that is used as binder in the coating can be an appropriate resin for stabilization of the pigment dispersions. Most conventional alkyds, polyesters, and thermosetting acrylics will stabilize the dispersion of most pigments. If there is a problem, the two most common changes to provide adequate stability are to increase the molecular weight of at least that portion of the resin used in the mill base or to increase the number of hydroxyl, amide, carboxylic acid, or other polar groups on at least that part of the resin used in the mill base. It has been shown that the highest molecular weight components of resins are selectively adsorbed [6].

If a monofunctional surfactant is used, the solvent is likely to win over most surfactant molecules owing to the large number of solvent molecules even though the surfactant might be somewhat more strongly adsorbed. This can be offset by increasing the concentration of surfactant molecules so that the equilibrium will be shifted in favor of the surfactant. However, this will leave excess surfactant in the final coating film and, thereby, probably reduce the performance properties of the final film. Thus, in organic media, monofunctional surfactants have not generally been a desirable choice for stabilization.

There are combinations of pigment/resin/solvent where stability against flocculation cannot be achieved and an additive such as a surfactant is required. Additives should be used only as a last resort. They may be effective in stabilizing the pigment dispersion, but they may also interfere with some other critical property or properties. For example, it is common to obtain poor adhesion to metal surfaces when surfactants are added to a coating. The most commonly used additive for stabilization is probably lecithin, a naturally occurring choline ester of phosphoglycerides. It is quite strongly adsorbed on the surface of many pigments and, therefore, lesser amounts are required than with most surfactants.

Because of the low molecular weight of the resins used, stabilization of dispersions in high solids coatings presents special problems. In many cases surfactants are required, but to avoid the problems of deleterious effects on film properties special surfactants or oligomeric surfactants with multiple adsorbable groups are being designed. Dispersions for high solids coatings are discussed in Section 20.3.

20.2. FORMULATION OF MILL BASES

The combination of resin/solvent/pigment used in making the pigment dispersion is called a mill base. The formulator must design a mill base for dispersing a pigment in the most appropriate dispersion equipment at the optimum efficiency. Pigment dispersion machinery is the most expensive machinery in the paint plant in terms of both capital and operating costs. It is, therefore, important to maximize the amount of pigment dispersed per unit time. Higher pigment loading means more efficient production; high loadings are possible when the viscosity of the vehicle (solvent plus resin) to be used in the mill base is low (low viscosity also gives faster wetting).

A properly stabilized pigment dispersion exhibits Newtonian flow, and its viscosity will follow the Mooney equation (Eq. 20.1) (see Section 19.5):

$$\ln \eta = \ln \eta_e + \frac{K_E V_i}{1 - \frac{V_i}{\phi}} \tag{20.1}$$

It can readily be seen that volume of pigment (internal phase) can be maximized by using the lowest possible viscosity (η_e) vehicle. Solvent alone would give low viscosity, fast pigment wetting, and high pigment content, but as we have seen, solvent alone cannot stabilize a dispersion against flocculation. Therefore, it is necessary to include some resin in the mill base. For maximum pigment loading, it is desirable to determine the minimum concentration of resin solution that will provide stability. This could be done by following the Rehacek procedure to determine the minimum concentration that still gives a point on the linear section of the curve. However, this procedure is very time consuming. Many years ago, Fred Daniel devised a simpler (although less accurate) method that is much faster [9].

20.2.1. Daniel Flow Point Method

The *Daniel flow point method* is a powerful tool for formulating mill bases efficiently, especially for dispersions to be made in ball and sand mills and related

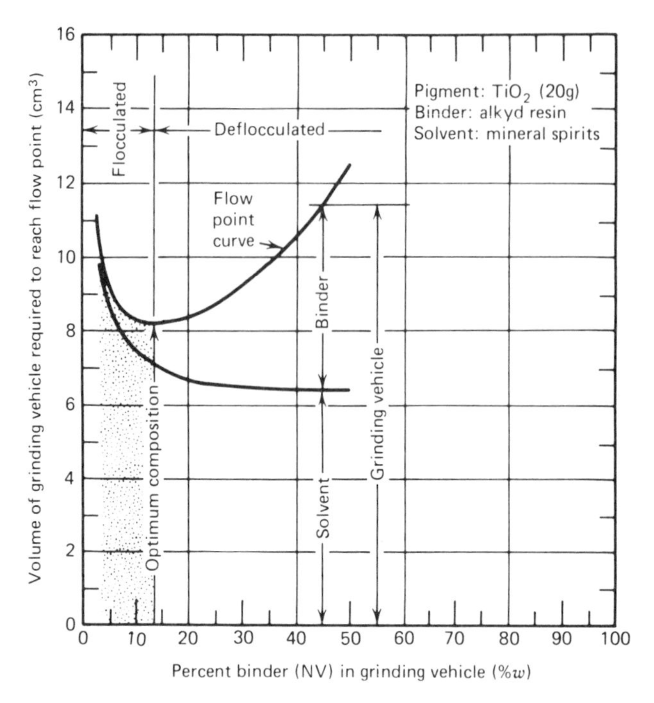

Figure 20.2. Daniel flow point plot: milliliters of solutions of an alkyd resin in mineral spirits per 20 g TiO_2 as a function of NVW resin in solution. (From Ref. [9], with permission.)

types of equipment [9]. It provides an estimate of the most appropriate resin concentration to use with a particular pigment. One makes a series of solutions of different resin concentration in the solvent. One determines the amount of each solution that must be added to a weighed amount of pigment so that when it is dispersed by rubbing vigorously with a spatula on a flat glass plate, the resulting dispersion will have a viscosity just low enough to flow readily off the spatula. One is using the spatula as both a dispersion machine and as a viscometer. Since the clearance between the spatula and the glass plate is small, the shear rate is high. A spatula is actually a fairly good dispersing machine. One plots the data obtained against the concentration of the solution used. This plot is an *isoviscosity* plot, that is, the viscosity of each of the dispersions is approximately the same as measured by whether they will just flow readily off the end of the spatula. (The actual low shear rate viscosity is approximately 10 Pa·s.) An example of such a plot is shown in Figure 20.2; in addition to the volume of resin solution, the volume of solvent in the resin solution is also plotted.

For every resin/solvent/pigment combination that can give a stable dispersion there is a minimum in the curve. Let us consider the significance of the minimum. To the right of the minimum, the viscosity of the external phase is increasing because the concentration of resin is increasing. As can be seen from Eq. 20.1, if η_e increases and viscosity is to stay constant, then V_i must decrease, that is, the amount of vehicle required per unit of pigment must increase.

To the left of the minimum, however, the amount of vehicle per unit of pigment is increasing even though η_e is decreasing. This is because the concentration of resin is insufficient to stabilize the dispersion. The pigment dispersion flocculates increasingly as the concentration of resin in the solutions decreases and the viscosity increases more and more steeply, so that additional resin solution must be added to reach the isoviscosity level.

The minimum point of the Daniel flow point determination corresponds to the minimum concentration of that resin in that solvent which permits preparation of a stable dispersion of that pigment. Since the determination is not highly accurate, it is usual to start experimental dispersions with a somewhat higher resin concentration. It is important to remember that if one adds more solvent to a stable dispersion with near the minimum resin concentration, the dispersion will flocculate.

In some cases, no minimum is found, and the amount of resin solution required per unit of pigment keeps decreasing as the concentration increases. This behavior signifies that a stable dispersion of that pigment cannot be made with that combination of resin and solvent.

Patton [9] gives many examples of data on the Daniel flow point method. He gives information expressed in volume units of pigment as well as the more conventional weight units.

20.2.2. Oil Absorption Values

In the early days of paint formulating, it was observed that the amount of pigment that could be incorporated with linseed oil in a mill base varied tremendously from pigment to pigment. To simplify the process of formulating mill bases to approximately equal viscosity, *oil absorption values* were determined. Linseed oil was slowly added to a weighed amount of pigment while rubbing with a spatula. Initially, balls of pigment wet with oil form in excess powdered pigment. When just enough oil has been added and worked into the pigment so that one coherent mass has been formed, the end point has been reached. The oil absorption value is calculated then as the number of pounds (grams) of linseed oil required to reach the end point with 100 lb (100 g) of pigment. Using the oil absorption value for a new pigment in comparison to that of another pigment for a mill base permits one to set a starting point for pigment content when formulating a mill base for the new pigment.

At the end point, there is just enough linseed oil to adsorb on the surface of all the pigment particles and fill the interstices between close-packed particles. Oil absorption values for different pigments vary over a wide range. The smaller the particle size, the higher will be the oil absorption. The small particle size pigment has more surface area; therefore, a larger amount of linseed oil is adsorbed on the surface. In some cases, such as some grades of carbon black, the pigment particles are porous. Some oil penetrates into these pores and this increases the amount of oil required and hence the oil absorption value. The ratio of surface area to volume can be affected by particle shape. For example, diatomaceous earth (see Section 18.3) has a very high surface area and hence very high oil absorption values. Pigment density has a major effect. High density pigments require less weight of oil to adsorb on the surface of a unit weight of pigment and to fill the interstices and hence have lower oil absorptions. There is little effect of changing vehicles from linseed oil. Therefore, data obtained with linseed oil can be used in helping formulate mill bases with any vehicle. Pigment suppliers provide oil absorption values for their pigments.

The precision of the oil absorption determination by the spatula method is not high. Operators working in different laboratories commonly will duplicate values only within ±15%. It has been said that with experience, deviations by a single operator can be reduced to ±2–3%. In spite of these error ranges, it is common to see oil absorption values given with three supposedly significant figures. A significant improvement in accuracy and precision has been achieved by using a mixing rheometer (see Section 19.3.4), such as a Brabender Plastometer, for carrying out the determination [10]. The mixer is loaded with a known amount of linseed oil followed by slow addition of pigment. The mixer imparts the necessary shear to separate pigment aggregates. The power required to turn the blades is recorded. As the amount of pigment is increased, the power requirement increases, when the oil absorption end point is passed, the dispersion mass breaks up into chunks leading to erratic readings with substantial dropoff in power requirements. The values obtained are more reproducible and are generally a few percent higher than those obtained by the spatula method. The relationship between oil absorption of pigments and the critical pigment volume concentration of films containing those pigments is discussed in Section 21.3.

20.3. PIGMENT DISPERSIONS FOR HIGH SOLIDS COATINGS

As high solids coatings have increased in importance, the occurrence of difficulties in making stable pigment dispersions has increased. Increasing the solids of organic solution coatings requires decreasing the molecular weight of the resins and reducing the number of functional groups per molecule. The reduced number of functional groups per resin molecule decreases the probability of the adsorption of resin molecules; there is a greater probability of solvent adsorption being favored and hence a greater likelihood of flocculation. Also, the low molecular weight results in thinner adsorbed layers of resin and associated solvent molecules. One can use a higher molecular weight resin having a higher functionality just for the pigment dispersion, with the rest of the resin being of low molecular weight and low functionality to reduce the viscosity. Although adsorption of the higher molecular weight, higher functionality resin will be favored at equilibrium, with a sufficient amount to assure stabilization of the pigment dispersion, some of the high molecular weight resin will also be in solution, leading to higher viscosity.

In many high solids coatings, it is not possible to make stable dispersions using the resins from the coating as the adsorbed layer. This has lead to the design of special dispersing aids sometimes called *hyperdispersants* [11–14]. Jakubauskas [11] describes the design parameters of such dispersants. He concludes that the most effective class of dispersant has a polar end with several functional groups and a less polar tail of sufficient length to provide for a surface layer with at least 10 nm thickness. Examples of such dispersants include polycaprolactone–polyethylenimine block copolymers [11], polycaprolactone capped with toluene diisocyanate post-reacted with triethylene tetramine [11], acrylic resins made by group transfer polymerization starting with methyl methacrylate, followed by glycidyl methacrylate, post-reacted with polyamines or

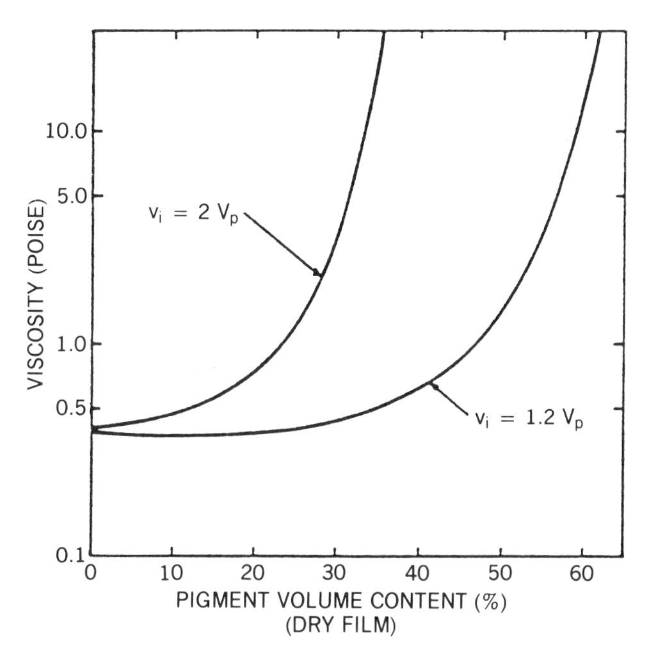

Figure 20.3. Calculations of the effect of PVC on the viscosity of two formulations both at 70 NVV but differing in thickness of the layer of adsorbed polymer solution. See text for assumptions. (From Ref. [15], with permission.)

polycarboxylic acids [12], low molecular weight polyesters from polyhydroxystearic acid [13], and proprietary hyperdispersants [14].

There is the further problem of the adsorbed layer thickness on the pigment surfaces in high solids coatings. In conventional coatings, the volume of pigment in the wet paint of even highly pigmented coatings is quite low so that differences in adsorbed layer thickness do not make a large difference in the viscosity of the wet coating (provided the dispersion is stable against flocculation). However, in high solids coatings, especially highly pigmented, high solids coatings, the adsorbed layer thickness can have a significant effect on the viscosity.

The effect of adsorbed layer thickness can be seen by considering a modification of the Mooney equation (Eq. 20.2), taking into account that the internal phase volume due to the adsorbed layer, V_a, in addition to that due to the pigment, V_p, is included in V_i. Using this equation a series of model calculations were made to illustrate the effect of adsorbed layer thickness on viscosity [15].

$$\ln \eta = \ln \eta_e + \frac{2.5\,(V_p + V_a)}{1 - \dfrac{(V_p + V_a)}{\phi}} \qquad (20.2)$$

The plot in Figure 20.3 shows the calculated dependence of viscosity of coatings with 70 NVV as a function of the pigment volume concentration in dry films to be prepared from the coatings. The possible level of pigmentation is much more limited with the thicker adsorbed layer than with the thinner one. The following principal assumptions were made in making the calculations: $V_i = 1.2\ V_p$ and $V_i = 2\ V_p$ (these are equivalent to an 8- and a 25-nm adsorbed layer respectively on a 200-nm pigment); $\phi = 0.65$; Solvent: $\rho = 0.8$, $\eta = 0.4$ mPa·s; Oligomer: $\rho = 1.1$, $\eta = 40$ mPa·s at 70 NVV, 4×10^5 mPa·s at 100 NVV. The viscosity dependence was assumed to follow Eq. 20.3, where w_r = weight fraction of resin:

$$\ln \eta = \ln \eta_s + \frac{w_r}{0.963 + 0.763\ w_r} \tag{20.3}$$

Surfactants have been designed that are so strongly adsorbed on a pigment that they are preferentially adsorbed even in the presence of a large amount of solvent; then little excess over the amount needed to saturate the pigment surface area is needed to stabilize a dispersion. For example, phthalocyanine blue modified by covalently attaching long aliphatic side chains has been used as a surfactant with phthalocyanine blue pigment, the phthalocyanine end of the molecules of surfactant in effect joins the crystal structure of the surface of the pigment particles so that little, if any, is in solution. The average adsorbed layer thickness needed to protect against flocculation was shown to be about 4.5 nm [5]. Specific, proprietary surfactants have been designed for other pigments.

A related approach is to bond long chains covalently to the surface of pigments. For example, the surface of silicon dioxide pigment particles has been reacted with trialkoxysilanes having long chain alkyl substituents [16]. Titanate orthoesters having three ethyl groups and one alkyl group with a long alkyl chain reduce viscosities of dispersions of some inert pigments. It may be that the ethyl groups exchange with hydroxyl groups on the pigment surfaces.

In many cases, pigment manufacturers are offering pigments that have been surface treated to provide stable dispersions even when the pigment is dispersed in solvent. See Ref. [17] for a review of surface treatment of pigments.

20.4. DISPERSIONS IN AQUEOUS MEDIA

The dispersion of pigments in aqueous media involves the same three factors as in organic media: wetting, separation, and stabilization. However, the unique properties of water add extra considerations. First, water's surface tension is high so that there is more likely to be a problem in wetting the surface of the pigment particles. Secondly, in some cases water interacts strongly with the surface of pigments; therefore, the functional groups on the stabilizers have to interact more strongly with the pigment surface to compete with the water. Furthermore, many applications of aqueous dispersions are in latex paints so that the systems have to be designed so that stabilization of the latex dispersion and the pigment dispersions do not adversely affect each other.

Inorganic pigments such as TiO_2, iron oxides, and most inert pigments have highly polar surfaces so that there is no problem of wetting them with water. The surfaces of inorganic pigments interact strongly with water, but the adsorbed layer of water does not by itself stabilize their dispersions against flocculation. Most organic pigments, however, require the use of a surfactant to wet the surfaces.

In contrast to dispersions in organic media, stabilization by charge repulsion can be the major mechanism (see Section 2.5.1). The stability of the dispersions depends on pH since that affects surface charges. For any combination of pigment, dispersing agent, and water there is a pH at which the surface charge is zero, this pH is called the *isoelectric point* (*iep*). At *iep*, there is no charge repulsion; above the *iep*, the surface is negatively charged; and below the *iep*, it is positively charged. The stability of dispersions is at a minimum at *iep* $\pm$ 1 pH unit [18]. The *iep* value for pigments varies over a fairly wide range, at least as wide as 4.8 for kaolin clay to 9 for $CaCO_3$.

Surface treatments [17] can have important effects on the stability of aqueous pigment dispersions. Special TiO_2 pigments have been developed for water-borne coatings. Sustantial differences in dispersion stability can result from differences in the composition and completeness of the surface treatment [19]. In some cases, the amount of surface treatment has been such that the TiO_2 content is as low as 75% of the pigment weight. Since hiding is related to the actual TiO_2 content, larger amounts of such highly treated pigments are needed to obtain equivalent hiding.

TiO_2 slurries in water are sold on a large scale. The dispersions are generally stabilized with an amine. Commercial stabilizing systems are proprietary. As an example, 2-amino-2-methyl-1-propanol (AMP) along with some di- and/or triethylene glycol can be used. The amine raises the pH above the *iep* of the pigment. The amino group is presumably strongly adsorbed on the pigment surface in competition with water, and the glycol ethers may associate with the aminoalcohol and in turn with water molecules adding to the adsorbed layer thickness. Such slurries have major advantages for manufacturers of latex paints compared to dry TiO_2 pigment. The supplier saves drying and packaging costs and the user saves handling and dispersing costs. The slurries are shipped in tank cars or tank trucks and are handled by pumping. Prices of slurry TiO_2 (on the basis of TiO_2 content) and of dry pigment are approximately equal.

In latex paints a combination of several agents is usually used to provide stabilization of the pigment dispersion. Most paint formulations contain several pigments. The *iep* values of the various pigments are different, which complicates the problem of charge stabilization. Commonly mixtures of surfactants are used. Anionic surfactants are frequently used as one component. Polymeric anionic surfactants (such as salts of acrylic copolymers in which acrylic acid and hydroxyethyl acrylate are used as comonomers) provide salt groups for strong adsorption on the polar surface of the pigment and hydroxyl groups for interaction with the aqueous phase and nonpolar intermediate sections to add adsorbed layer thickness. It seems probable that both charge repulsion and entropic repulsion are involved in the stabilization. Nonionic surfactants are frequently used along with an anionic surfactant. Generally, nonionic surfactants are less likely to give foaming. It is also common to use, as a further dispersion aid, potassium tripolyphosphate, the basicity of which may assure that the pH is above the *iep* of all pigments. Note that *potassium* tripolyphosphate is used in paints, not the sodium salt widely used in laundry detergents. The potassium salt is less likely to deposit on the surface of the film as a scum after being leached out of a dry paint film by water. See Tables 35.1 and 35.2 for examples of surfactant combinations in latex paints.

In contrast to the situation with most inorganic pigments in water, organic pigments, unless specially surface treated, generally have surface tensions much lower than the surface tension of water. Therefore, surfactants are needed to reduce the surface tension of the water to permit wetting that is an essential first step in pigment dispersion. Either anionic or nonionic monofunctional surfactants can be used. In the case of anionic surfactants, the dominant mode of stabilization is probably charge repulsion. In the case of nonionic surfactants, the relatively long polyether alcohol end is oriented out into the water and associated with multiple water molecules; stabilization is probably predominantly by entropic repulsion.

Many latex paint manufacturers buy, rather than manufacture, their color pigment dispersions. The wetting and stabilization methods used are proprietary. It seems logical to assume that polymeric surfactants with multiple nonpolar groups scattered along a polar backbone would be particularly appropriate for organic pigments. Owing to their lower sensitivity to polyvalent ions and usually reduced adverse effect on film properties, nonionic backbone polymeric surfactants are expected to be of the greatest interest.

Whenever possible, it makes sense not to dry the pigment but to simply add dispersant to the wet filter cake obtained in pigment manufacture. This saves the cost of drying and avoids cementing of aggregates and, hence, the need for the separation stage in making the pigment dispersion.

Additives can sometimes affect dispersion stability. For example, a hydrocarbon solvent used in a latex paint as a defoamer was shown to cause flocculation [20]. The paint formula had several different surfactants. It was found that the order in which the surfactants were added controlled whether or not the flocculation occurred.

There is need for more research on the factors leading to stability of dispersions in latex paints. There is reasonable understanding of the stabilization of single pigment systems. However, paints contain several pigments, commonly with both polar and nonpolar surfaces, and one or more latexes (see Chapter 35). Furthermore, the formulations contain one or more water-soluble polymers and/or associative thickeners (see Section 35.3) to adjust the rheological properties of the paint. As a result, the selection of appropriate dispersing agent combinations is done largely on an empirical basis. As in any other situation where empirical knowledge is the key to success, it is highly desirable to build a data base of combinations that work (and, particularly, combinations that do not work) to facilitate formulating with some new pigment combination or latex.

20.5. DISPERSION EQUIPMENT AND PROCESSES

A wide range of dispersion equipment is used in making pigment dispersions. An important difference is in the amount of shear stress that the equipment can exert on the pigment aggregates. Easily separated pigment aggregates can be dispersed in low shear stress equipment whereas some pigment aggregates require very high shear stress for separation. Discussion of all types of such machinery and their operation is beyond the scope of this text. Patton's book [9] provides more detailed discussion of some types and detailed engineering information is available from the machinery manufacturers. Reference [1] deals more fundamentally with the engineering aspects of some dispersion methods. We discuss a few of the most

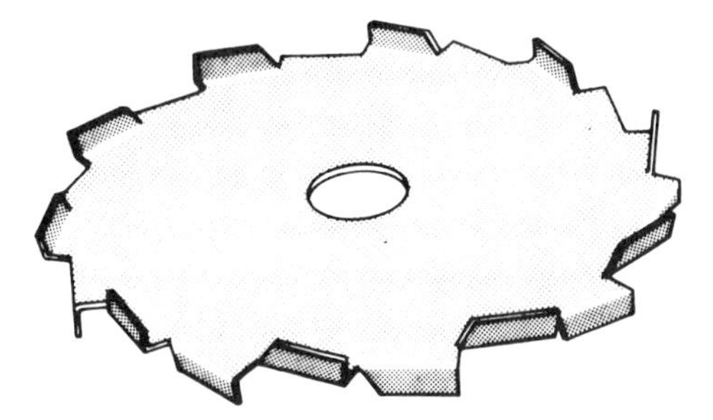

Figure 20.4. Schematic drawing of a high speed impeller disk. (From Ref. [9], with permission.)

important types with emphasis on the advantages and disadvantages of each. The coatings formulator must become familiar with the dispersion equipment available in his/her company's factories so that he/she can design formulations appropriate for production with that machinery.

From a processing point of view, there are three stages in making and using pigment dispersions: (1) premixing, that is, simply stirring the dry pigment into the vehicle and eliminating lumps; (2) imparting sufficient shear stress to separate the pigment aggregates; and (3) let down, that is, combining the pigment dispersion with the balance of the ingredients to make a coating. Some machines can carry out only the second step, some can do two of the three, and others can do all three.

Some pigment dispersions are made for specific batches of coating; after making the pigment dispersion, it is let down with the other components to make the final coating. Other pigment dispersions are made to be used in several related types of coatings or as *tinting pastes* for color matching a variety of coatings. In order to minimize inventory of tinting pastes, one tries to select vehicles for them that are compatible with a wide range of the types of coatings made by that company.

20.5.1. High-Speed Disk (HSD) Dispersers

High-speed disk (*HSD*) *dispersers* consist of a shaft with a disk that rotates at high speed in a vertical cylindrical tank. High-speed disk dispersers are also called high-speed impellers. A schematic drawing of a typical disk is shown in Figure 20.4. Increasingly, the disks are made from engineering plastics that provide greater abrasion resistance than steel.

The shearing action takes place by the differential laminar flow rates streaming out from the edge of the disk that is rotating typically at 4000–5000 rpm. The shear stress developed is relatively low and, therefore, HSD machines are appropriate only for relatively easily separated pigments. Obtaining predominantly laminar flow requires some minimum viscosity depending on the dimensions, but it is usually somewhat over 3 Pa·s. The higher the viscosity, the greater the shear stress that will be exerted on the aggregates and, therefore, the faster the process. The viscosity should be set such that the motor driving the shaft is running at peak power. It has been said that any formulator who has not stalled a 250-hp motor on a big tank disperser has never properly formulated a mill base for a high-speed disperser.

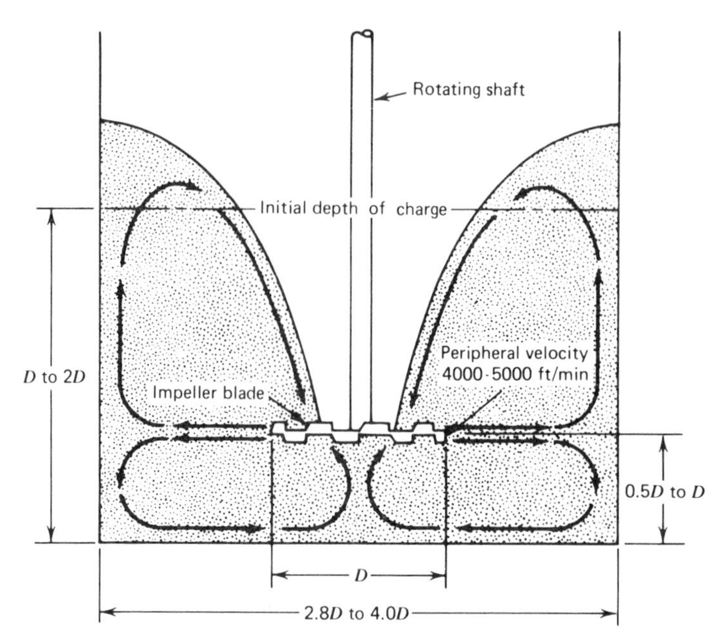

Figure 20.5. Diagram of a high-speed impeller disperser showing correct positioning of disk and optimum dimension ratios. (From Ref. [9], with permission.)

For maximum efficiency, the pigment loading should be maximized. In solvent-based coatings, this can be done by using as low a resin concentration vehicle solution as will provide stabilization against flocculation. The Daniel flow point method is an appropriate way of estimating this resin/solvent ratio. Then the ratio of this vehicle to pigment is set to achieve a viscosity at high shear rate such that the motor will draw peak amperage. In the case of dispersions in water for latex paints, water-soluble polymers are commonly included in the pigment dispersion stage of making a latex paint to increase the viscosity during dispersion.

The general overall flow diagram in a high-speed disperser is shown in Figure 20.5. Centrifugal force leads to flow up the sides of the tank. If the mill base is Newtonian and the dimensions and operating conditions are appropriate, the whole charge becomes intimately mixed and all portions of the mill base pass repeatedly through the zone of highest shear near the edges of the disk. However, if the mill base is shear thinning, the viscosity is high at the upper edge of the material on the sides of the tank where shear rate is low, resulting in hang up and incomplete mixing. Some mill bases contain a pigment designed to make the final coating shear thinning. The resulting hang up problem can be minimized by dispersing all of the pigments except the one that predominantly gives the shear thinning effect. After the balance of the pigments have been separated, the final pigment is slowly added. A desirable approach to the hang up problem is to design the dispersing tank with a slow-speed scraping blade that travels around the upper inside of the tank while the high-speed disk is spinning in the center of the tank.

The HSD dispersers are commonly used for premix, dispersion, and let-down operations. One initially loads the vehicle components, mixing at a low rpm, and then adds the dry pigment slowly near the shaft. After the pigment is loaded, the

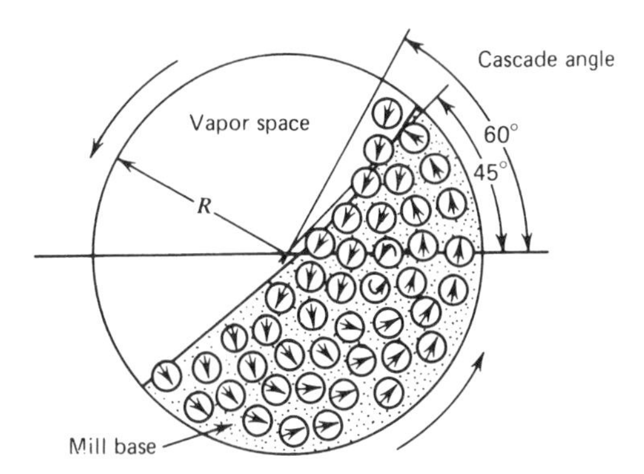

Figure 20.6. Schematic drawing of the cascading pattern in a ball mill. (From Ref. [9], with permission.)

speed is turned up to about 5000 rpm for the dispersion stage. In a properly formulated mill base, the dispersion stage requires about 15 minutes. The rpm value is reduced and the let down is carried out with the rest of the components in the formulation. In the case of latex paints, it is critical that the latex not be present during the dispersion stage since most latexes would coagulate when exposed to high shear. The latex is added in the let-down stage at low rpm.

Compared to other dispersion equipment, HSD dispersers generally have the lowest capital and operating costs. No separate premix is required, and the let down can be carried out in the same tank. Cleaning to change from one color to another is relatively easy. Solvent loss can be kept to a minimum by using a covered tank. The major limitation is that the shear stress imparted to pigment aggregates is relatively low so that the equipment can only be used for relatively easily separated pigments. However, pigment manufacturers have made substantial progress in making pigments with easily separated aggregates and HSD dispersers are probably used in making the largest share of coatings of any dispersion method. Laboratory disk dispersers are available, and results correlate well with production machines.

20.5.2. Ball Mills

A *ball* or *pebble mill* is a cylindrical container, mounted horizontally and partially filled with balls or pebbles. The mill base components are added to the mill and the mill is rotated at a rate such that the balls or pebbles are lifted up one side and then roll in a cascade to the lower side, as shown schematically in Figure 20.6. Intensive shear is imparted to pigment aggregates when the balls roll over each other with a relatively thin layer of mill base between them.

Two broad classes of such mills are used in the coatings industry. In one class, steel balls are used and the lining of the cylindrical mill is steel. These are generally called steel ball mills but also, commonly, just ball mills. In the other class, the balls are ceramic and the mill lining is porcelain. In the early days of the coatings industry, the balls were large round pebbles. These mills are commonly called pebble mills even though pebbles are no longer used and the balls are synthetic and smaller in size. One must be careful to understand how the terms are being used. Pebble mills never use steel balls. However, the term ball mill is used by some to mean either class of mill and by others to mean only steel ball mills. In this book, we use ball mill in the general sense and steel ball mill and pebble mill to specify each particular class.

Steel balls have the advantage that their density is higher and, therefore, shearing is greater and milling times can be shorter. There is always some ball wear. If one tries to disperse TiO_2 in a steel ball mill, one will obtain a gray rather than a white. (Dispersions for primers are sometimes run in steel ball mills, since they are commonly gray anyhow.) Pebble mills are used when discoloration is a disadvantage.

Ball mills operate most efficiently when their diameter is large so that the length of the cascade is long. The efficiency of operation is also very dependent on the loading of the mill. A mill should be loaded about half full of balls; this gives the longest cascade. The mill base volume should be just over enough to cover the balls when the mill is at rest. If the balls are all the same diameter spheres, the volume of balls in a half full mill is approximately 32% of the total volume of the mill. The volume occupied by the mill base when loaded for optimum efficiency is a little over 18%. If the balls are not uniform in diameter, the packing factor is higher and there is less space for mill base. If much more mill base is used than just enough to cover the balls, the time required for satisfactory separation increases.

The operation of a ball mill requires care in setting the rpm of the mill. If the rpm is too low, the balls are not carried up as high and the cascade is not as long. If the rpm is too high, the balls are carried past the 60° angle shown in Figure 20.6 and fall down when they get high enough rather than cascade. This reduces shear, can lead to ball breakage, and is more likely to lead to breaking the ultimate crystals of the pigment rather than just shearing the aggregates. If the rpm is still higher, the balls just centrifuge and little dispersion action is exerted. Experienced mill operators can tell whether the rpm setting is proper by the sound. Milling efficiency is also affected by the viscosity of the mill base. If the viscosity is too low, ball wear is high. If the viscosity is too high, the balls roll more slowly and efficiency is reduced. Optimum viscosity is dependent on ball size and density, the larger the ball size and the higher the density of the balls, the higher the optimum viscosity. Viscosities in the neighborhood of 1 Pa·s are commonly used.

The Daniel flow point method (see Section 20.2.1) was originally developed to assist in the formulation of mill bases for ball mills. It permits loading the maximum amount of pigment at the viscosity required for that mill. Proper mill base formulating and the use of proper volumes of balls and mill bases can make large differences in the time required for satisfactory separation. The time required depends on how easily the pigment aggregates are separated. The minimum time is usually on the order of 6 to 8 hours. Even difficult pigments should require no more than 24 hours. Sometimes one hears of 72 or more hours being required for dispersion. This almost always indicates poor formulation or mill loading.

Although the capital cost of ball mills is relatively high, operating cost is low. No premixing is required and the mill can run unattended. In some cases, let down is carried out in the mill, but is usually done separately. There is no volatile loss. Ball mills can be used to disperse all but the most difficult to separate pigments. They are difficult to clean and, therefore, most appropriate for making batch after batch of the same dispersion. The dispersion can be made, the batch emptied out, and the next batch started without cleaning the mill.

When the mill base is dumped at the end of a batch, a significant amount remains in the mill and must be rinsed; the rinse material is then added to the batch. This rinsing should not be done with solvent alone. If the mill base has been properly formulated with only a little higher concentration of resin solution than indicated by the minimum point on the Daniel flow point curve, the addition of solvent will flocculate the dispersion. Rinsing should be done with a resin solution at least as concentrated as that used in the mill base.

If a series of colors are to be made in the same mill, one should schedule to start with the lightest color and work toward the darkest color. This minimizes color contamination problems. Another problem is batch size. Coating manufacturers using ball mills usually have several different size mills but there is a limitation. The batch size cannot be less than covers the balls and cannot be much more than just covers the balls.

There is no directly comparable laboratory mill available. This is because the operation of a ball mill is so dependent on the diameter of the mill. Production mills range from 1.25 to 2.5 m (4–8 ft) in diameter. Some laboratories use so-called jar mills. They are usually less than 30 cm (1 ft) in diameter and are usually rolled at a much less than ideal rpm. Correlation with production operations is usually poor.

Quickee mills are more appropriate and much faster for making laboratory batches of dispersions at least roughly similar to production dispersions in a ball mill. A steel container is filled a little over half full with 30-mm steel balls and enough mill base to somewhat more than cover the balls. The container is then shaken on a paint shaker of the type used in paint stores. Easily separated pigments require 5–10 min. Difficult to separate pigments may take up to an hour of shaking. Jars with glass beads can be used in cases where discoloration from the steel balls is excessive.

20.5.3. Sand Mills

Sand mills were invented by Hochberg and developed to get around the batch size limitations of ball mills. A schematic drawing of a sand mill is shown in Figure 20.7. There is a high-speed rotor in the center, and the space between the rotor and the cylinder is partially filled with sand. A premixed mill base is pumped into the bottom of the mill. It flows up through the mill, being exposed to shear as it passes between the rapidly moving sand particles, and out through a screen that keeps the sand in the mill. Horizontal mills have also been developed. The original mills were charged with sand with an average diameter of about 0.7 mm. While sand is still used, small ceramic balls are used more frequently. Small steel balls are also used; then the mill is commonly called a *shot mill* because the steel balls originally used were the shot used in shotgun shells. Shot mills are generally used to disperse carbon black or other dark pigments.

A typical sand mill operates with a peripheral speed of the rotor of 10 m s^{-1}. This impels the sand particles at high speed so that even though their size is small, the shear rate between them is high and, of course, in passing through the mill, a pigment aggregate passes between many pairs of sand particles. Mill bases should be formulated using slightly higher resin concentrations than the minimum in the

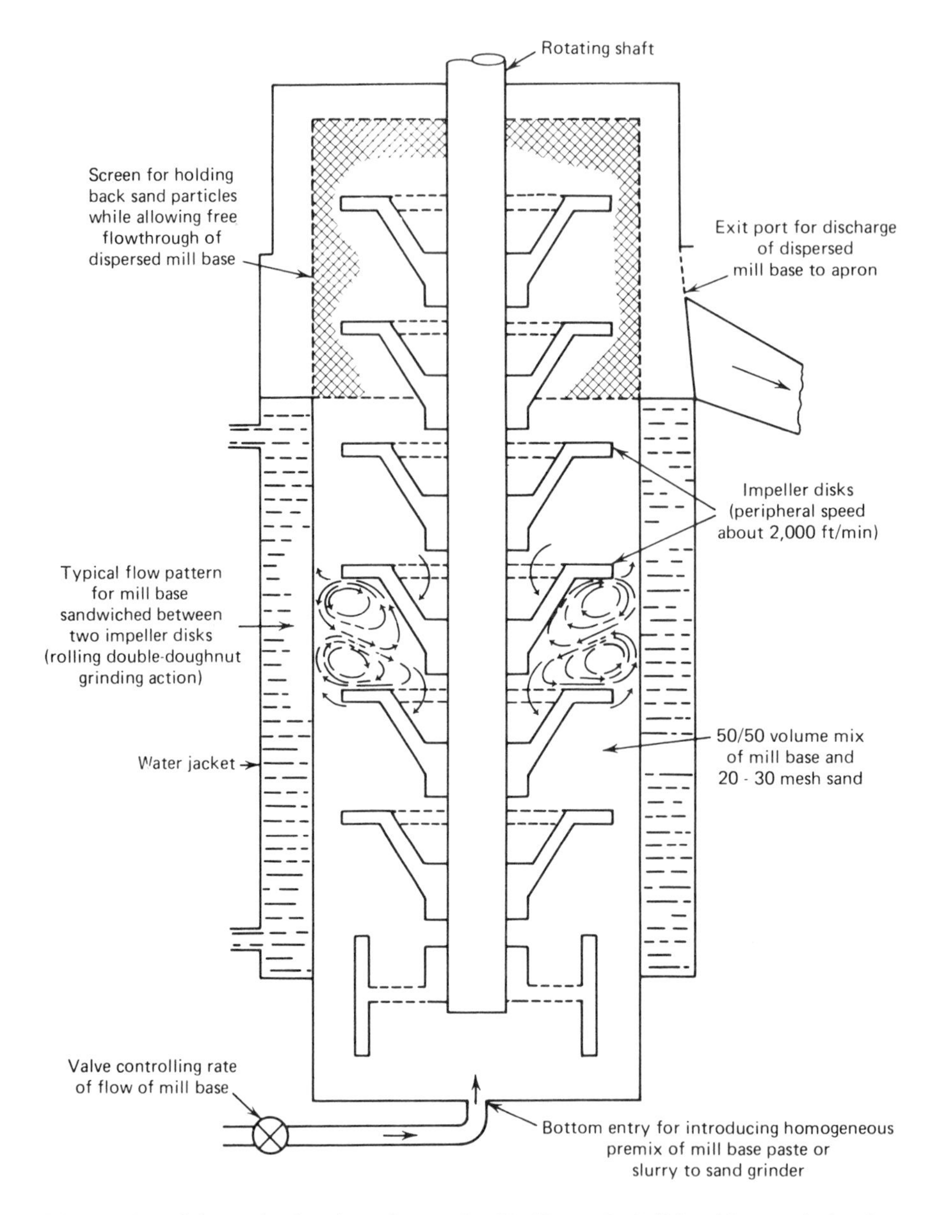

Figure 20.7. Schematic drawing of a sand mill. (From Ref. [9], with permission.)

Daniel flow point curve. The higher the viscosity of the mill base, the longer the average residence time in the mill and, hence, the greater the degree of shear. For easily separated pigments, it is desirable to use lower viscosity to achieve greater flow rate through the mill. The range of viscosities used is 0.3–1.5 Pa·s. For more difficult to separate pigments, it is sometimes necessary to pass through the mill (or a battery of mills in series) two or three times.

The capital investment and operating costs of the mills are relatively low. Premixing is required and, of course, the let down must be done separately. Batch

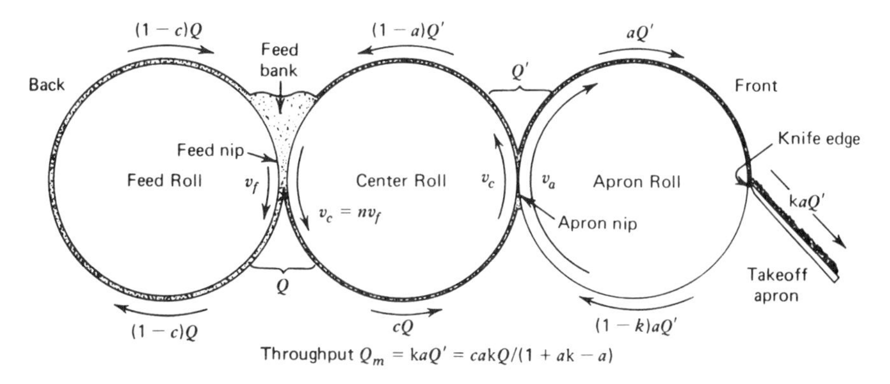

Figure 20.8. Schematic diagram of a three-roll mill. (From Ref. [9], with permission.)

size is flexible. Cleaning is a problem; generally, mills are reserved for similar colors to minimize cleaning. Sand mills are effective at separating pigment aggregates provided the aggregates are small compared to the sand particles. Laboratory sand mills are available that give quite good correlation with production mills. Generally instead of pumping in the bottom of the mill, they are designed to simply pour the premix into the top of the mill. In terms of coatings volume, the use of sand mills, and related equipment, is probably second only to HSD dispersers.

20.5.4. Three-Roll and Two-Roll Mills

Although formerly widely used in making dispersions for coatings, the use of *three-roll mills* is much more limited today. Figure 20.8 shows a schematic drawing of a three-roll mill. Aggregate separation results from the shear developed as the mill base passes through the nip between each pair of rolls. The viscosity of the mill bases is higher than the other methods discussed thus far—5 to 10 Pa·s or even higher. Since the mill base is exposed on the rolls, solvents must have low vapor pressures to minimize evaporation.

Three-roll mills have comparatively high capital and operating costs. Premixes are required and let down is a separate operation. Skilled operators are required. On the positive side, shear rates are relatively high so that difficult to separate pigment aggregates can be processed, batch size is versatile, and cleanup is relatively simple. Production usage today is limited almost entirely to small batches of dispersions with no solvent, or only low-volatility solvent, where one must operate at high viscosity. Three-roll mills are convenient for laboratory use.

Two-roll mills exert even greater shear rates than three-roll mills. They are usually used with solvent-free high molecular weight polymer/pigment systems. They can separate even the most difficult to separate aggregates. They are most commonly used for dispersing very expensive pigments. The high capital and operating costs are justified when the economic value of achieving the last 10–20% of potential color yield means substantial difference in product cost. Another use is for dispersing certain carbon black pigments when the desired "jetness" is only attained with virtually complete separation. Two-roll mills are particularly appropriate when the dispersion is to be used in a transparent coating. This means that

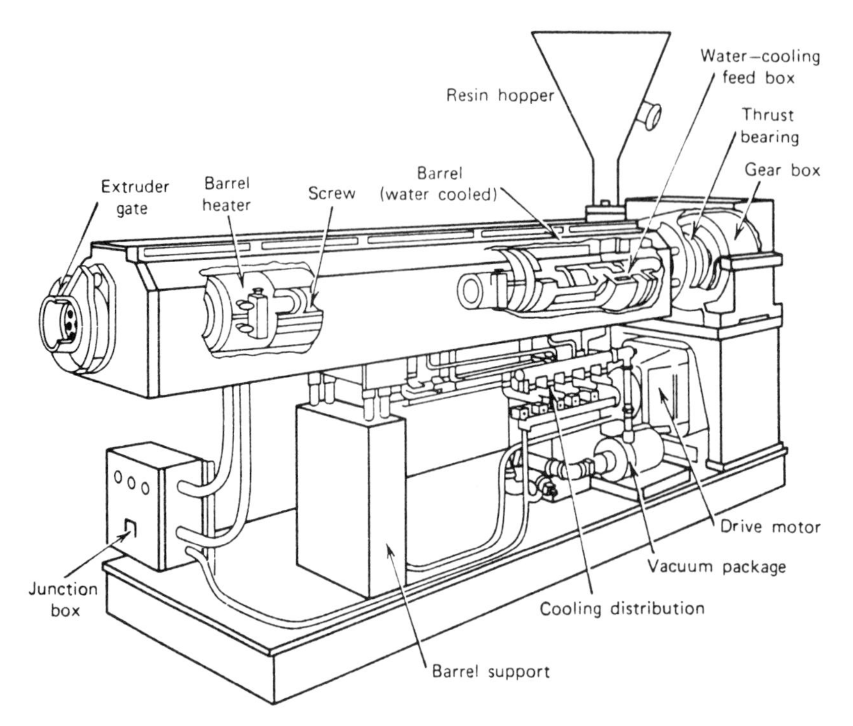

Figure 20.9. Sectional view of a typical extruder. (From Ref. [21], with permission.)

essentially all of the pigment aggregates should be broken down to ultimate particle size to eliminate (or at least minimize) light scattering.

Two-roll mills have very high capital and operating costs. Volatile loss is complete. Premixes are required and further processing of the dispersion is required to make a liquid dispersion that can be incorporated into a coating. Few coatings companies operate two-roll mills; most purchase pigment dispersions of the type for which two-roll mills are appropriate from companies that specialize in making pigment dispersions.

20.5.5. Extruders

Extruders are being increasingly used for pigment dispersion. Pigment dispersion for almost all powder coatings is done in extruders (see Section 31.4.11); also, more and more high-viscosity liquid dispersions are being processed in extruders. An extruder has a screw feeding through a cylinder and pressing the material out the end through a die. The barrel of the extruder can be operated in a wide temperature range. The shear impressed in the screw is very high and any pigment aggregates can be broken down in an extruder. The most difficult to disperse pigments require the longest residence times. In the case of solids, the product is usually chopped into small pieces and then pulverized either for use as a powder or for ease of dissolving. Figure 20.9 shows a schematic diagram of an extruder.

20.6. EVALUATION OF DEGREE OF DISPERSION

Assessment of *degree of dispersion* is a critical need for establishing original formulations and optimizing processing methods as well as for quality control. Generally speaking, the coatings industry does a poor job in this critical evaluation. Differences in degree of dispersion come from two factors: incompleteness of separation of the original aggregates into individual crystals and flocculation after separation.

For white and colored pigments, the most effective evaluation method is by determination of *tinting strength* in comparison to a standard. In the case of a batch of white dispersion, one weighs out a small sample of the white dispersion and then carefully mixes into it a small weighed amount of a standard color dispersion, say blue for the sake of illustration. At the same time, one weighs out a standard white sample with exactly the same ratio of the same blue standard. After thorough mixing, one puts a small amount of each tint mixture adjacent to each other on a piece of white paper and draws down both samples with a stiff flat-ended spatula so that the edges of the two samples touch each other. One can then compare the color of the two draw downs. If the batch is a darker blue than the standard, the tinting strength of the batch of white is low, meaning it is not equally dispersed. To test a color dispersion, say blue for illustration, one would carry out the same procedure except one would use the standard white in both samples and the standard blue in one and the batch of blue in the other.

One can check for flocculation by pouring some of the tint mix made for the draw down onto a piece of tin plate. Immediately after making the pour down, one rubs the wet coating with a forefinger. If the color changes, the dispersion is flocculated. For example, if the mix of blue with white becomes bluer where it is rubbed, the blue pigment dispersion is flocculated. If, on the other hand, the rubbed portion becomes a lighter blue, the white is flocculated.

Flocculation can also be detected by examining the flow of the dispersion. Well-stabilized dispersions have Newtonian flow properties. If a dispersion is shear thinning (and does not contain a component designed to make it shear thinning), it is flocculated. The degree of shear thinning is proportional to the extent of flocculation.

A further method of assessing pigment dispersion is by settling or centrifugation experiments. The rate of settling is governed by the particle size and the difference in density of the dispersed phase from the medium. A well-separated, well-stabilized dispersion settles or centrifuges slowly, but when settling is complete the layer thickness is small. A well-separated but poorly stabilized dispersion settles quickly to a bulky sediment. The floccules settle more quickly because of their large size and they form a bulky sediment because continuous phase is trapped within the floccules. Flocculated sediment is also readily stirred or shaken back to a uniform suspension in contrast to the sediment formed from the nonflocculated dispersion. If the pigment settles (or centrifuges) relatively quickly to a compact layer, the separation step is incomplete. The larger aggregates settle more quickly but the volume is not large because less internal phase is trapped in the sediment than in a flocculated system.

Settling or centrifuging tests can provide qualitative or semiquantitative information which is sufficient for development work and quality control purposes. For

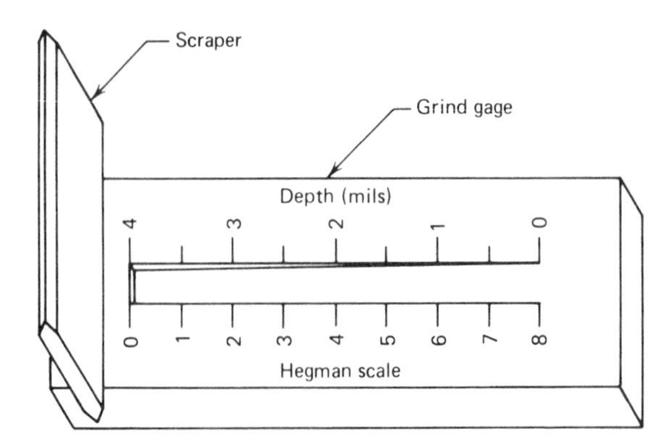

Figure 20.10. Sketch of a grind gauge and scraper for measuring "fineness of grind" of pigment dispersions. (From Ref. [9], with permission.)

research purposes, it may be important to determine more quantitative data, for example, the extent of flocculation. The degree of flocculation can be calculated from rates of centrifugation [5].

One can also examine the dispersion microscopically. One must use caution in preparing the samples for examination by microscopy. In general, it is necessary to dilute the sample. If the sample is diluted with solvent, there is a high probability of flocculation. In such a case, one may report that the dispersion is flocculated when actually the flocculation was a result of sample preparation. Electron microscopic studies of the surfaces of etched dry coating films can be useful in assessing variations in dispersion [22].

A technique called the *flocculation gradient technique* is probably the most rapid and accurate method for quantitative study of the degree of pigment dispersion in both liquid coatings and dry films. The method was originally developed by Balfour and Hird [23] to study systems containing TiO_2 pigments. The amount of 2500-nm infrared radiation scattered by a film as a function of film thickness is measured. (Particles, generally, scatter longer wavelength radiation more effectively than they do visible light.) There must be a significant difference in the refractive index at 2500 nm between the pigment and the vehicle. A plot of backscatter against film thickness gives a straight line whose gradient increases with increasing flocculation. The technique has proven useful in evaluating flocculation of a variety of pigments [24]. Infrared scattering measurements are also useful in determining the degree of flocculation in liquid dispersion samples [24]. Other examples of the use of infrared back scattering measurements are given in Refs. [8] and [25].

Positronium annihilation studies of free volume in pigment dispersions show promise of being the basis for a sensitive measurement of the degree of dispersion [26].

The most widely used method of testing for *fineness of grind* in the coatings industry is the use of a draw down gauge, commonly the *Hegman gauge*. A sketch of one such gauge is shown in Figure 20.10. A sample of the dispersion to be tested is placed on the steel block before the zero reading and then drawn down by the steel bar scraper. One then lifts the block up and quickly looks across the drawn down sample to see at which graduation one can start to see particles projecting or streaks caused by particles being dragged along. It is said that the higher the scale reading, the "better" the dispersion.

Unfortunately, the device is not capable of measuring degree of dispersion. First, we should remember that the major problem in making satisfactory dispersions is avoiding flocculation. However, the gauge cannot detect flocculation at all, since the drawing down step breaks up any floccules. Next, the particle sizes of properly dispersed pigments are small compared to the depth of the groove on the gauge. The depth on some gauges ranges from 0 to 10 mil (250 μm) in graduation units of 1.25 mil (approx. 30 μm). The one shown in Figure 20.10 ranges from 4 mil (100 μm) to 0 in steps of 0.5 mil (12.5 μm). The TiO_2 pigment particles have an average size of about 0.23 μm, about 2 orders of magnitude smaller than the graduation steps of the gauge. Even aggregates with 10^4-10^5 particles could escape detection. Many color pigment particles are even smaller and carbon black particles can be as small as 5 nm. Some inert pigment particles are as large as 2 or 3 μm, still an order of magnitude smaller than the groove depths. Obviously, the gauge cannot test whether or not all or most of the particles are less than the groove depths. Blakely has demonstrated in the case of TiO_2 dispersions that only approximately 0.1% of the total pigmentation of a paint system was responsible for an unacceptable fineness of grind rating [27].

Why are such devices used? They do give some idea of whether the big aggregates are getting broken up. They do give some idea of the presence of dirt particles. The determinations are fast, taking about a half minute. A tinting strength determination by an experienced person requires 2 or 3 minutes; but, after spending this time, one has an assessment of the dispersion instead of an essentially meaningless number. In the 19th century when very coarse, difficult to separate, pigments were all that were available, the gauge might have had some value. There will be no excuse for using it in the 21st century.

GENERAL REFERENCES

G. D. Parfitt, *Dispersions of Powders in Liquids*, 3rd ed., Applied Science Publishers, London, 1981.

T. C. Patton, *Paint Flow and Pigment Dispersion*, 2nd ed., Wiley-Interscience, New York, 1979.

REFERENCES

1. J. Winkler, E. Klinke, and L. Dulog, *J. Coat. Technol.*, **59** (754), 35 (1987); J. Winkler, E. Klinke, M. N. Sathyanarayana, and L. Dulog, *J. Coat. Technol.*, **59** (754), 45 (1987); J. Winkler and L. Dulog, *J. Coat. Technol.*, 59 (754), 55 (1987).
2. K. Rehacek, *Ind. Eng. Chem., Prod. Res. Dev.*, **15**, 75 (1976).
3. A. Saarnak, *J. Oil Col. Chem. Assoc.*, **62**, 455 (1979).
4. L. Dulog and O. Schnitz, *Proc. XVIIth FATIPEC Congress*, Vol. II, 409 (1984).
5. R. B. McKay, *Proc. VIth Int. Conf. Org. Coat. Technol.*, 1980, Athens, p. 499.
6. J. Lara and H. P. Schreiber, *J. Coat. Technol.*, **63** (801), 81 (1991).
7. A. Brisson and A. Haber, *J. Coat. Technol.*, **63** (794), 59 (1991).
8. S. Dasgupta, *Prog. Org. Coat.*, **19**, 123 (1991).
9. T. C. Patton, *Paint Flow and Pigment Dispersion*, 2nd ed., Wiley-Interscience, New York, 1979.
10. T. K. Hay, *J. Paint Technol.*, **46** (591), 44 (1974).

11. H. L. Jakubauskas, *J. Coat. Technol.*, **58** (736), 71 (1986).
12. J. A. Sims and H. J. Spinelli, *J. Coat. Technol.*, **59** (752), 125 (1987).
13. J. D. Schofield and J. Toole, *Polym. Paint Colour J.*, Dec 10/24, 170 (1980).
14. J. D. Schofield, *J. Oil Colour Chem. Assoc.*, **74**, 204 (1991).
15. L. W. Hill and Z. W. Wicks, Jr., *Prog. Org. Coat.*, **10**, 55 (1982).
16. K. Hamann and R. Laible, *Proc. XIV FATIPEC Congress*, 17 (1978).
17. J. Schroeder, *Prog. Org. Coat.*, **16**, 3 (1988).
18. W. H. Morrison, Jr., *J. Coat. Technol.*, **57** (721), 55 (1985).
19. T. Losoi, *J. Coat. Technol.*, **61** (776), 57 (1989).
20. R. E. Smith, *J. Coat. Technol.*, **60** (761), 61 (1988).
21. J. Y. Oldshue and D. B. Todd, *Kirk-Othmer Encyclopedia of Chemical Technology*, 3rd ed., Vol. 15, Wiley, New York, 1985, p. 635.
22. A. Brisson, G. L'Esperance, and M. Caron, *J. Coat. Technol.*, **63** (801), 111 (1991).
23. J. G. Balfour and M. J. Hird, *J. Oil Colour Chem. Assoc.*, **58**, 331 (1975).
24. J. E. Hall, R. Benoit, R. Bordeleau, and R. Rowland, *J. Coat. Technol.*, **60** (756), 49 (1988).
25. J. E. Hall, R. Bordeleau, and A. Brisson, *J. Coat. Technol.*, **61** (770), 73 (1989).
26. B. Mayo, J. P. Pfau, and R. E. Sharpe, *J. Coat. Technol.*, **59** (750), 23 (1987).
27. R. R. Blakely, *Proc. XIth FATIPEC Congress*, 187 (1972).

CHAPTER XXI

Pigment Volume Relationships

Traditionally, coatings formulators have worked with weight relationships, but volume relationships are generally of more fundamental importance and practical significance. While there had been a few isolated earlier examples of recognition of the importance of volume considerations in the performance of coatings, credit for the full realization of this importance belongs to W. K. Asbeck and M. Van Loo [1]. They looked at a series of performance variables as a function of the *pigment volume concentration* (PVC), that is, the volume percent of pigment in a *dry* paint film. (Some authors call PVC pigment volume content.) The term PVC should never be used to specify the volume of pigment in a wet paint film. This has been done occasionally in the literature and has been responsible for serious misinterpretations. Although PVC is universally expressed as a percentage value, there are some authors who express PVC as a pigment volume fraction in equations instead of a percent without bothering to tell the reader.

Asbeck and Van Loo observed that many properties of paint films changed abruptly at some PVC as the PVC was increased in a series of formulations. They designated the PVC where these changes occurred, the *critical pigment volume concentration*, CPVC. They also defined CPVC as that PVC where there is just sufficient binder to provide a complete adsorbed layer on the pigment surfaces and to fill all the interstices between the particles in a close-packed system. Below CPVC, the pigment particles are not close packed and binder occupies the "excess" volume in the film. Above CPVC, the pigment particles are close packed, but there is not enough binder to occupy all of the volume between the particles; in other words, above CPVC there are voids in the film. Slightly above CPVC, these voids are air bubbles in the film; but as PVC increases, the voids interconnect and film porosity increases sharply.

When films are prepared from coatings with PVC near to CPVC, there may well not be uniform distribution of pigment through the dry film, so that some parts of the film may locally be above CPVC and other parts below CPVC. Also some properties start to change as soon as PVC increases to the extent that there are air voids in films and others change when the PVC is sufficiently greater than CPVC that the film begins to be porous. Coatings with flocculated pigment clusters result in films with nonuniform distribution of pigment particles and, therefore, the CPVC of films prepared from coatings with flocculated pigment particles is

MECHANICAL PROPERTIES

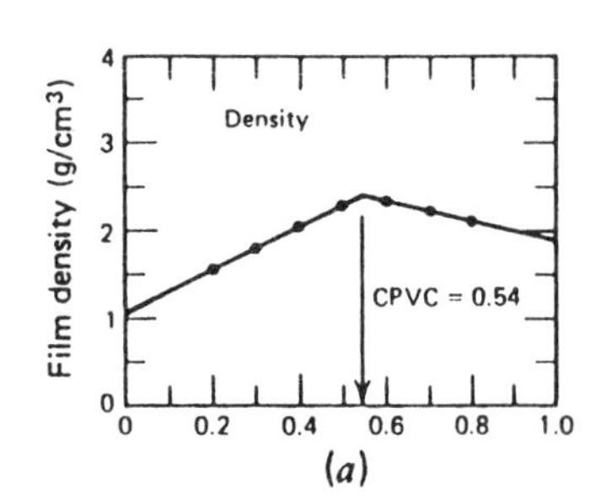

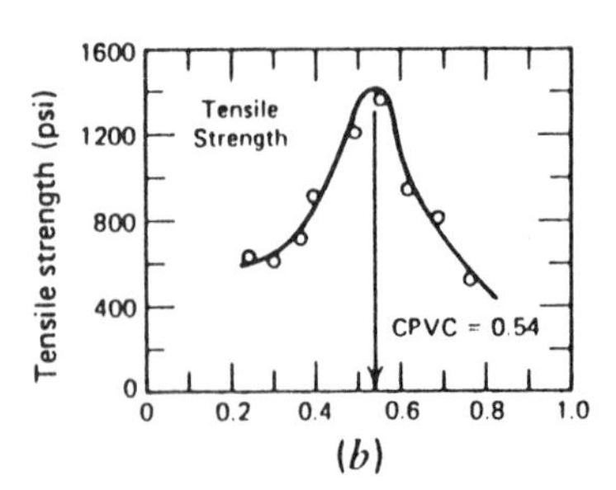

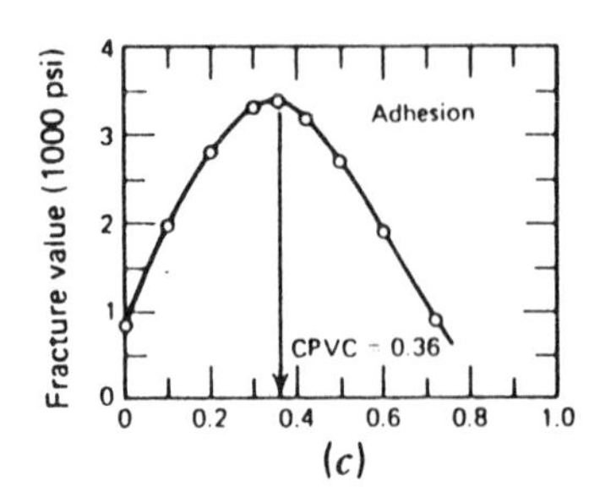

Figure 21.1. Relationship between PVC and (*a*) density, (*b*) tensile strength, and (*c*) "adhesion." (From Ref. [2], with permission.)

lower than the CPVC of films with the same pigment combination that are not flocculated.

The CPVCs of different pigments vary over a wide range, at least 18–68 (see Section 21.2 for a discussion of the factors affecting CPVC). The precision and accuracy of determinations of CPVC can vary substantially depending on the method of determination (see Section 21.3 for discussion of determination of CPVC).

21.1. RELATIONSHIPS BETWEEN FILM PROPERTIES AND PVC

As PVC is increased in a series of coatings made with the same pigments and binders, the density increases to a maximum when PVC equals CPVC and then decreases, as shown schematically in Figure 21.1*a*. The increase reflects that pigments (with very few exceptions) have higher densities than binders. However, above CPVC the lower density of air reduces the film density.

As shown schematically in Figure 21.1*b*, tensile strength of a series of films generally increases with PVC to a maximum at CPVC, but then decreases above CPVC. Below CPVC the pigment particles serve as reinforcing particles and increase the strength. It can be considered that polymer molecules adsorb on the surface of multiple pigment particles providing the equivalent of cross-links. Accordingly, more force is required to break this physical network with increasing pigment levels. However, above CPVC, the air voids weaken the film. Generally, abrasion and scrub resistance of films also drop as PVC is increased beyond CPVC.

Plots of adhesion as a function of PVC are sometimes reported as in Figure 21.1*c*. It appears that adhesion goes through a maximum at CPVC, but the relationship may not be that simple. In adhesion testing, failure can be cohesive or adhesive (see Section 26.6). In cohesive failure the coating film breaks; in adhesive failure there is separation at the interface between coating and substrate. In many, if not all, cases where results such as shown in Figure 21.1*c* are obtained, the failures at low PVCs are cohesive failures, as are those above CPVC; only those near CPVC, if any, are adhesive failures. Such results merely reflect the changes in the cohesive tensile strength of the film, not changes in adhesion between the substrate and the coating. If the cohesive strength of a film is lower than its adhesive strength, film failure does not provide any information about adhesive strength.

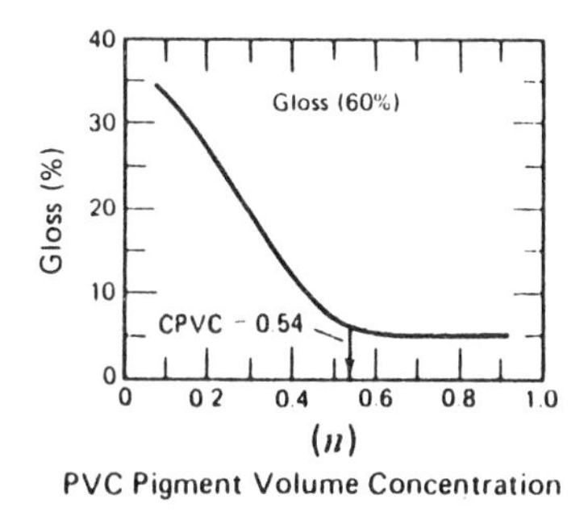

Figure 21.2. Relationship between gloss and PVC. (From Ref. [2], with permission.)

Stain resistance decreases sharply above CPVC since staining liquids can penetrate into the pores leaving color behind that is difficult to remove. Porosity also affects other properties. If one applies a single coat of a paint with PVC above CPVC to steel and then exposes the panel to humidity, rapid rusting can occur since the pores permit water and oxygen to get to the surface of the steel with little interference. On the other hand, if one applies an alkyd-based paint with PVC above CPVC to a wood substrate, one is less likely to get blistering than with a similar paint with PVC below CPVC. When water gets into the wood behind the paint, it can escape through the pores of the alkyd paint when PVC is above CPVC but not when the PVC is below CPVC.

Gloss is related to PVC. In general, unpigmented films have very high gloss. The initial few percent of pigment has little effect on gloss; but above about 6–9 PVC, gloss drops until PVC approaches CPVC (see Section 17.3). Figure 21.2 shows a typical relationship between gloss and PVC.

It is almost always desirable to make primers with a high PVC since the rougher, low gloss surface leads to better intercoat adhesion than a smooth, glossy surface. It is sometimes desirable to design a primer with PVC greater than CPVC. Adhesion of a top coat to such a primer is enhanced by mechanical interlocking resulting from penetration of vehicle from the top coat into the pores of the primer. Presumably many of the pores in the primer are filled with binder from the top coat, which decreases the PVC of the top coat. If the PVC of the primer is much above the CPVC, there can be an undesirable loss of gloss of the top coat; the primer is said to have poor *enamel hold out*. The term means that the gloss of an enamel applied over such a coating will be lower than when applied over a nonporous substrate.

Hiding is also affected by PVC; as pigmentation increases, generally hiding increases. Initially hiding increases rapidly, but then tends to level off. In the case of rutile TiO_2, as shown in Figure 16.9, hiding goes through a maximum, gradually decreases with further increase in PVC, and then increases sharply above CPVC. This increase in hiding above CPVC results from the air voids left in the film when PVC is above CPVC. The refractive index of air (1.0) is less than that of the binder (approximately 1.5) so that there is light scattering by the air interfaces in addition to interfaces between the pigment and the binder. The effect becomes very large as the interfaces between air and pigment increase with increasing PVC. For example, if rutile TiO_2, with a refractive index of 2.76, is in the formulation, the

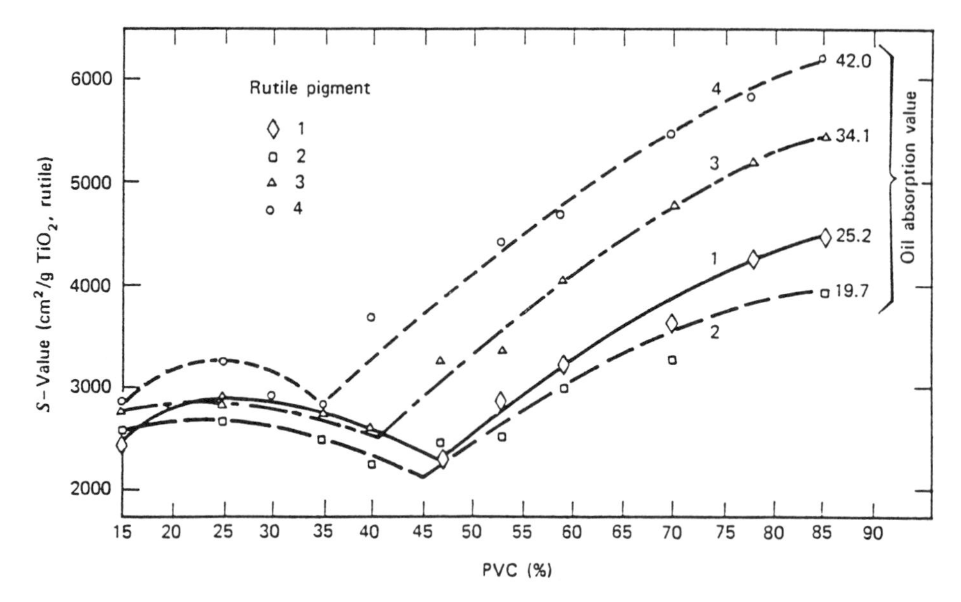

Figure 21.3. Scattering coefficient as a function of PVC of flat paints made with different oil absorption grades of rutile TiO_2 and Lorite inert pigment. (From Ref. [3], with permission.)

refractive index difference between TiO_2 and air (1.76) is much larger than between TiO_2 and binder (1.26).

Owing to the high cost of TiO_2, paints are not generally formulated with a PVC of TiO_2 greater than about 18% since incremental hiding at higher PVC is not cost efficient. (This value is dependent on the actual TiO_2 content of the TiO_2 pigment and on the stability of the TiO_2 dispersion.) Rather than using large quantities of TiO_2 in high PVC coatings, lower-cost inert pigments are used to occupy the additional volume. The effect is illustrated in Figure 21.3. The scattering efficiency versus PVC data are based on a series of paints made with four TiO_2 pigments having differing oil absorption values and a clay pigment (Lorite). Maximums in scattering efficiencies, below CPVC, occur at about 25 PVC. Above CPVC, there is a sharp increase in scattering coefficient and hence hiding. Note that the CPVC generally increases as the oil absorption of the TiO_2 decreases, but that scattering efficiency increases with oil absorption. These changes presumably reflect decreasing particle size as oil absorption increases.

The scattering efficiency of TiO_2 is also affected by the particle size of inert pigments used with it [4]. Inclusion of some inert pigment with a particle size smaller than that of TiO_2 (i.e., less than 0.2 μm) increases the efficiency by acting as a so-called *spacer* for the TiO_2 particles.

The increase in hiding above CPVC can be useful. For example, the hiding of white ceiling paints can be substantially improved by formulating to PVC above CPVC. This permits hiding with one coat, which is particularly desirable in ceiling paints. The stain resistance and scrub resistance of the paint are inferior to similar paints with PVC less than CPVC, but stain and scrub resistance are not important in ceiling paints. Of course, the high PVC paints have the further advantage that their cost is lower.

Tinting strengths of white paints also increase significantly as the PVC of a series of paints is increased beyond CPVC. The air voids above CPVC increase light scattering so that a colored paint dries with a lighter color than one with the same amount of color pigment but with PVC below CPVC.

As demonstrated above, changes in volume relationships almost always control the physical properties of films. Thus, there is a major advantage in formulating coatings on the basis of volume rather than weight relationships. For any particular application there is a ratio of PVC/CPVC most appropriate for the combination of properties needed for that application. Once this ratio has been established, changes in pigment combinations for that application should be made such that this PVC/CPVC ratio is maintained. This important concept is developed in more detail by Bierwagen [5]. He also emphasizes that one must exercise particular care when formulating with PVC near CPVC since relatively small changes in pigment ratios or packing, flocculation, or film formation could affect film properties substantially [6].

While it would be desirable to have accurate determinations of CPVC available, even reasonable estimates can be useful since they permit one to focus a series of experiments within a range approximating the desired PVC/CPVC ratio. The decision as to the proper pigment loading of a particular formulation should be based on actual experimental data, not on the theoretical best PVC value. Even without CPVC data, the concept is of great value to the formulator. It is critical to recognize that performance properties vary with volume not with weight relationships and that CPVC increases with increasing particle size distribution. Qualitative use of oil absorption and density values for individual pigments, together with the recognition that CPVC increases when using mixtures of pigments with different particle size distributions, permit one to start formulating in a reasonable range of compositions. On the other hand, when using weight relationships one is working blindly.

21.2. FACTORS CONTROLLING CPVC

There are large variations in CPVC depending on what pigment or pigment combination is used in the coating, and the extent, if any, of pigment flocculation.

With the same pigment composition, the smaller the particle size, the lower is the CPVC. The ratio of surface area to volume of small particle size pigments is greater than that of larger particle size pigments. Hence, a higher fraction of binder is adsorbed on the surface of the smaller pigment particles so the volume of pigment in a close-packed final film is smaller. One can consider this effect from the standpoint of a modification of the Mooney equation (Eq. 21.1). The volume fraction of internal phase is the sum of the volume of pigment, V_p, and the volume of the adsorbed layer, V_a. In a solvent-free system, V_p equals PVC and V_p is equal to CPVC when $V_p + V_a = \phi$. Thus, if V_a is larger, the V_p corresponding to CPVC is smaller. Note that at CPVC, viscosity of a solvent-free system approaches infinity.

$$\ln \eta = \ln \eta_e + \frac{K_E(V_p + V_a)}{1 - \dfrac{(V_p + V_a)}{\phi}} \tag{21.1}$$

The CPVC depends on particle size distribution; the broader the distribution, the higher the CPVC. As discussed in Section 19.5, broader particle size distribution of spherical, dispersed-phase systems increases packing factors. In low-gloss coatings, the least expensive component of the dry film of the coating is the inert pigment; in order to minimize cost, it is desirable to maximize the inert pigment content. Since the properties should remain constant, one does not want to change the PVC/CPVC ratio; the lowest cost systems, therefore, are those with the highest CPVC. This means minimizing the amount of very fine particle size pigment but at the same time maximizing the particle size distribution. Obviously, compromises are needed, but broad particle size distribution is advantageous.

Pigment dispersion affects CPVC; CPVCs of films from paints in which the pigment is flocculated are lower than CPVCs from corresponding paints with nonflocculated pigment dispersions. Films prepared from coatings with flocculated pigment clusters have less uniform distribution of pigment and hence are more likely to have portions where there are local high concentrations of pigment. Solvent containing resin is trapped inside these clusters of pigment particles. When the coating dries, the solvent diffuses out of the resin solution trapped in the floccules, in some cases leaving behind insufficient binder to fill the spaces. The CPVC is defined as the PVC above which voids are present in a film; there are likely to be voids in films from a flocculated system at a lower PVC than in a nonflocculated system. In one example, it is reported [7] that CPVC decreased from 43 to 28 with increasing flocculation. Asbeck [7] suggests using the term UCPVC, *ultimate CPVC*, to describe the CPVC with a nonflocculated pigment combination. It seems to us better to recognize that flocculation is one of the many variables that can affect CPVC, especially since it may be difficult to determine UCPVC experimentally.

Even beyond the possible effect of the vehicle of the coating on flocculation, one might expect CPVC to be dependent on the binder composition as it affects the thickness of adsorbed layers. Thicker adsorbed layers might be expected to lead to lower CPVC. Actually, CPVC with a given pigment or pigment combination seems to be essentially independent of the binder composition (except in latex paints, which are discussed in Section 21.4). It may be that the differences are too small to detect with the relatively imprecise methods for determining CPVC. It may also be that the shrinkage forces encountered when films are formed press the particles against each other to the extent that only some minimal layer of binder is left between particles regardless of the original adsorbed layer thickness. No basic studies of the phenomenon have been reported.

21.3. DETERMINATION OF CPVC

Critical pigment volume concentration has been determined by many different procedures [5,8]. In many cases the precision, that is, the reproducibility of the value, is relatively poor and the accuracy, that is, nearness to the "true" value, is sometimes also poor or indeterminate. Bierwagen [6] has emphasized that in the formation of films of coatings with PVC at or near CPVC, local fluctuations within a film of volume fractions of binder and pigment are possible. Thus, there could be parts of a film with PVC > CPVC while the average composition might be less than CPVC. In view of these uncertainties, one must be careful in assessing the importance of small differences in PVC and CPVC values.

Many changes in film properties have been used as a means of determining CPVC, however, the one generally considered to be the most accurate is tinting strength. A series of white paints with increasing PVC are prepared and tinted with the same ratio of color to white pigment. At CPVC there is a relative abrupt increase in the white tinting strength of the paint resulting from the "white" air bubbles above CPVC. The technique is most easily applicable to white paints but can be applied to colored paints.

The CPVC can be determined by filtering a coating and measuring the volume of the pigment filter cake. Asbeck [7] recommends a specially designed filter that he calls a *CPVC cell.*

The CPVC for any pigment or pigment combination can be calculated from oil absorption (OA), as shown in Eq. 22.2, provided the OA value is based on either a nonflocculated dispersion or determined at a sufficiently high shear rate that any floccules are separated. The definitions of both OA and CPVC are based on close-packed systems with just sufficient binder to adsorb on the pigment surfaces and fill the interstices between the pigment particles. Oil absorption is expressed as grams of linseed oil per 100 g pigment, CPVC is expressed as milliliters of pigment per 100 mL of film; ρ = density of the pigment(s) and 93.5 is 100 times the density of linseed oil (note that both OA and CPVC are expressed as percentages, not as fractions).

$$\text{CPVC} = \frac{1}{1 + \dfrac{(\text{OA})(\rho)}{93.5}} \tag{21.2}$$

The significance of the interrelationship depends on the observation that OA and CPVC are approximately independent of the binder, provided the pigment particles are not flocculated. This independence on binder seems surprising; nevertheless, from the limited data available, it appears to be a reasonable approximation (except for latex paints). Since the accuracy of calculated CPVCs depend on the accuracy of OA determinations, OA values determined by using a mixing rheometer, such as a Brabender Plastometer (see Sections 19.3.4 and 20.2.2), are preferable to values determined by the spatula rub-up method [9]. Such an intensive mixer provides greater shear rate for separation of pigment aggregates than spatula rubbing and also the data points are taken while the dispersion is under high shear so that, even if the binder does not stabilize the pigment against flocculation, the volume fractions represent a nonflocculated system.

Effects of variation in the procedures for determining OA, including the use of other liquids besides linseed oil, are reported in Ref. [10]. The discussion of OA determination in this paper is interesting and useful; however, the conclusions drawn by the author about CPVC are erroneous. The author failed to recognize that when the pigment dispersions were diluted with solvent, the pigment flocculated.

The accuracy of the CPVC value obtained depends on the accuracy of the OA values. Since many OA values are not accurate, Asbeck [7] recommends against such calculations but many workers have found the calculations useful.

Although the CPVC for individual pigments and for specific combinations of pigments can be calculated from oil absorptions, the CPVC values of pigment combinations cannot be calculated from these values alone, since the differences in particle size distribution with pigment combinations affects the packing factor. The OA values for each combination of pigments must be determined experimentally. A variety of equations have been developed to calculate CPVC from data on individual pigments [8]. The most successful use OA values, densities, and average particle sizes of the individual pigments [11,12]. The equations assume that the particles are spheres, a fair assumption for many but not all pigments. It was shown that the calculated values correspond reasonably well to experimentally determined CPVC values.

21.4. CPVC OF LATEX PAINTS

There is considerable controversy about the applicability of CPVC to latex paints. Some maintain that one cannot apply the concept. Others maintain that there is no fundamental difference in the CPVC concept between solvent-borne and latex paints. Bierwagen [13] emphasizes that experimental errors in determining the CPVC of latex paints are even greater than with solvent-borne paints so that one must use caution in reaching conclusions. Many workers in the field, however, maintain that the concept is useful in latex paints but that CPVC must be thought of quite differently in latex paints than in solvent-based paints.

A study of the effect of PVC on the hiding of latex paints led to the conclusion that the increase in hiding at CPVC is a convenient method of determining the CPVC of latex paints [14]. The authors concluded also that the CPVC was lower in latex paint than in solvent-based paint. Patton [15] has recommended that the term *latex CPVC*, LCPVC, be used to distinguish from CPVC in solvent-borne paints.

It has been found that LCPVC is smaller than the CPVC of a solvent-borne paint with the same pigment combination. Furthermore, although CPVC is approximately independent of the binder in solvent-borne paints, LCPVC varies with the latex and some other components of latex paints. It has been found that LCPVC increases as the particle size of the latex decreases. Also, LCPVC increases as the T_g of the latex polymer decreases, and a coalescing agent increases LCPVC. Since LCPVC is smaller than CPVC, the ratio of volume percent of binder in the film of a solvent based paint, V_s, to that of a latex based paint, V_l will always be less than 1 with the same pigment combination. This ratio (Eq. 21.3), has been called the *binder index*, e [14].

$$e = \frac{V_s}{V_l} \tag{21.3}$$

It has been proposed, and demonstrated with a limited number of examples, that this ratio is independent of the pigment combination. It follows that, if one knows the binder index for some latex, one can calculate the LCPVC for paints made with that latex and other pigments from the CPVC values calculated from oil absorptions (or, for mixtures, by calculation from OA, density, and particle size data).

The difference in CPVC between latex and solvent-based paints follows from the difference in film formation. For the sake of comparison, let us consider a highly idealized pair of systems. In the solvent-based paint, let us assume that all the pigment particles are spheres with the same diameter and that the CPVC is 50. In the latex paint, we will use the same pigment along with a latex whose particles are all spheres with the same diameter as that of the pigment particles.

In the solvent-based paint, there is a layer of resin swollen with solvent on the surface of all the pigment particles. Ideally, when a solvent-borne paint with PVC = CPVC is applied, the solvent evaporates and the "resin-coated" pigment particles arrange themselves in a random close-packed order with binder filling in all the spaces between the pigment particles.

At the same ratio of pigment to binder as in a solvent-based paint where PVC = CPVC, the idealized latex paint would contain equal numbers of latex particles and pigment particles. When we apply a layer of the latex paint, the water evaporates and we get a close-packed system of spheres. But this time, some of the spheres are pigment particles and some are latex particles. There would not, however, be a uniform arrangement of alternating latex and pigment particles in a three-dimensional lattice. Rather, there would be at best a statistical distribution of particles. In some areas there would be small clusters of pigment particles, and in other areas there would be small clusters of latex particles. As the film forms, the latex particles coalesce, flowing around the pigment particles. However, the viscosity of the coalesced binder is high and it is difficult for the polymer to penetrate into the center of clusters of pigment particles. As film formation proceeds, the water left inside the pigment clusters diffuses out of the film, leaving behind voids. Remember that the definition of CPVC is that PVC above which there are voids in the film. Although the PVC of 50 in the solvent-based paint equaled CPVC, the same PVC of 50 in the latex paint results in a film with PVC > LCPVC. The probability of having clusters of pigment particles can be decreased by increasing the number of latex particles until void-free films are obtained, but this necessarily reduces PVC. In other words, LCPVC is lower than CPVC.

If the T_g of the polymer is lower, the viscosity of the polymer at the same temperature is lower and the distance that the latex can penetrate into clusters of pigment particles can increase. Thus, the LCPVC of paint made with lower T_g latex is higher, although still lower than the CPVC with the solvent-based paint. Analogously, the reduction of viscosity of the polymer by coalescing agent increases LCPVC. Incidentally, it follows that the higher the temperature during film formation, the higher the LCPVC should be, although no experimental test of this hypothesis has been published. It also follows, although again it is untested, that LCPVC should depend on time. Although the viscosity of the polymer is high and inhibits flow between the particles, the viscosity is not infinite and perhaps in time voids would be filled.

If a smaller particle size latex is used while maintaining the same ratio of pigment to binder volume, the number of latex particles in our idealized paint would be larger than the number of pigment particles. Now the probability of clusters of pigment particles forming would be reduced. As a result, one would expect, as is found experimentally, that LCPVC increases with decreasing particle size of the latex. Particle size distribution of a latex affects its packing factor [16], and presumably affects LCPVC.

Not only does larger particle size latex decrease LCPVC, flocculation of the latex particles would also be expected to drastically decrease LCPVC. If one formulates a latex paint so that the PVC is slightly less than LCPVC with an unflocculated latex, but the latex flocculates, the PVC would be greater than LCPVC. Again, there has been no data published on the effect of latex flocculation on LCPVC. The need for further research in pigment volume relationships in latex paints is obvious.

GENERAL REFERENCE

G. P. Bierwagen and T. K. Hay, *Prog. Org. Coat.*, **3**, 281 (1975).

REFERENCES

1. W. K. Asbeck and M. Van Loo, *Ind. Eng. Chem.*, **41**, 1470 (1949).
2. T. C. Patton, *Paint Flow and Pigment Dispersion*, 2nd ed., Wiley-Interscience, New York, 1979, p. 172.
3. P. B. Mitton in T. C. Patton, Ed., *Pigment Handbook*, Vol. 3, Wiley-Interscience, New York, 1973, p. 321.
4. J. Temperley, M. J. Westwood, M. R. Hornby, and L. A. Simpson, *J. Coat. Technol.*, **64** (809), 33 (1992).
5. G. P. Bierwagen and T. K. Hay, *Prog. Org. Coat.*, **3**, 281 (1975).
6. G. P. Bierwagen, *J. Coat. Technol.*, **64** (806), 71 (1992).
7. W. K. Asbeck, *J. Coat. Technol.*, **64** (806), 47 (1992).
8. R. W. Braunshausen, Jr., R. A. Baltrus, and L. De Bolt, *J. Coat. Technol.*, **64** (810), 51 (1992).
9. T. K. Hay, *J. Paint Technol.*, **46** (591), 44 (1974).
10. H. F. Huisman, *J. Coat. Technol.*, **56** (712), 65 (1984).
11. G. P. Bierwagen, *J. Paint Technol.*, **44** (574), 46 (1972).
12. C. R. Hegedus and A. T. Eng, *J. Coat. Technol.*, **60** (767), 77 (1988).
13. G. P. Bierwagen and D. C. Rich, *Prog. Org. Coat.*, **11**, 339 (1983).
14. F. Anwari, B. J. Carlozzo, K. Choksi, M. Chosa, M. DiLorenzo, C. J. Knauss, J. McCarthy, P. Rozick, P. M. Slifko, and J. C. Weaver, *J. Coat. Technol.*, **62** (786), 43 (1990).
15. Ref. [2], pp. 192–204.
16. K. L. Hoy and R. H. Peterson, *J. Coat. Technol.*, **64** (806), 59 (1992).

CHAPTER XXII

Application Methods

Many methods for applying coatings are used depending on the circumstances. Capital costs, operating costs, film thickness and appearance requirements, and the structure of the object to be coated are among the factors that affect the choice of application method for each particular case. Reduction of VOC emissions is a driving force in the development of new modifications of application methods. The amount of solvent required and the possibility of recovering the solvent are affected by the application method. Several of the important application methods are considered in this chapter. Further information and discussion of some additional application methods can be found in the general references provided at the end of the chapter. Electrodeposition of coatings and the application of powder coatings are discussed in Chapters 30 and 31, respectively.

22.1. BRUSHES, PADS, AND HAND ROLLERS

Brushes, pads, and hand rollers are frequently used for application of architectural coatings. They are used by both do-it-yourselfers and professional painters. Although the same paints can usually be applied by spray gun, few do-it-yourselfers use spray guns. On the other hand, professional painters use spray guns for applying architectural paints whenever possible, to save time.

22.1.1. Brush Application

A wide variety of brushes is available: narrow and wide, long- and short-handled, and nylon, polyester, or hog bristle [1]. Hog bristles are appropriate for solvent-borne paints but not for water. Nylon bristles are appropriate for water-borne paints but are swollen by some solvents. Polyester bristle brushes can be used with either. Brushes all have in common a large number of bristles that hold paint in the spaces between the bristles. When the paint is applied, pressure forces paint out from between the bristles. The forward motion of the brush splits the layer of paint so that part is applied to the surface and part remains on the brush. The same principles apply to brushes where an open-cell polyurethane foam replaces the bristles in a conventional brush.

Viscosity characteristics are critical in using brushes. Pick-up of paint on the brush is controlled by paint viscosity at a relatively low shear rate, around 15–30 s^{-1}, the shear rate of dipping and removing a brush from a paint can. If the viscosity is too high, too much paint will be brought out of the container with the brush; if the viscosity is too low, too little paint will be on the brush.

Ease of brushing requires a low viscosity. The shear rate between the brush and the surface to which the paint is being applied is relatively high, estimated to be in the range of 5000–20,000 s^{-1} [2]. The viscosity at high shear rate controls the ease of brushing; high viscosity leads to high "brush drag." The film thickness applied is importantly affected by the viscosity at high shear rate; applied film thickness increases with increasing viscosity. For most applications, a viscosity of 0.1–0.3 Pa·s is appropriate. In general, solvents with relatively slow evaporation rates must be used to slow the increase in viscosity of the paint on the brush. At least some degree of shear thinning is essential, and it is usually desirable for the coating to be thixotropic.

When paint is applied by brush, the surfaces of the newly applied wet film shows lines of thin and thick wet film called brush marks. The origin of these furrows is not clearly understood. They do not result from the individual bristles of the brush, as is evident from comparing the size and number of brush marks with those of the bristles of the brush. When urethane foam "brushes" are used, brush marks still result even though the brush has no bristles. The brush marks probably result from the irregularities in wet film thickness that result from splitting the wet film between the brush and the substrate as the brush is applying the paint. Whenever layers of liquids are split, the surfaces are initially irregular. These irregularities are apparently made into lines by the linear movement of the brush. It is desirable to formulate a coating so the brush marks can flow out nearly level before the film dries. As discussed in Sections 23.1 and 23.2, low viscosity promotes leveling but increases the probability of sagging. Thixotropic flow properties are generally desirable because this delays the increase in viscosity after brushing, permitting a compromise between leveling and sagging. Formulation problems are particularly difficult in the case of latex gloss paints as discussed in Section 35.3.

22.1.2. Pad Application

Pad applicators are replacing brushes to some extent for the do-it-yourself market [1]. The most common type of pad consists of a sheet of nylon pile fabric attached to a foam pad that is attached to a flat plastic plate with a handle. For low-viscosity coatings like stains and varnishes, a lamb's wool pad is used.

Pads have a number of advantages compared to brushes. Pads hold more paint than a similar width brush and can apply paint up to twice as fast. In general, pad application leaves a smoother layer than brush application. Extension handles can be used with pads, thus reducing the need for moving ladders. Pads, especially refills, are less expensive than brushes. On the other hand, pads require the use of trays, which results in some paint loss and solvent evaporation. Cleanup is more difficult with pads than with brushes.

22.1.3. Hand Roller Application

Hand rollers are the fastest method of hand application and are widely used in applying architectural paints to walls and ceilings. A wide variety of rollers and roll coverings are available [1]. There are rollers with built-in wells to minimize the need for dipping in the tray, and there are also power-filled rolls.

The viscosity requirements are similar to those for brush application. When paint is applied by roller there is also film splitting. At the nip, the coating is under pressure; as the roller moves on, the pressure is released resulting in *cavitation*, that is, formation of small bubbles. As the roller continues to move, the bubbles expand and the walls between them are stretched. On further extension, the walls between the bubbles begin to resemble fibers; still farther from the nip, as the roller moves, the fibers are drawn longer and longer and eventually break. After breaking, the ends of the fiber are pulled by surface tension to minimize the surface area of the applied coating but, if there is not sufficient time at low viscosity for leveling to occur, track marks are left in the coating.

A further complication arises with longer fibers that may break in two places; as a result, droplets of loose paint fly off to land on the painter or the floor. This is called *spattering*. If spattering could be eliminated, painting speed could be increased, and time spent on masking and laying drop cloths could be reduced. However, all current formulations spatter; some much more than others. The phenomenon is not completely understood. Glass has pointed out that as the fibers are drawn out, extensional flow rather than shear flow is involved [3,4] (see Section 19.6.3). It follows that extensional viscosity rather than shear viscosity affects the development of fibers from roller application and, in turn, the roughness of the film and the degree of spattering. Spattering can occur with any kind of paint but is particularly difficult to minimize in latex paints. Formulation variables affecting spattering of latex paints are discussed in Section 35.2.

22.2. SPRAY APPLICATION

Spraying is a widely used method for applying paints and coatings in architectural and especially in industrial applications. In general, spraying is much faster than application by brush or hand roller. Spraying is used on flat surfaces but is particularly applicable to coating irregularly shaped articles. The principal disadvantages are the greater difficulty in controlling where the paint goes and the inefficiency of application, since only a fraction of the spray particles are actually deposited on the object sprayed. One must mask areas where paint is not desired, and there is a likelihood of contaminating the general area with spray dust.

Many different kinds of equipment are used for spraying; all atomize the liquid paint into droplets. Droplet size depends on the type of spray gun and coating; variables include: air and fluid pressure, fluid flow, surface tension, viscosity, and, in the case of electrostatic application, voltage. The choice of spray system is affected by capital cost considerations, efficiency of paint utilization, labor costs, size and shape of objects to be coated, among other variables. Coating formulations must be established for the particular spray equipment and conditions.

Table 22.1. Typical Baseline Transfer Efficiencies

Type of Spray Gun	Transfer Efficiency (%)
Compressed air	25
Airless	40
Air-assisted airless	50
High volume, low pressure air	65
Electrostatic air	60–85
Electrostatic rotary	65–94

Source: Ref. [5].

Not all the droplets are deposited on the surface being coated. Some of the droplets approach the surface and *bounce back*, carried by eddy currents of air. The higher the pressure, the higher the forward velocity of the air, and, as a result, the greater the percentage of bounce back. Some droplets miss the object being coated—this is called *overspray*. Some droplets may fall out of the spray pattern under the force of gravity. *Fall out* is higher with large spray patterns and longer distances between the spray gun and the surface being sprayed. The sum of all this waste determines *transfer efficiency*; transfer efficiency is defined as the percent of coating solids leaving the gun that is actually deposited on the coated product.

Transfer efficiency has always been an important cost factor, since low transfer efficiency means that more coating must be used for the same surface area. Now, it is especially critical to achieve high transfer efficiency to reduce VOC emissions by utilizing less coating.

Transfer efficiency is affected by many variables. The size and shape of the product being coated is a major variable. Low transfer efficiency can be expected when spraying a chain link fence; high transfer efficiency can be expected when spraying a large wall. Less obvious, but still important, are variables such as conveyor line speed and how objects being painted are hung on a conveyor line. The method of application has a major effect on the transfer efficiency—whether the application is manual or automated, whether compressed air, airless, or rotary spray guns are used, and whether or not an electrostatic system is used. In manual systems, the skill of sprayers is very important. In the case of automated systems, system design is critical. Some factors involved are: distance between the gun and the surface, angle of the spray gun, stroke speed uniformity, extent and uniformity of overlapping, and precision of triggering.

To compare the transfer efficiency of different spray application methods, the spray gun industry has developed standard procedures permitting comparison of what is called *baseline transfer efficiency* [5]. Typical percentages are given in Table 22.1. It must be emphasized that these percentages are strongly affected by many variables. For example, transfer efficiency in coating a chain link fence is low at best, and transfer efficiency in coating large wall expanses is high in any case.

22.2.1. Compressed Air Spray Guns

Compressed air spray guns cause atomization of the paint by fine streams of compressed air. It is the oldest spray method in use today and still is selected by a large number of applicators. Figure 22.1 shows a cross section of a compressed air spray gun. A stream of paint is driven through the nozzle orifice by relatively low pressure, 1–5 kPa (1.5–7 psi), or, in other types of guns, by the suction caused by the rapid air flow past the outside of the orifice. The stream of paint coming out of the orifice is atomized into small droplets by fine streams of compressed air at pressures of 25–50 kPa (35–70 psi). The degree of atomization is controlled by: (1) the viscosity of the paint (the higher the viscosity at the high shear rate encountered going through the orifice, the larger the particle size), (2) the air pressure (higher air pressure, smaller particle size), (3) the diameter of the orifice (smaller orifice, smaller particles), (4) the pressure forcing or pulling the paint through the orifice (higher pressure, smaller size), and (5) surface tension (lower surface tension, smaller particle size). The outer jets of compressed air shown in the drawing adjust the shape of the stream of atomized particles coming from the gun. If these jets were not present, a cross section of the stream would be roughly circular. Generally, an elliptical cross section pattern permits more efficient application. Looked at from the side, the pattern is a flattened cone, often called a *fan*. Guns can be hand held or attached to robots.

In general, compressed air guns are less expensive than other types of guns. Atomization can be finer than with other methods. The system is versatile, and virtually any sprayable material can be sprayed with air spray. The level of control can be high if the operators are skilled. However, as noted in Table 22.1, baseline transfer efficiency is lowest of all the spray methods.

A substantial improvement in transfer efficiency of compressed air spraying has been made by use of *high volume, low pressure* (HVLP) air guns. The guns are designed to operate at lower air pressures, 2–7 kPa (3–10 psi), but with higher air volumes. The guns are designed with large unrestricted air passages to handle a large volume of air. Because of the low pressures, bounce back is reduced and overall transfer efficiencies of 65% or even higher can be achieved. High pressure, low volume spray guns are being used increasingly in such applications as automobile repair shops. California South Coast VOC regulations now require 65% transfer efficiency and air pressures of 7 kPa or less. An alternative approach to HVLP is low volume, low pressure (LVLP). In LVLP guns the air pressure is also less than 7 kPa and the air volume is substantially reduced by mixing the air and paint inside the gun.

22.2.2. Airless Spray Guns

In airless guns the paint is forced out of an orifice at high pressure, 5–35 MPa. As the paint comes out of the orifice, the pressure is released, resulting in cavitation that leads to atomization. The atomization is controlled by the viscosity (higher viscosity, larger particle size), pressure (higher pressure, smaller particle size), and

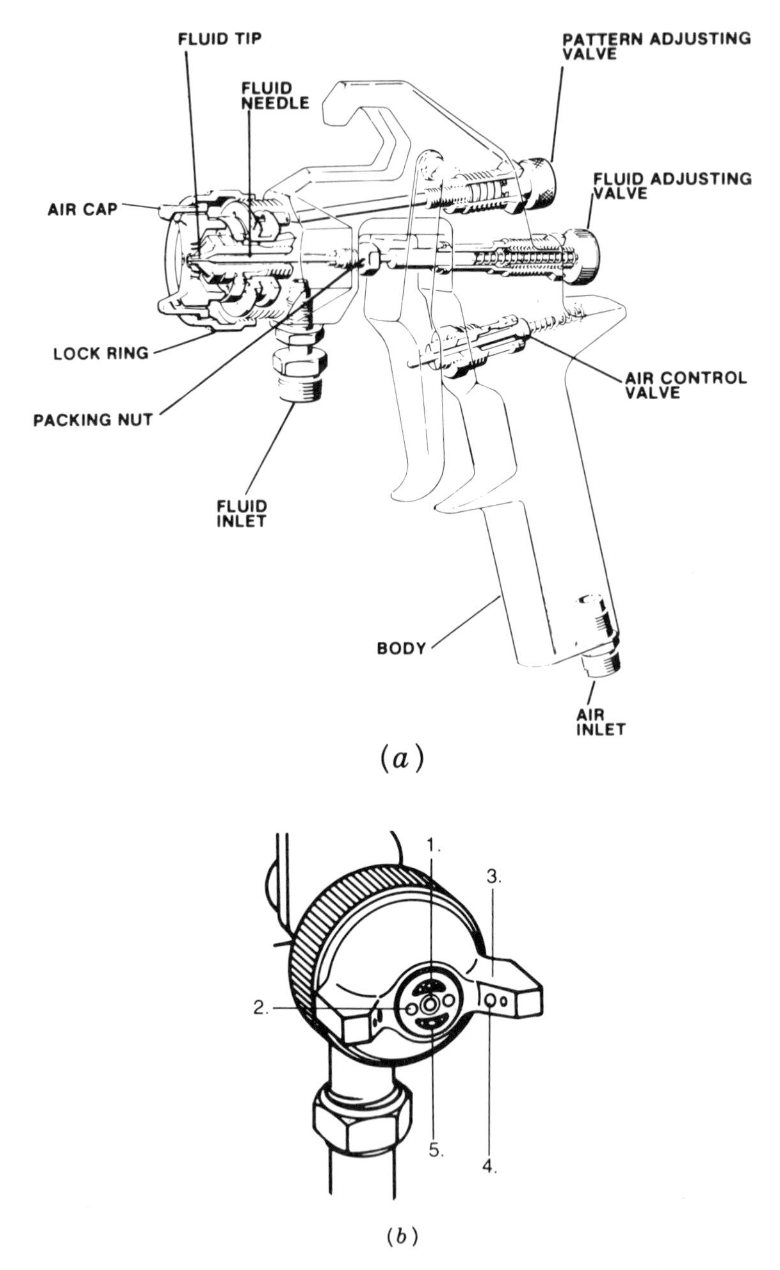

Figure 22.1. (*a*) Schematic cross section of a compressed air spray gun. (*b*) Schematic nozzle of a spray gun; 1. Annular ring around the fluid nozzle tip. 2. Containment hole. 3. Wings, horns, or ears. 4. Side-port holes. 5. Angular converging holes. (From Ref. [1], with permission.)

surface tension (lower surface tension, smaller particle size). The shape of the fan or pattern of the spray is controlled by the orifice size and shape. Air-assisted, airless spray guns are also available; the atomization is airless but there are external jets to help shape the fan pattern [6]. Both hand-held and robot airless guns are available.

The particle size of the droplets from airless spray guns is larger than with air spray: 70–150 μm as compared with 20–50 μm [7]. Also airless guns give a so-called *fishtail spray*; that is, there is a relatively sharp edge to the spray droplet fan with quite uniform droplet distribution within the fan. In contrast, the fan from air guns is *feathered* at the edge; that is, the number of droplets drops off at the edge of the fan with some being quite widely spaced. As a result of these differences, one can generally achieve more uniform film thickness with air spray; air-assisted airless application gives intermediate results.

Paint can be applied more rapidly by airless than by compressed air guns, an important advantage in permitting more rapid production. However, as the application rate increases, the likelihood of applying excessive paint thickness also increases, particularly in painting objects with complex shapes. Excess paint thickness is not only wasteful but may lead to sagging.

Since there is not a stream of compressed air accompanying the paint particles and because the droplet size is generally larger, there is less solvent evaporation from the atomized particles from airless guns than from compressed air guns. Solvents with higher relative evaporation rates are generally required in formulating paints for airless spray application.

The absence of the air stream reduces the problem of bounce back. This is a particularly important advantage of airless guns in spraying into recesses in irregularly shaped objects. On the other hand, spraying down a recessed section that is open on the opposite end is easier with a compressed air gun since the air stream helps to carry the paint particles along.

An aerosol paint container is a type of airless spray unit. A liquefied gas, commonly propane, supplies the pressure to force the paint out of the orifice. Since the pressure is relatively low, the viscosity of the paint must be low to get proper atomization.

22.2.3. Electrostatic Spraying

As shown in Table 22.1, transfer efficiency can be substantially higher with electrostatic spray units. In the simplest case, a wire is built into the orifice of the spray gun. An electric charge of the order of 50–125 kV is impressed on the wire. At the fine end of the wire, an electric discharge leads to ionization of the air. As the atomized paint particles pass through this zone of ionized air, they pick up a negative charge. The object to be painted is electrically grounded. When the paint particles approach the grounded surface of the object, the differential in charge attracts the particles to the surface. Taking the example of a chain link fence, an increased fraction of the paint particles deposit on the metal fence even to the extent that particles that had passed through the holes are attracted back to the back side of the fence. This *wrap around* effect permits painting both sides of the fence by spraying from only side. There is a lower, but still fairly high, overspray loss even

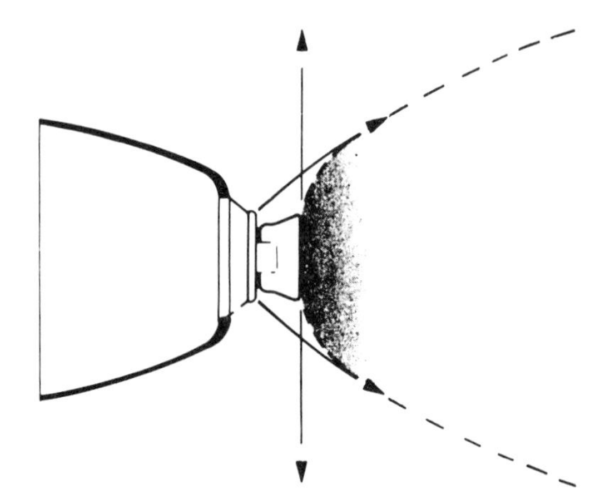

The Bell: The geometry of the spray pattern is varied according to the object to be painted by adjusting an annular compressed air shroud

Figure 22.2. Bell electrostatic spray equipment. (From Ref. [1], with permission.)

with electrostatic spray. With objects like automobiles or appliances overspray losses can be reduced by over 50%, resulting in transfer efficiencies around 80%.

The presence of the electrostatic field promotes atomization. As a result, completely different types of devices can be used for spraying. Figures 22.2 and 22.3 show disk- and bell-type spray units; they are so-called rotary atomizers. In both cases, coating is pumped through the tube leading to the middle of the disk or bell; the unit is rotated at a fairly high speed, on the order of 900 rpm (or higher depending on the diameter); and the coating flows out to the edge of the unit. The edge of the disk or bell is charged with a potential on the order of 60–150 kV. As the coating is thrown off the edge of the unit it is atomized and picks up a charge.

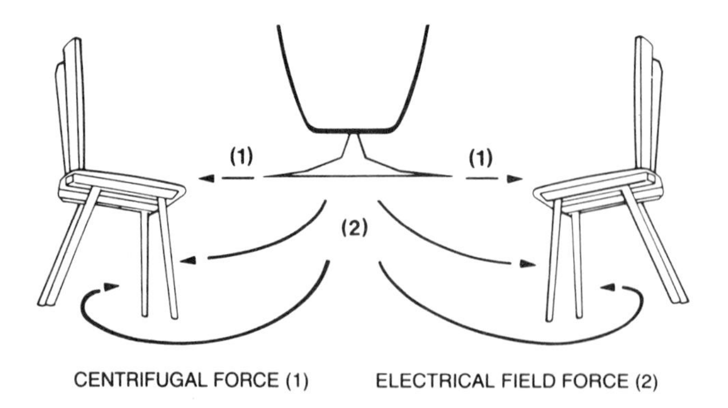

The Disc: The combined centrifugal and electrostatic forces of the ultra-high speed disc provide an extremely uniform finish and excellent penetration. Transfer efficiencies are high

Figure 22.3. Disk electrostatic spray equipment. (From Ref. [1], with permission.)

Particle size is controlled by coating viscosity and the peripheral speed of the unit. Very high speed bells, up to 60,000 rpm, are now being used that permit application of coatings with viscosities as high as 1.5–2 Pa·s compared to 0.05–0.15 Pa·s characteristically used with conventional speed disks and bells as well as spray guns. This ability to handle higher viscosity coating permits use of less solvent and hence higher application solids or higher molecular weight at the same application solids. However, it has been reported that application of pigmented coatings by high-speed electrostatic bells tends to give films with lower gloss [8]. It is suggested that the centrifuging effect leads to differences in pigment content among atomized particles leading to formation of uneven films. Uniformity of application and transfer efficiency can often be improved by compressed air shaping of the cloud of atomized coating, analogously to the effect of the shaping air in conventional air guns.

Low overspray losses and good wrap around using electrostatic spray depend on the charge pickup by the atomized paint particles, which is controlled by the conductivity of the paint. If the conductivity is too low, the particles do not pick up sufficient charge from the ionized air. This is most likely to be a problem with coatings having only hydrocarbon solvents, especially aliphatic hydrocarbons. Nitroparaffin or alcohol solvents can be substituted for a portion of the hydrocarbon solvents with good results. For coatings with free carboxylic acid groups on the resin, addition of a small amount of a tertiary amine like triethylamine will increase the conductivity sufficiently to give good charging. If the conductivity is too high, there is increased danger of electrical shorting. Generally, conductivity is measured but the results are expressed as resistivity; optimum resistivity varies with equipment and the coating operation, ranging from 0.05 to 20 megohms.

Conductivity of water-borne coatings is much higher than that of solvent-borne coatings; resistivities of the order of 0.01 megohms are reported [9]. Addition of slow evaporating water-miscible solvents with nonpolar ends, such as 2-butoxyethanol, somewhat reduces surface conductivity. However, spray equipment must be specially designed so that the paint line is grounded to minimize the possibility of electricity feeding back through the fluid line to ground [9].

Electrostatic spray is the most important method for applying powder coatings. See Section 31.5.1 for discussion of spraying powder coatings.

Electrostatic spray application is not devoid of difficulties. The substrate must be electrically conductive so that the necessary charge differential can be set up by grounding the object to be sprayed. It also can be difficult to get coating into recessed areas even with airless electrostatic spray application due to a *Faraday cage effect.* A Faraday cage results from the pattern of field lines between the electrode on the gun and the grounded object. The strong electrical field induced by the difference in voltage establishes field lines that the atomized particles follow between the gun and the object. However, areas surrounded by grounded metal, as in the corner of a steel case, are shielded from the electric fields by the metal: a Faraday cage is established and few particles enter such a shielded area.

22.2.4. Hot Spray

Since the viscosity of coatings must be relatively low, generally 0.05–0.15 Pa·s, for proper atomization, the solids, especially for lacquers that are comprised of relatively high molecular weight resins, must be quite low. An important approach for increasing solids (decreasing solvent) is the use of hot spray. A hot spraying system is designed with a heat exchanger to heat the coating to temperatures from 38 to 65°C. The system must be designed so that the coating is recirculated when the gun is turned off, even temporarily. At the elevated temperature, the viscosity is reduced sufficiently to permit a significant increase in solids.

While these systems were originally designed for use with quite low solids coatings like lacquers, they are useful for high solids coatings. For example, it is reported that the solids of a top coat for appliances can be increased from 55 to 65% using a paint heater before a disk electrostatic spray gun [2]. The viscosity of high solids coatings generally decreases more sharply with increasing temperature than that of conventional coatings—a desirable characteristic for hot spray application. There is the further advantage that the temperature drops between leaving the spray gun orifice and arrival on the work, leading to a correspondingly large viscosity increase. This characteristic is desirable for reducing the sagging problem that can be very serious with high solids coatings (see Section 23.2). Yet another advantage is that the temperature is controlled, reducing variations in production conditions.

22.2.5. Supercritical Fluid Spray

A new approach to spray application that is entering the pilot demonstration stage and initial commercial use as this is written, is *supercritical fluid spray* [7]. The supercritical fluid of choice is carbon dioxide with a critical temperature of 31.3°C and a critical pressure of 7.4 MPa. In the supercritical state, carbon dioxide exhibits solvency characteristics similar to aromatic hydrocarbons, but is not counted as VOC. Thus, it is possible to replace a fraction of the solvents in a coating with CO_2, resulting in reduction in the VOC. A dual feed gun is used with a low solvent content coating as one feed and with supercritical CO_2 as the other. Of course, the temperature must be controlled and the pressure must be above 7.4 MPa. The process is said to be applicable to most types of solvent-borne coatings; VOC can be reduced by 30–70% without changing molecular weight of the resins.

The high pressure means that airless guns must be used. Fortunately, the supercritical fluid spray system minimizes some of the problems of utilizing airless spray with conventional coatings [7]. When the coating leaves the orifice of the spray gun, the very rapid vaporization of the CO_2 breaks up the atomized droplets, further reducing particle size. The droplet size is comparable to that obtained with compressed air spray guns and significantly smaller than obtained with airless guns. Furthermore, the fan pattern is more similar to that obtained with compressed air guns (feathered) rather than the usual pattern for airless guns discussed in Section 22.2.2. The loss of the CO_2 is apparently complete before the droplets reach the surface so that the viscosity of the applied film on application is relatively high, thereby minimizing sagging.

While impressive results have been obtained on a laboratory scale and initial commercial applications, substantial testing will be required to see how widely useful the approach will be. Supercritical fluid spray is an excellent example of the value of creative thinking in developing new approaches to difficult problems.

22.2.6. Formulation Considerations for Spray-Applied Coatings

Formulation of the solvent mixture for spray-applied coatings and effective use of spray guns requires taking into consideration the very large ratio of surface area to volume of the atomized coating droplets and the flow of air over the surface of those droplets. As discussed in Section 15.1, these two factors exert major influences over the rate of solvent evaporation. In applying lacquers by spray, it has been shown that over half of the solvent in the coating may evaporate between the orifice of the spray gun and the surface of the substrate. If the solvent mixture is balanced properly and the spray gun is used properly, a sag-free relatively smooth lacquer surface can be achieved. If either is not proper, sagging or rough surfaces or, in extreme cases of poor spraying, both sagging and rough surfaces on the same substrate are obtained. As discussed in more detail in Sections 23.1 and 23.2, low viscosity of the coating after it arrives on the surface generally facilitates leveling but also increases sagging.

Proper control requires a careful balance of solvent evaporation rate with the particular spraying equipment and procedure. The greater the distance from the spray gun orifice to the work, the greater the fraction of solvent lost. Coatings are formulated to work best at a specific distance between the gun and the surface. This distance should be kept as constant as possible throughout the spraying operation. The problem can be illustrated by the result of trying to spray lacquer on a flat vertical surface by holding one's arm in a constant position and bending one's wrist to spray a wider area of the surface. When the gun is aimed perpendicular to the substrate surface, the distance is at a minimum. When the wrist is bent to the furthest degrees, the distance is at a maximum. If the solvent mixture was properly balanced for an intermediate distance, the lacquer sprayed perpendicularly would be likely to sag; the lacquer film at the extreme distance would tend to be rough, resulting from poor leveling.

The rate of solvent loss is affected by the degree of atomization. If the average particle size is smaller, the surface/volume ratio will be higher and the extent of solvent loss will be greater. The rate of solvent evaporation is also affected by air flow over the surface of the droplets. In general, more solvent will evaporate when compressed air guns are used rather than airless guns or spinning bells.

The rate of air flow through the spray booth can affect the degree of solvent loss. Temperature in the spray booth can be an important factor; during hot weather it is common to change the solvent mixture to slow the rate of evaporation.

In formulating a solvent mixture for a coating in the laboratory, the same type of spray gun at approximately the same distance from the work that will be encountered in the customer's factory should be used. A final adjustment must be made in the customer's plant under regular production conditions. If those conditions change, the solvent mixture will have to be changed. This is one reason why industrial coatings are almost always shipped at higher concentrations than the customer will actually use. This permits modifications of both solvent levels and solvent composition (by changing the reducing solvent) to accommodate temperature changes in the spray booth and other variables.

Viscosity of the coating to be sprayed must be adjusted to obtain appropriate atomization for the spray gun being used. The critical viscosity for atomization is that at the high shear rates, $10^3-10^6\ s^{-1}$ [2], encountered as the paint passes through the orifice of the gun. Shear thinning systems with relatively high viscosity at low shear rate can be sprayed successfully. The upper limitation on viscosity at low shear rate is the need to have a satisfactory flow rate of coating through the tubing to the spray gun which varies from gun to gun. Generally, architectural coatings, which must not sag significantly when heavy films are applied to large wall expanses, are shear thinning paints. Generally, the same paint that would be used for brush application can be used for spray; in the case of latex paints with some spray equipment, the viscosity at low shear rate may have to be reduced by diluting with a small amount of water.

Most conventional industrial coatings exhibit Newtonian flow; however, many high solids coatings and water-borne coatings are shear thinning. For spray application, the evaporation rate of the solvent combination is adjusted so that it is slow enough to permit leveling but rapid enough to minimize sagging. Viscosity for application is generally checked with an efflux cup (see Section 19.3.6). However, efflux cups should be used with care since they will not detect the presence of shear thinning or thioxtropic flow properties. They should only be used for control purposes. The proper way to establish viscosity for production spraying is by using the production spray gun under the conditions of use. Then, having found the proper degree of thinning with solvent, one can establish an efflux cup time range as the standard for that coating for use in that gun under those circumstances.

Many water-reducible coatings are slightly shear thinning after being reduced to spray viscosity with water. This means that the efflux cup time of the reduced coating will be longer than with most solvent-borne coatings to give the appropriate atomization. Remember that atomization is controlled by viscosity at high shear rate and efflux cup time is controlled by viscosity at low shear rate.

As discussed in Section 23.2, the control of sagging of some spray-applied high solids coatings cannot be done by adjustment of solvent evaporation rates. In such cases, the coating has to be formulated so that it is shear thinning and thixotropic. Efflux cup standard times for spray application of such coatings are different than for conventional Newtonian flow coatings and may be of limited utility even for quality control.

22.2.7. Dual Feed Spray Systems

Very reactive systems with short pot lives, such as two-package polyurethane coatings, can be spray applied. In the most common type of equipment, the two packages from separate reservoirs are metered into a small, efficient mixing chamber just before the spray gun orifice. The average dwell time in the mixing chamber is a fraction of a second and the average residence time in the gun is not much longer. The system is designed so that, when spraying is interrupted, the mixing chamber and the gun are automatically flushed with solvent to prevent clogging. Frequent checking and maintenance are required to assure that the proper ratio of components are being applied. In other types of equipment, the components are sprayed from two orifices; mixing occurs after atomization, so the coating cannot gel in the equipment.

22.2.8. Overspray Disposal

In industrial production, the advantage of decreased overspray is not just savings in the cost of the lost paint and lower VOC emissions. Any overspray has to be trapped so that it will not contaminate the surrounding area. This is generally done by using a *water-washed spray booth*, that is, a spray booth where the wall behind the work being sprayed is a continuous waterfall that is recirculated during spraying. The overspray is collected as a sludge. While it is sometimes possible to rework the sludge into low-grade paints, generally the sludge must be disposed of properly in approved hazardous solid waste disposal landfills. Such disposal has become very expensive; increased transfer efficiency with electrostatic systems can substantially reduce waste disposal cost.

Although water-wash spray booths work very efficiently with solvent-borne coatings, the separation of sludge is less efficient when some water-borne coatings are sprayed. The overspray does not completely coagulate when it strikes the water as it does with solvent-borne coatings. This makes separation more difficult and can limit recirculation of the water. It has been found that froth flotation methods adapted from ore recovery processes permit relatively rapid separation of the sludge [10]. It has also been reported [11] that separation of sludge from water-borne coatings can be improved by addition to the water tank of an emulsion of a melamine-formaldehyde resin for paint detackification along with water-soluble cationic and/or nonionic acrylamide polymers for paint flocculation.

22.3. DIP COATING

Dip coating can be a very efficient procedure for applying paint; it features both relatively low capital cost equipment and low labor requirements. The principle is simple: the object to be painted is dipped into a tank full of coating and pulled out, and excess coating drains back into the dip tank. In practice, satisfactory paint application by dipping is more complex. While the excess paint is draining off the object, a gradation of film thickness develops. The thickness at the top of the object is thinner than at the bottom of the object. As the draining is occurring, solvent is evaporating. The differences in film thickness can be minimized by controlling the rate of withdrawal of the object from the dip tank and the rate of evaporation of the solvent. If the object is withdrawn slowly enough and the solvent evaporates rapidly enough, film thickness approaching uniformity on vertical flat panels can be achieved. In actual production, the rate of withdrawal is usually faster than optimum so there is some thickness differential between the top and bottom of the coating.

Care must be exercised in selecting, as well as in changing, volatile solvents for dip coating because of flammability hazards and changes in viscosity of the dip tank that result from evaporation of solvent. Changes in viscosity result in changes in film thickness, which increases with increasing viscosity. Achieving consistent results requires maintaining the viscosity of the coating constant, which becomes more difficult as the volatility of the solvent is increased. Solvent can, of course, be added to replace solvent lost from the tank.

Successful use of dipping in production lines requires that the coating be very stable. Viscosity can increase not only by loss of solvent but also by chemical reactions of coating components. It must be remembered that only a relatively small fraction of the paint in the tank is removed each time an object is dipped. Fresh paint is added frequently to make up for these removals, but the paint in the tank will be a mixture of old and newly added paint. It is evident that some of the original charge of paint into the tank will be present for a long time. If cross-linking reactions occur in the bath, viscosity will increase. While the viscosity can be reduced by adding more solvent, this will reduce the solids so that dry film thickness will be reduced. In other words, the extent of such reactions must be minimal. When oxidizing alkyds are the vehicle for a dip coating, oxidation must be avoided because this will lead to cross-linking. Stabilization requires an antioxidant, but the antioxidant must be sufficiently volatile that it will escape during the early stages of the baking cycle or it will inhibit the cross-linking of the dry film. On the other hand, it must not be so volatile that it is rapidly lost from the dip tank. Isoeugenol is an example of a widely used antioxidant for alkyds in dip coatings.

A major advantage of dip coating is that all surfaces are coated with paint, not just the outer surfaces accessible to spray. However, there are difficulties in dipping irregularly shaped objects. Paint may be held in pockets or depressions giving pools of paint that do not drain. To minimize this problem, the point(s) of hanging of the object on the hooks of the conveyor line that carries it down into and out of the tank must be carefully designed and selected. Objects to be dip coated must be designed with drain holes that minimize the paint pooling but do not interfere with the performance or appearance of the product. Lower edges, and especially lower corners, build up high film thickness. To some degree, this buildup can be reduced by electrostatic *detearing*, that is, by passing the object over a highly charged electrode that causes a charge concentration at such points, resulting in the pulling off of the drops.

Water-borne coatings are supplanting solvent-borne dip coatings in many applications. They reduce flammability hazards and VOC emissions.

22.4. FLOW COATING

Flow coating and dip coating are related methods. Objects to be flow coated are carried on a conveyor through an enclosure in which streams of coating are squirted on the object from all sides. The excess coating material runs off and is recirculated through the system. A major advantage is that the volume of paint required to fill the lines of a flow coater is substantially less than that required for a dip tank for objects of the same size. This reduces inventory cost and increases paint turnover, hence reducing somewhat the problem of bath stability. It is common to design a flow coater so that the atmosphere in the enclosed area is maintained in a solvent saturated condition. In this way, evaporation of solvents is minimized until paint flow and leveling have been obtained. There is still a gradation of film thickness from the top to the bottom of the object, but usually less than from dip coating.

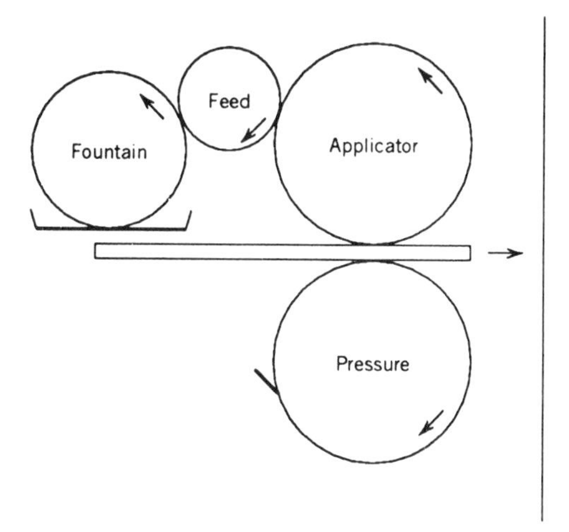

Figure 22.4. Direct roll coater. (From Ref. [1], with permission.)

Highly automated flow coating lines have been used for applying coatings to major appliances. The design permits more rapid line speeds than are possible with conventional dip tanks.

22.5. ROLL COATING

Roll coating is widely used and is efficient, but it is only applicable to uniform, generally flat or cylindrical, surfaces. Relatively slow evaporating solvents must be used to avoid viscosity buildup on the rolls of the coater. The pot life of the coating must be relatively long since the rate of turnover of coating through the system is relatively low. There are many types of roll coating procedures; the two most common are *direct roll coating* and *reverse roll coating*.

In direct roll coating, the sheets or stock to be coated pass between two rollers, an applicator roller and a backup roller, that are being rotated in opposite directions. The rollers pull the material being coated between the rolls as illustrated in Figure 22.4. Direct roll coating is used for coating sheet stock and sometimes for coil stock. In direct roll coating, the *applicator* rollers are generally covered with a relatively hard polyurethane elastomer. Coating is fed to the applicator roll by a smaller *feed* or *doctor* roll that is, in turn, fed by a *pickup* roll. The pickup roll runs partially immersed in a tray (called a *fountain*) containing the coating. Film thickness is dominantly controlled by the clearance between the feed roll and the applicator roll.

Several variations on direct roll coating are possible by use of different types of applicator rolls. The applicator rolls can have cut out sections so that they do not coat the entire substrate surface. The applicator roller can be engraved with small recessed cells over its whole surface. The cells are filled with coating, the surface is scraped clean with a *doctor blade*, and the coating remaining in the cells is transferred to the substrate being coated. Such so-called *precision coaters* apply a carefully controlled amount of coating only to the areas of the substrate that contact

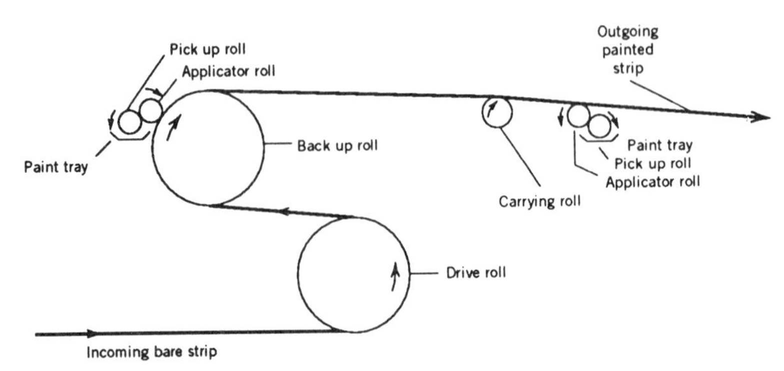

Figure 22.5. Reverse roll coater for coating both sides of coil stock. (From Ref [1], with permission.)

the roller. In another variation, the applicator roller is a brush roller; it is used to apply thick coatings to relatively rough surfaces.

In direct roll coating, as the coated material comes out of the nip between the rollers, the wet layer of coating is split between the roller and the substrate. As a result, the coating has a highly ribbed surface as it emerges from the coater—a phenomenon called *roll tracking*. The coating must be designed to level so this tracked appearance disappears, or at least become less obvious, before the coating stops flowing. To minimize solvent loss on the rolls and to keep the viscosity low to promote leveling, slow evaporating solvents are used and coatings are formulated so that they exhibit Newtonian flow.

In reverse roll coating the two rollers are rotating in the same direction and the material being coated must, therefore, be pulled through the nip between the two rolls, as shown in Figure 22.5. It is not generally feasible to coat sheets by reverse roll coating, but the process is widely used in coating coil stock. Reverse roll coating has the major advantage that the coating is applied by wiping rather than film splitting. A smoother film is formed, and the problems of leveling are minimized.

22.6. CURTAIN COATING

Curtain coating is widely used in coating flat sheets of substrate material such as wall panels. A coating is pumped through a slot in the *coating head* so that it flows out as a continuous *curtain* of coating. The material to be coated is carried under the curtain on a conveyor belt. The curtain should always be wider than the substrate to be coated so as to avoid edge effects on the film thickness. A recirculating system is required to return to the overflow from the sides and from between sheets to the coating head. A schematic diagram is shown in Figure 22.6.

Film thickness is controlled by the width of the slot, the pressure from pumping, the viscosity of the coating, and the rate of passage of the substrate being coated. The faster the line is running, the thinner the coating. Where applicable, curtain coating is an excellent method. No film splitting is involved so the film is laid down

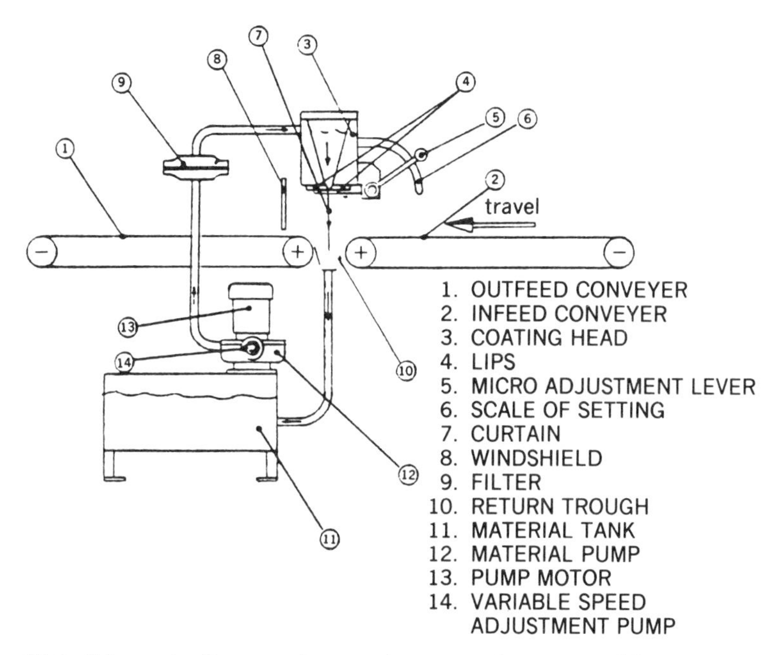

Figure 22.6. Schematic diagram of a curtain coater. (From Ref. [1], with permission.)

essentially smooth. Film thickness can be very uniform. The coating must be very stable and the evaporation of solvent must be closely controlled and solvent added to make up for loss of solvent. If, as explained in Section 23.3, a particle of low surface tension lands on the flowing curtain, surface tension differential driven flow can lead to a hole in the curtain that in turn leads to a gap in the coating being applied to the substrate. As the surface tension of the coating is reduced, particles in the air are less likely to have lower surface tensions than the coating, and the probability of holes is reduced.

GENERAL REFERENCE

S. B. Levinson, *Application of Paints and Coatings*, Federation of Societies for Coatings Technology, Blue Bell, PA, 1988.

REFERENCES

1. S. B. Levinson, *Application of Paints and Coatings*, Federation of Societies for Coatings Technology, Blue Bell, PA, 1988.
2. C. K. Schoff, *Rheology*, Federation of Societies for Coatings Technology, Blue Bell, PA, 1991.
3. J. E. Glass, *J. Coat. Technol.*, **50** (641), 72 (1978).
4. R. H. Fernando and J. E. Glass, *J. Rheology*, **32**, 199 (1988).
5. J. Adams, "Spray Applications Processes for Environmental Compliance," FSCT Symposium, Louisville, KY, May 1990.
6. M. G. Easton, *J. Oil Colour Chem. Assoc.*, **66**, 366 (1983).

7. K. A. Nielsen, D. C. Busby, C. W. Glancy, K. L. Hoy, A. C. Kuo, and C. Lee, *Polym. Mater. Sci. Eng.*, **63**, 996 (1990).
8. K. Tachi, C. Okuda, and K. Yamada, *J. Coat. Technol.*, **62** (791), 19 (1990).
9. M. J. Diana, *Products Finishing*, July, 54 (1992).
10. E. W. Fuchs, G. S. Dobby, and R. T. Woodhams, *J. Coat. Technol.*, **60** (767), 89 (1988).
11. S. F. Kia, D. N. Rai, M. A. Shaw, G. Ryan, and W. Collins, *J. Coat. Technol.*, **63** (798), 55 (1991).

CHAPTER XXIII

Application Defects

During the application and drying of paint films many kinds of defects or imperfections can develop in the film. In this chapter, we deal with some of the most important defects and, to the extent possible, discuss the causes of the defects and possible approaches for eliminating, or at least minimizing, their occurrence. Unfortunately, the nomenclature for many of these defects is not standardized. It is critical for people describing a problem and people trying to solve it to use the terms in the same way. For example, as will be described, *popping* can be called *cratering* and vice versa, and the word *telegraphing* has been applied to several different defects.

Many of the defects are related to surface tension phenomena. Surface tension is discussed in Section 2.4, but its importance in relation to many defects is so great that it seems appropriate to review the principles here.

The forces at an interface of a liquid differ from those within the liquid because of the unsymmetrical force distributions on the surface molecules. The surface molecules possess higher free energy, equivalent to the energy required to remove the surface layer of molecules per unit area. Nature strives to minimize such free energy. One way is by reducing the surface area of the liquid. Since a sphere encloses a maximum ratio of volume to surface area, surface tension forces continually try to shape liquids into droplets. For the same reason, surface tension drives the flow of a rough or uneven liquid surface toward becoming a smooth surface. The smooth surface has less interfacial area with air than the rough surface, hence there is a reduction in surface free energy as the surface becomes smoother. Surface tension can be expressed as an amount of force in the surface perpendicular to a line of unit length. Units are newtons per meter or, more commonly, $mN\ m^{-1}$. In older literature and to a degree still today, one sees dynes per centimeter: $1\ dyne\ cm^{-1} = 1\ mN\ m^{-1}$.

Another aspect of this drive toward the lowest surface free energy is the equilibrium orientation of molecules at the surface of a liquid. Segments of the molecules that minimize surface tension tend to orient at the surface. The lowest surface tension results from perfluoroalkyl groups at the surface. The next lowest are methyl groups. Progressively higher surface tensions result from aliphatic chains, aromatic rings, esters and ketones, alcohols, and, finally, water. (Mercury has a still higher surface tension.)

Polyfluorosubstituted aliphatic hydrocarbon chains have the lowest surface tensions of any of the materials we deal with in the coatings field. Next lowest is poly(dimethylsiloxane). The very flexible (easily rotatable) backbone of siloxane bonds permits orientation of a large population of methyl groups at the surface. The surface tension of linear aliphatic hydrocarbons increases as chain length increases, reflecting the larger ratio of methylene to methyl groups. In general also, as the chain length of aliphatic esters, ketones, and alcohols increases, the surface tension increases.

Water has the highest surface tension of the volatile components used in coatings. Addition of small amounts of surfactants to water gives low surface tension with the hydrocarbon chains on the surface.

In general, if more than one type of molecule is present, the segments of those molecules that lead to the lowest surface energy come to the surface of a liquid. When a liquid is stirred or otherwise agitated, the molecules at the surface are mixed in with the rest of the liquid. When the agitation is stopped, reorientation to give the lowest surface tension occurs. The equilibrium surface composition is not reestablished immediately. When coating films are applied, they are subjected to considerable agitation. As Bierwagen [1,2] has pointed out, the surface tension of importance in governing some aspects of coating behavior may not be the *equilibrium surface tension* but rather a *dynamic surface tension*.

The amount of time to establish equilibrium after agitation has ceased varies depending on the composition [3]. Unfortunately, there has not been adequate quantitative study of the rates with different systems of importance in the coatings field. Qualitatively, one can say that equilibrium is established most rapidly when dealing with small, flexible molecules and where there are large differences in the polarity of components in the system. Reaching equilibrium takes longer when the molecules with the lowest potential surface tension groups are polymers. But, if the polymers have moderate molecular weight and flexible backbones, they can apparently reach the surface relatively rapidly. [Poly(dimethylsiloxane) mentioned above is an example.] Low molecular weight octyl acrylate copolymers are widely used as additives to reduce surface tension of coating films as they are forming. In water-borne coatings, it has been shown that different surfactants differ in their rates of reaching equilibrium surface tension [3].

It is useful to remember that surface tension increases with decreasing temperature. It is also a quite valid generalization that solvents have lower surface tensions than coatings resins. Therefore, surface tension increases as solvent evaporates from a film of resin solution owing both to the change in concentration and temperature.

If two liquids with different surface tensions are placed in contact with each other, the liquid of low surface tension flows to cover the liquid with higher surface tension since this results in a lower overall surface free energy. Such flow is a *surface tension differential driven flow*; some authors prefer the terminology *surface tension gradient driven flow*. Flows of this type have been observed for millenia but Carlo Marangoni, a 19th century Italian physicist, is credited with providing a sound scientific understanding of the phenomena [4].

A commonly observed example of the Marangoni effect is the flow that takes place when a clean glass containing wine or brandy is tipped, wetting the side of the glass with the liquid, and then returned to an upright position. Liquid then flows up the side of the glass, forming a bead of greater film thickness along the upper edge of the wetted area. In many cases, the amount of liquid collecting in the bead becomes so great that the force of gravity leads to droplets flowing back down the side of the glass. These "tears of wine" have been known since biblical times, but why does the

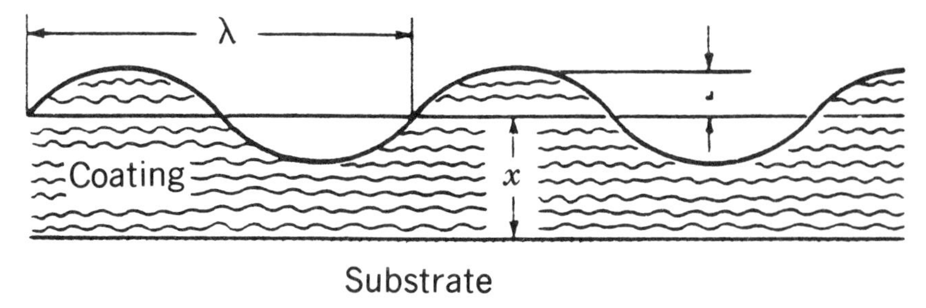

Figure 23.1. Schematic diagram of the cross section of brush marks. (From Ref. [6], with permission.)

phenomenon occur? Ethyl alcohol has a higher relative evaporation rate and a lower surface tension than water. Evaporation occurs most rapidly along the edge of the layer of wine or brandy on the side of the glass. This leads to a reduction in the alcohol concentration along the edge compared to the concentration in the bulk of the liquid. The lower concentration alcohol solution at the edge has a higher surface tension. To minimize surface free energy, the low surface tension (higher alcohol concentration) liquid in the bottom of the glass flows upwards to cover the higher surface tension liquid at the edge of the liquid. Evaporation continues at that edge, reducing the alcohol concentration and leading to a continuation of the flow of the lower surface tension liquid up the glass. The surface tension differential is also affected by the temperature change. As ethyl alcohol evaporates, temperature decreases, further increasing the differential in surface tension.

In summary, there are two important types of flow resulting from surface tension effects. Surface tension driven flows occur to minimize the surface area of a liquid. Surface tension differential driven flows occur to cover a liquid, or any other surface, of higher surface tension with a liquid of lower surface tension.

23.1. LEVELING

Most methods of application of coatings lead initially to formation of a rough wet film. It is generally desirable for both appearance and performance to have the irregularities level out. The most widely studied leveling problem has been the leveling of brush marks. While a person unacquainted with the field might first say that leveling results from gravitational effects, this is clearly not, at least to any significant degree, the case. If gravity were a significant factor, paints applied to ceilings should level much more poorly than paints applied to floors, but that is not the case. Based on studies of the flow of mineral oil, Orchard [5] proposed that the driving force for leveling is surface tension and established mathematical models for the variables that he proposed would control the rate of leveling.

Orchard's treatment has been widely applied to flow of coatings. Patton [6] illustrates the model with an idealized cross section of a wet film exhibiting brush marks that follow a sine wave profile as shown in Figure 23.1. He gives several forms of what is called the Orchard equation and shows their derivations. A convenient form, Eq. 23.1, relates the change in amplitude of the sine wave to time.

$$\ln \frac{a_0}{a_t} = \frac{5.3\ \lambda\ x^3}{\lambda^4}\frac{dt}{\eta} \tag{23.1}$$

where

a_0 = initial amplitude (cm)
a_t = amplitude at time t (cm)
x = average coating thickness (cm)
λ = wavelength (cm)
γ = surface tension ($mN\ m^{-1}$)
η = viscosity (Pa·s)
t = time (s)

Most rapid leveling occurs when wavelength is small, viscosity is low, surface tension is high, and film thickness is large. Unfortunately, the paint formulator has little or no control over most of the variables. Wavelength is determined by application conditions; in brushing it increases as pressure on the brush increases and also as the thickness of the coating increases. High surface tension increases the rate of leveling; however, the formulator is limited in optimizing this factor since, as will be seen, high surface tension can lead to other defects. Thicker films promote leveling; however, increasing film thickness increases cost of painting and, as seen in the next section, increases the probability of sagging on vertical walls. The principal means of control left to the formulator is viscosity.

In most cases viscosity changes during the time in which leveling occurs. The Orchard equation does not adequately take this variability into consideration. As solvent evaporates, viscosity increases. Further, if the system is thixotropic, the viscosity is reduced by the high shear rate exerted during brushing and subsequently increases with time as this shearing force is removed at the very low shear rate involved in the leveling process. Another potential shortcoming in the Orchard treatment is the assumption of constant surface tension; as noted earlier, dynamic surface tension may be critical.

The Orchard model provides satisfactory correlation between experimental data and predictions when the liquid film has Newtonian flow properties and sufficiently low volatility so that the viscosity does not change during the experimental observations. The model does not provide satisfactory correlations in at least some coatings where the properties of the coating change during leveling. Overdiep [7,8] devised methods of observation that permitted following the location of the ridges and valleys. He found with two alkyd coatings that brush marks leveled to an essentially smooth film, but then ridges grew where there had been valleys and valleys formed where there had been ridges. Although surface tension can and does cause a ridged film to level, it cannot cause a level film to become ridged because that creates more surface area. Overdiep proposed that flow driven by surface tension differential is the major driving force.

As can be seen in Figure 23.1, the wet film thickness in the valleys of the brush marks is less than in the ridges. When the same amount of solvent evaporates per unit area of surface, the fraction of solvent that evaporates from the coating in the valleys is larger than from the coating in the ridges. As a result, the concentration of the resin solution in the valleys is higher than that in the ridges. Correspondingly, the surface tension in the valleys is higher than on the ridges. Following the Mar-

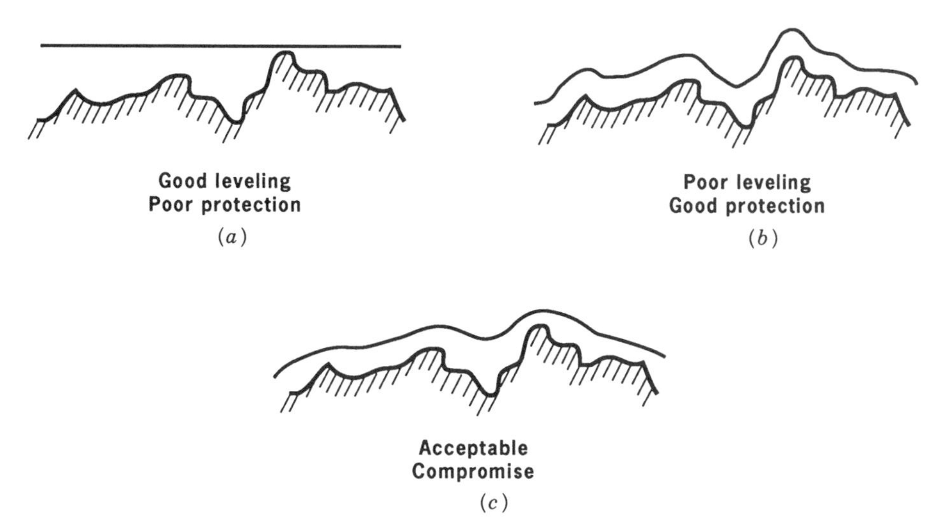

Figure 23.2. Alternate leveling results (*a–c*) after applying a coating to a rough surface. (Adapted from Ref. [7], with permission.)

angoni effect, coating flows from the ridges into the valleys. In other words, Overdiep proposed, and was able to demonstrate experimentally, that with volatile solvents, the primary driving force for leveling was not surface tension but surface tension differential. In some cases, as he showed, these differentials can lead to overshooting of the smoothest stage and lead to the growth of ridges. The extent of the flow driven by surface tension differential depends on the rate of evaporation of the solvent.

Overdiep [7] was particularly interested in what happens with uneven paint films over a rough substrate. He reasoned that surface tension driven flow might give the smoothest film. However, as illustrated in Figure 23.2*a*, this may be undesirable because protection in thin areas might be limited. On the other hand, as shown in Figure 23.2*b*, surface tension differential driven flow would tend to yield equal film thickness but with the surface of the film following the roughness of the substrate rather than being level. Overdiep suggests it might be best to adjust a coating so that both types of flow are significant so as to achieve a compromise with reasonable film smoothness and without very thin thickness sections, as shown diagrammatically in Figure 23.2*c*. The balance could be controlled by the volatility of the solvent; with very low volatility, leveling would be surface tension driven; with relatively high volatility, the major effect would be surface tension differential; with intermediate volatility, both phenomena could be important.

In spray application, surface roughness consists of bumps surrounded by valleys rather than ridges and valleys. Since the effect is somewhat reminiscent of the appearance of orange skins, it is commonly called *orange peel* (Fig. 23.3). The bumps are generally much larger than spray droplets.

A variety of factors can affect orange peel. Most commonly orange peel is encountered in spraying coatings having solvents with high evaporation rates. It is common for people to conclude that because of the fast evaporation, the viscosity

Figure 23.3. Typical orange peel pattern (15X). (From Ref. [9], with permission.)

of the coating on the substrate builds up so rapidly that the leveling is poor; in some situations, that is probably the case.

However, in the late 1940s, it was found that leveling of sprayed lacquer films could frequently be substantially improved by addition of very small amounts of silicone fluid [low molecular weight poly(dimethylsiloxane)]. Contrary to the common wisdom that all leveling was surface tension driven, that is, promoted by high surface tension, here was a case where adding a material known to reduce surface tension substantially improved leveling.

Hahn [10] provided an explanation for the phenomenon. When one sprays a lacquer, initially the surface is fairly smooth and, as you watch, orange peel grows. Hahn proposed that the growth of the orange peel results from a surface tension differential driven flow. The last atomized spray particles to arrive on the wet lacquer surface have traveled for a longer distance between the spray gun and the surface, hence have lost more solvent, have a higher resin concentration and, therefore, have a higher surface tension than the main bulk of the wet film. The lower surface tension wet lacquer flows up the sides of these last particles so as to minimize overall surface free energy. That is, the surface tension differential driven flow grows the orange peel. If one adds a surface tension depressant, such as a silicone fluid, that orients very rapidly to the surface, the surface tensions of the wet lacquer surface and of the last atomized particles to arrive are all uniformly low so there is no differential to promote growth of orange peel. The leveling that does occur in such a film is surface tension driven by the rather low surface tension after addition of silicone fluid. While this may be a small driving force, it is more important that the growth of irregularities due to differentials in surface tension is avoided. Octyl acrylate copolymers can also give an overall low surface tension and minimize orange peel growth.

It would be of interest to see what the effect of silicone fluid or octyl acrylate copolymers would be on the leveling of brush applied coatings. To the extent that the leveling is surface tension differential driven as found by Overdiep, the leveling should be made poorer, not improved. The results of the experiment are not given in the literature, but apparently these additives are not used to promote leveling in brush applied coatings, only in spray applied coatings.

Electrostatically sprayed coatings are likely to show surface roughness greater than a corresponding nonelectrostatically sprayed surface. As a result, the final passes in the case of automotive top coats are generally applied without electrostatic charging even though the major part of the top coat may be applied using electrostatic spray to reduce overspray. It has been speculated that the greater surface roughness obtained with electrostatic spray results from arrival of the last charged particles onto a coated surface that is quite well electrically insulated from the ground. These later arrivals may retain their charge sufficiently long to repel each other and thereby reduce the opportunity for leveling.

It has also been suggested [11] that when coatings are applied by high-speed bell electrostatic spray guns, differentials in the pigment concentration within the spray droplets may result from the centrifugal forces. These pigment concentration differentials lead to rougher surfaces and reduction in gloss of the final films; it is reported that gloss reduction increases as the rpm of the bell increases.

There have not been enough intensive studies of the factors controlling level of a wide variety of coatings. The leveling of powder coatings is discussed in Section 31.3. Leveling problems are particularly severe with latex paints. Latex paints, in general, exhibit a greater degree of shear thinning and more rapid recovery of viscosity after exposure to high shear rates than do paints made with solutions of resins in organic solvents. Owing to their much higher dispersed phase content, the viscosity of latex paints changes much more rapidly with loss of volatile materials than does the viscosity of solvent-borne paints. No experimental work has been reported on the relative importance of surface tension and surface tension differentials in leveling of latex patients; however, it seems probable that the leveling is primarily surface tension driven. The surface tension of water is high, but the presence of surfactants imparts low surface tension to latex paint. Furthermore, it is probable that this low surface tension is established rapidly, although not instantaneously, after the agitation of application stops. Perhaps more importantly, the surface tension is uniformly low since it is almost unchanged as water evaporates. Thus the generally poor leveling of latex paints may result in part from the absence of surface tension differentials to promote leveling. The low surface tension may not provide adequate driving force for leveling in a film whose viscosity is increasing rapidly with time. The problems of leveling of latex paints are discussed further in Section 35.3.

23.2. SAGGING

When a wet coating is applied to a vertical surface, the force of gravity causes it to flow downwards (*sag*) to some extent. Differences in film thickness at various places lead to differing degrees of sagging, resulting in curtains or drapes of paint. Patton [12] gives Eq. 23.2, showing the variables that affect the volume of paint that sags as a function of time.

$$V_s = \frac{x^3 \rho g t}{300\eta} \tag{23.2}$$

where

x = Initial film thickness (cm)
V_s = Volume of coating that has sagged after time t
ρ = density (g cm^{-1})
g = Gravitational constant (s cm^{-2})
t = time (s)
η = viscosity (Pa·s)

The driving force in sagging is gravity (g). Density of the coating (ρ) is a factor; in some cases high density inert pigments can be avoided, but generally the formulator has little latitude to control density. Thick films should be avoided, but hiding generally dictates some minimum film thickness. Therefore, viscosity is the major variable available for controlling sagging. Unfortunately, higher viscosity to control sagging reduces leveling. Overdiep [13] has developed equations that take into consideration the changes in viscosity after application.

The tendency to sag can be evaluated by observing the behavior of films applied under conditions simulating field use. However, this does not provide a numerical basis for evaluating the extent of sagging. Various tests have been developed. The most commonly used are *sag-index blades*, a straight edge applicator blade with a series of ¼-in. gaps of different depths at 1/16-in. across the blade [14]. A drawdown, which is a series of stripes of paint of various thickness, is made on a chart and immediately the chart paper is placed in a vertical position. When the paint sags, a stripe may sag down to the edge of the next stripe; if the paint is very subject to sagging, a thin stripe will sag down to the next stripe and, if the paint is more resistant to sagging, only thicker stripes will sag down to the next stripe. For research purposes, Overdiep has developed a more sophisticated method, the sag balance as described in Ref. [13].

In spray applied solvent solution coatings, sagging can generally be minimized while still achieving adequate leveling by a combination of proper use of the spray gun and control of the rate of evaporation of solvent. The goal is to manipulate viscosity so that it is initially low for leveling but builds up before severe sagging has been encountered. In brush and hand roller applied coatings where fast evaporating solvents cannot be used, thixotropic systems have to be devised that permit leveling to occur before the viscosity recovers, but where the recovery of the viscosity occurs soon enough that sagging is not serious. As would be expected, latex paints in general are less likely to exhibit sagging than are solvent solution paints since they are almost always thixotropic.

In very high solids coatings, especially when spray applied, sagging can be a serious problem. This is contrary to what most formulators expected. In order to obtain the same dry film thickness as with a conventional coating, less wet film need be applied. Since sagging decreased with the cube of film thickness, one might expect less problem controlling the sagging of high solids coatings than conventional coatings. Furthermore, the increase in viscosity resulting from the loss of the same amount of solvent from a high solids coating is greater than the increase in viscosity of the conventional coating. This factor would also lead one to expect that sagging should be easier to control with a high solids coating. However, it has been commonly found in practice that sagging of high solids coatings is more difficult to control.

Although other factors may also be involved, one reason for the unexpected problems is that substantially less solvent is lost during spraying (i.e., after leaving the gun and before arriving at the substrate surface) of high solids coatings [15,16]. This lower loss of solvent leads to less increase in viscosity for the high solids coating as compared to the conventional coating with the resultant greater likelihood of sagging.

The reasons for the lower solvent loss have not been clearly established. It has been proposed that, because of higher surface tension, high solids coatings atomize to give larger particle size droplets than conventional coatings. The lower ratio of surface area to volume would lead to lower solvent losses. However, one should be able to adjust the spray gun pressures and so forth to obtain equivalent atomization. The differences may result from a colligative effect on solvent evaporation. Resins in high solids coatings are lower in molecular weight and concentrations are higher. For both these reasons, the ratio of the number of solvent molecules to resin molecules is lower in the case of high solids coatings than conventional coatings. This would lead to a decrease in rate of solvent loss (see Section 15.1.5 for a model calculation). However, it seems doubtful that this difference could account entirely for the large differences in solvent loss, such as reported by Wu [15].

In conventional coatings, solvent evaporation from the spray droplets is controlled by the rates of evaporation of solvent in the coating. As discussed in Section 15.1.4, at later stages of solvent evaporation from a film, the rate of diffusion of solvent molecules to the surface becomes the factor limiting the rate of evaporation of solvent. It has been suggested that the stage of diffusion control of the rate of solvent loss is reached after less loss of solvent from high solids than from conventional coatings, so that solvent evaporation from spray droplets of high solids coatings is markedly reduced [17,18]. In this regard, it may be advantageous to use linear rather than branched backbone solvents in high solids coatings. Although the relative evaporation rate of branched solvents is higher than isomeric linear chain solvents, linear molecules can diffuse through a solution with limited free volume more rapidly than branched ones (see Section 15.1.4). Further research is needed to elucidate the reasons for slower solvent loss but, meanwhile, the coatings formulator must control sagging of high solids coatings.

As discussed in Section 22.2.4, hot spraying may help control sagging. When the applied coating cools on striking the object, the viscosity increase reduces sagging. Use of carbon dioxide under supercritical conditions may be particularly useful in controlling sagging since the CO_2 flashes off almost instantaneously when the coating leaves the orifice of the spray gun leading to an increase in viscosity (see Section 22.2.5). High-speed electrostatic bell application permits application of coatings at higher viscosity which also helps (see Section 22.2.3).

However, sagging of many high solids coatings cannot be adequately controlled by adjustment of solvent composition of the coating and application variables. It is then necessary to make the systems thixotropic. For example, a dispersion of a fine particle size silicon dioxide, precipitated silicon dioxide, bentonite clays treated with a quaternary ammonium compound, or polyamide gels can be added to impart thixotropy. One tries to formulate so that the recovery to high viscosity is slow enough to permit reasonable leveling and rapid enough to control sagging. However, such agents increase the high shear viscosity somewhat and hence require higher solvent levels. They also tend to lower gloss and are frequently not effective at higher temperatures.

The problem of sagging in high solids automotive metallic coatings can be particularly severe (see Section 33.1.3.1). Even a small degree of sagging, that might not be noticeable in a white coating, will be very evident in a metallic coating since it affects the orientation of the metal flakes. Furthermore, use of SiO_2 to impart thixotropy is undesirable since even the low scattering efficiency of SiO_2 is enough to reduce the metallic flop in the coatings. Therefore, acrylic microgel compositions have been developed that can impart thixotropic flow using swollen gel particles [19]. In the final film, the index of refraction of the polymer from the microgel is essentially identical with that of the cross-linked acrylic binder polymer so there is no light scattering to interfere with metallic flop. It has been reported that there is also an improvement in the strength of the final film when microgels are incorporated [20].

Another problem that can be encountered with high solids coatings is *oven sagging* [17]. The coating appears to be fine until it is put into the oven; then sagging occurs. Oven sagging results from the strong temperature dependence of the viscosity of high solids coatings. As compared to conventional coatings, there is a much steeper drop in viscosity as the coated product enters the hot oven which promotes sagging. Oven sagging can be somewhat controlled by zoning the oven. A lower temperature in initial zone gives more time for solvent loss and perhaps some cross-linking so that viscosity increases as a result of the higher solids or higher molecular weight before the film is subjected to high temperature.

Water-reducible coatings are less likely to give sagging problems than high solids coatings, but there are circumstances where they will exhibit delayed sagging. The viscosity of these coatings is very dependent on the ratio of solvent to water as well as to solids content (see Section 7.3). As the water and solvent evaporate, the residual water to solvent ratio can give lower viscosity in spite of higher solids and sagging can result. Such behavior can depend on the relative humidity during the flash-off period after spraying. It has been found with a water-reducible acrylic enamel that sagging occurred above but not below a critical relative humidity [21] (see Sections 15.1.3 and 15.1.6 for the definition and discussions of critical relative humidity).

23.3. CRAWLING, CRATERING, AND RELATED DEFECTS

If one applies a coating with a relatively high surface tension to a substrate with a comparatively low surface tension, the coating will not wet the substrate. The mechanical forces involved in the application may spread the coating on the substrate surface but, since the surface is not wetted, surface tension forces will tend to draw the liquid coating back toward spherical shape. Meanwhile solvent is evaporating and, therefore, viscosity is increasing so that before the coating can pull up into spheres the viscosity is high enough that flow essentially stops. The result is a very uneven film thickness with areas having little, if any, coating ad-

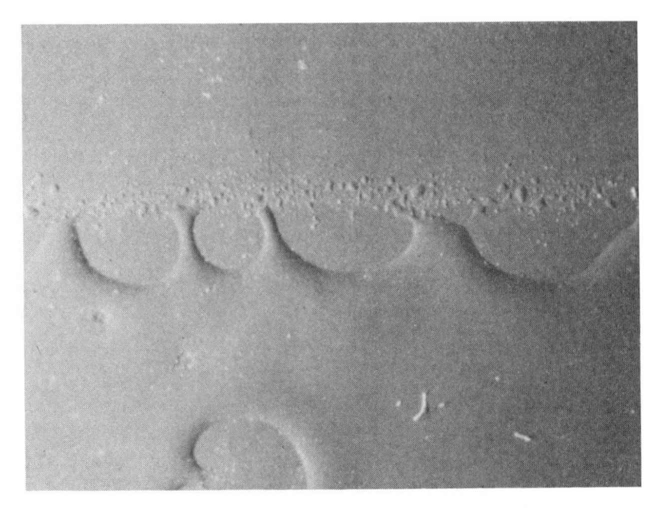

Figure 23.4. Crawling of a topcoat applied over a low surface energy primer (7X). (From Ref. [9], with permission.)

joining areas of excessive film thickness. This behavior is commonly called *crawling*; it is also called *retraction*. Figure 23.4 shows an example of crawling. In waterborne coatings, it has been shown that crawling can depend on the rate of establishment of equilibrium surface tension with different surfactants [3].

Crawling can result from applying a coating to steel with oil contamination on the surface. It is especially common in coating plastics. In some cases, crawling results from the failure to remove a mold release agent completely from a plastic molded part. Application of a high surface tension top coat to a low surface tension primer can lead to crawling. If a coating contains silicone fluids or fluorocarbon surfactants, there is likely to be a crawling problem when a subsequent coat is applied. To avoid crawling, the surface tension of the coating must be lower than the surface tension of the substrate.

If one handles a primer surface with bare hands and then applies a relatively high surface tension top coat, it is likely that the top coat will draw away from the oils left behind in fingerprints. This result of crawling which copies a pattern of low surface tension areas on the substrate has been called *telegraphing*. Care is needed with the term telegraphing since this is only one of several phenomena called telegraphing.

Crawling can also result from the presence in the coating of surfactant-type molecules that can orient rapidly on a highly polar substrate surface. In such a case, even though the surface tension of the coating is lower than that of the substrate, it could be higher than the surface tension of the substrate after the surfactant oriented on the surface. This could occur if the polar group of the surfactant associated with the substrate and the long nonpolar end becomes the surface that the coating must wet.

If one adds excess silicone fluid to a coating in order to correct a problem like excessive orange peel, small droplets of insoluble fractions of the poly(dimethylsiloxane) can migrate to the substrate surface and spread on it, leaving a new substrate surface that the coating cannot wet, resulting in crawling. A little

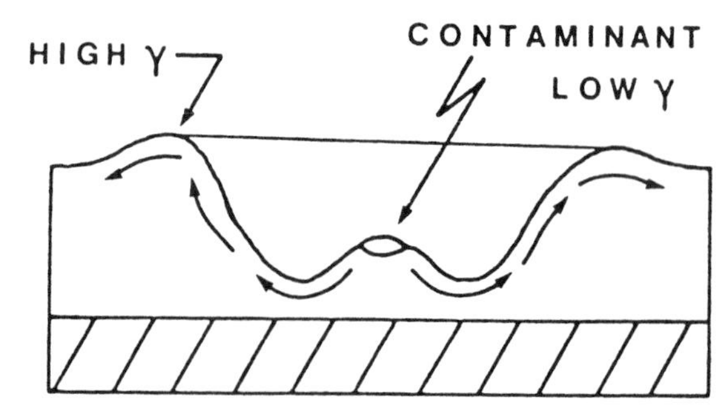

Figure 23.5. Schematic diagram of a crater. (From Ref. [9], with permission).

silicone fluid can solve some defect problems but even a small excess can cause what might be a worse problem. It has been reported that higher molecular weight fractions of poly(dimethylsiloxane) are insoluble in many coating formulations [22]. Modified silicone fluids, such as polysiloxane/polyether block copolymers, have been developed that are compatible with a wider variety of coatings and are less likely to cause undesirable side effects.

High solids coatings are likely to have higher surface tensions than conventional coatings. To achieve the high solids, lower molecular weight resins with lower equivalent weights must be used. This means that the concentration of polar functional groups such as hydroxyl groups is higher and hence surface tension will generally be higher. Also, the solvents that give the lowest viscosity systems are likely to be relatively high surface tension solvents. Therefore, there is a greater likelihood of crawling problems with high solids coatings. The effect of a long series of additives on crawling and other film defects has been reported [23].

Cratering is the appearance of small round defects that look somewhat like volcanic craters on the surface of coatings. A schematic drawing and a photograph are shown in Figures 23.5 and 23.6, respectively. Cratering should not be confused with popping, which is discussed in the next section. Cratering results from a small particle of low surface tension contaminant which is in the coating or lands on the wet surface of a freshly applied paint film [10]. Some of the low surface tension

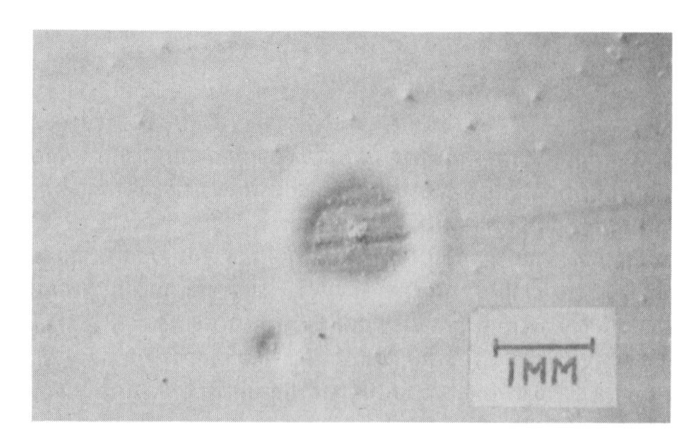

Figure 23.6. Typical crater. (From Ref. [9], with permission.)

material dissolves in the adjacent surface of the coating film, thereby creating a localized surface tension differential. As a result of the Marangoni effect, this low surface tension part of the coating film flows away from the particle to try to cover the surrounding higher surface tension liquid coating. Since as the flow occurs, solvent evaporates, the differential in surface tension increases and the flow continues. However, loss of solvent causes an increase in viscosity which impedes flow, leading to formation of a characteristic crest around the pit of the crater. Commonly, as shown in Figure 23.5, a particle of contaminant can be seen in the center of a crater.

Craters must be avoided in applying coatings. The user applying the coating should try to minimize the probability of low surface tension contaminants arriving on the wet coating surface. For example, the spraying of lubricating oils or silicone fluids on or near the conveyor carrying freshly painted parts is an almost sure way to cause craters. However, in most factories the presence of some contaminating small particles cannot be avoided; therefore, the formulator must design coatings that minimize the probability of cratering. Lower surface tension coatings are less likely to form craters since a smaller number of the contaminating particles will have still lower surface tension. Alkyd coatings have low surface tensions and seldom give cratering problems (or crawling problems either). In general, solvent-based polyester coatings are more likely to give cratering problems than acrylic coatings, which tend to have lower surface tensions. High solids coatings, because of their generally higher surface tensions, are more likely to give cratering problems than conventional coatings. Some water-borne coatings are also vulnerable to cratering. In one case it was found that trace particles from cosmetics worn by production workers caused cratering of a water-borne automotive base coat.

Additives can be used to minimize cratering. Small amounts of silicone fluid generally eliminate cratering but, as noted before, caution is required in selecting the amount and type of silicone fluid used to avoid crawling problems. Octyl acrylate copolymer additives usually substantially reduce cratering. All of these additives operate by reducing the surface tension to a level lower than that of most contaminants that could cause cratering. If the entire surface has a uniformly low surface tension, there will be no surface tension differential driven flow. A comparison of the effects of a range of additives on the control of defects, such as cratering, in a range of coatings has been reported [24]. The report also includes possible side effects such as reduction of gloss, as well as loss of adhesion of coatings applied over the surface of the coatings containing the additives.

There are many other examples of film defects resulting from surface tension differential driven flows. In coating tin plate sheets, the coating is applied by roller and the coated sheets are passed on to warm wickets that carry the sheets approximately vertically through an oven. In some cases, one can see a pattern of the wicket as a thin area on the final coated sheet. The heat transfer to the sheet is fastest where it is leaning against the metal wicket. The surface tension of the liquid coating on the opposite side drops locally because of the higher temperature. This lower surface tension material flows toward the higher surface tension surrounding coating leaving an area of thin coating. This defect has also been called telegraphing.

In spraying flat sheets, one can get an effect called *picture framing*; the coating is thickest at the edges and, just in from the edge the coating is thinner than average. The contrast in hiding of the substrate can make the differences in film thickness very evident. Solvent evaporates most rapidly from the coating near the edge where the air flow is greatest. This leads to an increase in resin concentration at the edge and to a lower temperature. Both factors increase the surface tension there, causing the lower surface tension coating adjacent to the edge to flow out to cover the higher surface tension coating.

Surface tension differential driven flow can also result when overspray from spraying a coating lands on the wet surface of a different coating. If the overspray has lower surface tension then the wet surface, cratering occurs. If the overspray has high surface tension compared to the wet film, local orange peeling results.

In applying coatings by curtain coating (see Section 22.6), it is critical that the curtain of flowing paint remain intact. If a particle of contaminant of lower surface tension than that of the coating lands on the surface of the flowing curtain, surface tension differential driven flow will cause a thin area in the curtain which can lead to a hole in the curtain. When this part of the curtain is deposited on the panel being coated, an uncoated area results. The problem is minimized by using coatings of the lowest possible surface tension. Since the curtain is flowing, the surface tension of the coating is not that measured at equilibrium; Bierwagen [2] discussed this potentially difficult problem.

23.4. FLOATING AND FLOODING; HAMMER FINISHES

The film defect called *floating* is most easily seen in coatings pigmented with at least two pigments. For example, a light blue gloss enamel panel can show a mottled pattern of darker blue lines on a lighter blue background. The pattern tends to be hexagonal, but seldom perfectly so. Alternatively, with a different light blue coating, the color pattern might be reversed: the lines could be very light blue with the background being a darker blue.

These effects result from pigment segregations that occur as a consequence of convection current flows driven by surface tension differentials while a film is drying. Rapid loss of solvent from a film during drying leads to considerable turbulence. Convection patterns are established whereby coating material flows up from the lower layers of the film and then circulates back down into the film. As the fresh material flows across the surface before it turns down, solvent evaporates, concentration increases, and temperature drops. As a result, surface tension increases. The resultant surface tension differential drives continuation of the convection current. The flow patterns are roughly circular, but as they expand they encounter other flow patterns and the convection currents are compressed. If the system is quite regular, a pattern of hexagonal *Bénard cells* is established. The cells are named after a 17th century French scientist who pointed out the commonness of hexagonal flow patterns in nature. As the solvent evaporation continues, the viscosity increases, and it becomes more difficult for the pigment particles to move. The smallest particle size, lowest density particles continue moving for the longest time and the largest particle size, highest density particles stop moving sooner. The observed segregated pattern of floating results.

Floating is particularly likely to occur if one of the pigments if flocculated and the other is present as a nonflocculated dispersion of fine particle size. The fine particle size pigment keeps moving the longest and is trapped in larger amounts

Figure 23.7. Typical Bénard cell pattern. (From Ref. [9], with permission.)

where the convection current turns back into the film at the border between adjacent cells. The border between the cells has a higher concentration of the finer particle size material whereas the center of the cells is more concentrated in the coarser pigment. If, in the example of the light blue paint, the white pigment is flocculated and the blue is not, one would find darker blue lines on a lighter blue background. If, on the other hand, the blue were flocculated and not the white, there would be lighter blue lines on a darker blue background. Figure 23.7 shows a photograph of a Bénard cell pattern in an industrial coating; Figure 23.8 shows a schematic diagram of convection patterns in Bénard cell formation.

Obviously, floating can be reduced by properly stabilizing pigment dispersions so that neither pigment is flocculated. However, one can get floating even without flocculation. This results from using pigments with very different particles sizes and densities. An obvious example is the use of fine particle size, high-color carbon black with titanium dioxide to make a gray paint. Not only is the particle size of

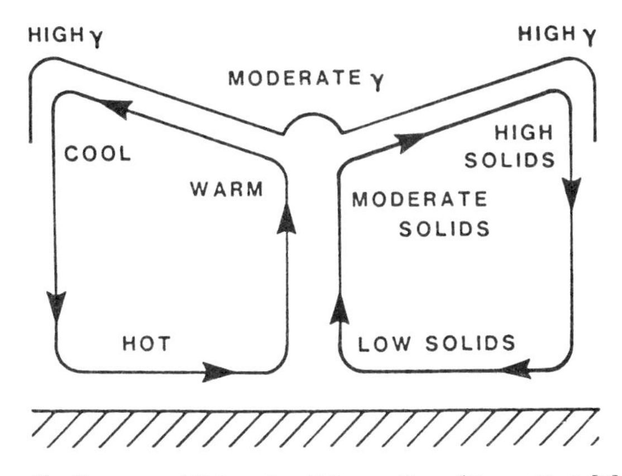

Figure 23.8. Schematic diagram of Bénard cell formation. (From Ref. [9], with permission.)

the TiO_2 several times that of the carbon black, the TiO_2 also has about a fourfold higher density. The problem can be minimized by avoiding such combinations; a much larger particle size, weaker black, such as lamp black, can be used to make a gray with substantial reduction in the probability of floating.

Floating is reduced if slower evaporating solvents are used. Surface tension differentials are then less likely to be established and, if they are not present, Marangoni flow and floating will not occur. If one has a coating that exhibits floating, the formulator should first reformulate to eliminate the problem: most commonly, selective pigment flocculation is the cause and should be corrected. The second choice is to use an additive. As in the case of other flow phenomena driven by surface tension differentials, floating can be prevented by adding a small amount of a silicone fluid. To avoid crawling, the amount of silicone must be kept small. The probability of problems can be reduced by using a very dilute solution of the silicone fluid in solvent. Since slow evaporating solvents also can help reduce floating, it is common to use slow evaporating solvents in making such additive solutions.

A somewhat related phenomenon is called *flooding*. When flooding occurs a mottled appearance is not present, but the color of the surface is different than should have been obtained from the pigment combination involved. For example, one might have a uniform gray coating but the gray would be darker than would be expected from the ratio of black to white pigments. Flooding results from surface enrichment by one or more of the several pigments in the coating [9,25]. The stratification is thought to occur as a result of different rates of pigment settling within the film caused by differences in pigment density and size or flocculation of one of the pigments. Flooding is accentuated by thick films, low vehicle viscosity, low evaporation rate solvents—anything that tends to keep the film at low viscosity longer and allow more pigment settling. The remedies are to avoid flocculation and low-density fine particle size pigments if possible, and to use faster evaporating solvents and higher viscosity vehicles. Floating, flooding, and related color defects are discussed in Ref. [25].

While floating is usually undesirable, ingenuous coatings formulators have taken advantage of the problem by purposely inducing floating to make attractive coatings. The coatings are called *hammer finishes* because they look a little like the pattern one would get striking a piece of metal with a ball peen hammer. Hammer finishes were used on a very large scale, especially for coating cast iron components where it was desirable to hide the surface roughness. Such coatings contain large particle size, nonleafing aluminum pigment and dispersions of transparent fine particle size pigments—commonly phthalocyanine blue.

One way of getting a hammer effect is to spray a metallic blue coating and then spray a small amount of solvent on the wet film. The surface tension is lowest where the drops of plain solvent land and surface tension differential driven convention current flow patterns are set up, leading to floating where the lines have more blue and the centers of the patterns have more aluminum with less blue.

There are also so-called *self-hammer* coatings, formulated to give a hammer finish pattern without the need for the spatter spray of solvent. The effect is accomplished by using fast evaporating solvents in the coating together with a resin, such as a styrenated alkyd resin that will give fast drying. Although hammer finishes are still sold, their usage has dropped off because smooth plastic molded parts are now used that do not need a coating to hide a rough surface.

23.5. WRINKLING; WRINKLE FINISHES

The term *wrinkling* refers to the surface of a coating that looks shriveled or wrinkled into may small hills and valleys. In some cases, the wrinkle pattern is so fine that to the unaided eye the film appears to have low gloss rather than to look wrinkled. However, under magnification, the surface can be seen to be glossy but wrinkled. In other cases, the wrinkle patterns are broad or bold and are readily visible to the naked eye.

Wrinkling results when the surface of the film becomes very high in viscosity while the bottom of the film is still relatively fluid. It can result from rapid solvent loss from the surface, followed by later solvent loss from the lower layers. More commonly, it results from more rapid cross-linking at the surface of the film than in the lower layers of the film. The subsequent solvent loss or cure results in shrinkage in the lower layers which pulls the surface layer into the wrinkled pattern. Wrinkling is more apt to occur with thick films than with thin films because the possibility of different reaction rates and differential solvent loss within the film increase with thickness.

The earliest examples of wrinkling were with drying oil films, especially if all or part of the oil was tung oil and cobalt salts were used as the only drier. Tung oil cross-links relatively rapidly when exposed to oxygen of the air, and cobalt salts are very active catalysts for the autoxidation reaction but are poor through driers (see Section 9.2.2). These factors favor differential surface cure. After there has been significant cross-linking of the surface, the lower layers of the film cross-link, leading to shrinkage resulting in wrinkling of the surface layer. Depending on the ratio of tung oil to other drying oils or, in the case of alkyd systems, the oil length of the alkyd, and the ratio of cobalt to driers such as lead (or zirconium) salts that promote through dry, the wrinkle pattern can be fine or bold. While in many cases wrinkling is undesirable, ingenious paint people turned the disadvantage into an advantage and for many years *wrinkle finishes* were sold on a large scale for applications such as office machinery. Somewhat analogous to hammer finishes, wrinkle finishes covered unevenness in cast metal parts. Their usage has dropped markedly since plastic molded parts have supplanted many metal castings.

Today, wrinkling is usually an undesired defect. It is most commonly encountered in improperly formulated or applied MF cross-linked coatings in which amines are used to neutralize acidity in the coating formulation and/or amine blocked sulfonic acid catalysts are used for package stability. The probability of encountering wrinkling in such coatings increases as the volatility of the amine increases. For example, triethylamine leads to wrinkling under conditions where dimethylaminoethanol neutralized coatings do not wrinkle. Increasing catalyst concentration leads to increasing probability of wrinkling. Greater film thickness increases the probability of wrinkling [26].

Another type of coating where wrinkling commonly occurs in the laboratory is in UV curing of pigmented acrylate coatings with free radical photoinitiators (see

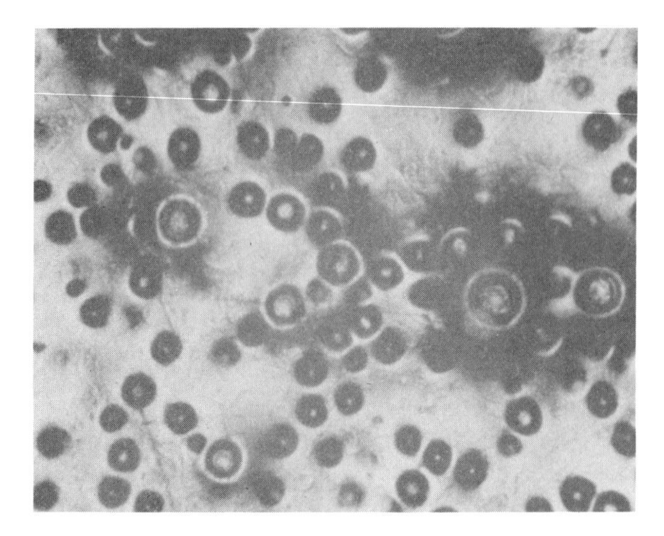

Figure 23.9. An extreme example of popping. (From Ref. [9], with permission.)

Section 32.1.3). High concentrations of photoinitiator are required to compete with absorption by the pigment. Penetration of UV through the film is reduced owing to its absorption by the pigment as well as the photoinitiator. There is very rapid cross-linking at the surface and slower cross-linking in the lower layers of the film resulting in wrinkling. Wrinkling in such systems is likely to be more severe if the curing is done in an inert atmosphere rather than in air. In the latter case, cure differential is reduced by oxygen inhibition of surface cure. For this reason, UV curing by cationic polymerization, which is not air inhibited, is even more prone to surface wrinkling.

23.6. POPPING

Popping refers to formation of blisters in the surfaces of films that break but do not flow out before viscosity increase prevents leveling. Figure 23.9 shows an extreme case of popping. Sometimes, the surface layer increases in viscosity to the extent that the blisters do not break; this condition is also called popping, but more usually blistering or bubbling.

Popping results from rapid loss of solvent at the surface of a film during initial flash-off. The surface develops a high viscosity relative to solvent-rich lower layers of the film. When the coated object is put into an oven, solvent volatilizes in the lower layers of the film, creating bubbles that do not readily pass through the high-viscosity surface. As the temperature increases further, the bubbles expand, finally bursting through the top layer. The viscosity of the film meanwhile has increased so much that it is no longer possible for the coating to flow back together and heal the eruption.

Popping can also result from, or at least be made worse by, entrapment of air bubbles in a coating. If the surface of the film has a high viscosity, the bubbles

Table 23.1. Critical Film Thickness for Popping

	Critical Dry Film Thickness (μm)	
Copolymer T_g (°C)	Water	Solvent
−28	50	120
−13	30	>70, <95
−8	20	>70, <95
14	10	55
32	5	25

may remain in the film until the coating goes into the oven. The air expands with higher temperature and the bubbles may burst through the surface. Air bubbles are more likely to be entrapped during spray application than other means of application.

A third potential cause of popping can be the evolution of volatile byproducts of cross-linking after the surface viscosity has increased to the extent that the bubbles of volatile material cannot readily escape through the surface.

The probability of popping increases with film thickness since there is a greater chance of developing a large differential in extent of solvent loss as film thickness increases. A means of evaluating the relative likelihood that a series of coatings will show popping is to determine the maximum film thickness of each coating that can be applied without popping when the films are prepared, flashed off, and baked under standardized conditions [27]. This thickness is called the *critical film thickness for popping*.

Popping can be minimized by spraying more slowly in more passes, by longer flash-off times before the object is put into the oven, and by zoning the oven so that the first stages are relatively low in temperature. The probability of popping can also be reduced by having a slow evaporating, good solvent in the solvent mixture. This tends to keep the surface viscosity low enough for bubbles to pass through and heal before the viscosity at the surface becomes too high.

Popping can be particularly severe with water-reducible baking enamels, as shown in Table 23.1. The pairs of enamels were identical except that, in one set, they were reduced for application with only solvent and in the other set with water [27]. As can be seen, the critical film thickness for popping was consistently lower for the water-reduced coatings.

The data in Table 23.1 also illustrate another variable that affects the probability of popping. Critical film thickness for popping in solution coatings decreases as the T_g of the acrylic resin in the coating increases. This is true in both the solvent- and water-reduced coatings, but the effect is particularly large in the water-reduced compositions.

It is common experience that popping is more difficult to control in water-reduced baking finishes than in solvent-borne coatings. There are probably many reasons for this. A wide variety of solvents with different evaporation rates is available for adjusting the formulations of solvent-borne coatings, but water only has one vapor pressure/temperature curve and the curve is steeper than for any organic solvent. Water can be retained by forming relatively strong hydrogen bonds with polar groups on the resin molecules at room temperature; these hydrogen bonds break at higher temperature, releasing the water. The heat of vaporization of water, 2260 J/g, is higher than that of organic solvents—373 J/g for 2-butoxyethanol as an example. This higher heat of vaporization slows the rate of temperature increase of water-reduced coatings, further increasing the probability of popping [27].

In latex baking coatings, the T_g effect on popping is the reverse of that with water-reducible coatings. Popping is more likely to occur with lower T_g latex polymers. With the lower T_g polymer, coalescence of a surface film before the water has completely evaporated is more likely than is the case with the higher T_g polymer.

23.7. FOAMING

During application, paint is subjected to substantial agitation and mixing with air, creating the opportunity for foam formation. In the last section, the problem of entrapping air in spray application of baking coatings was referred to, but the problem of foaming is most severe in room temperature applied latex paints. Spray application of latex paints can lead to air entrapment, but hand roller application is even more likely to cause foaming.

Formation of a foam involves the generation of a large amount of surface area; it follows, therefore, that the lower the surface tension, the less the energy required to generate a given amount of foam. However, foam bubbles in pure low-viscosity liquids are not stable and break essentially instantaneously. There must be something else present to stabilize the foam. Although water has a high surface tension and therefore might be more difficult to foam, foam bubbles in water are easier to stabilize since a wider variety of components can be put in water that will rapidly migrate to the surface of a bubble to stabilize it. A surfactant not only reduces the surface tension of water, facilitating foam formation, but also migrates to the surface of the droplets to give an oriented surface layer with a high viscosity, stabilizing the foam bubbles. In formulating a latex paint, an important criterion in the selection of any surfactants or water-soluble polymers as thickeners is their effect on foam stabilization [3].

A variety of additives can be used to break foam bubbles. They all depend on creating surface tension differential driven flow on the surface of the bubble. If the surface tension of a spot on the surface can be lowered, liquid from that area will flow away to try to cover neighboring higher surface tension areas. But the wall thickness of bubbles is thin and, as material flows away, the wall gets still thinner and therefore weaker, so that this spot on the surface of the bubble breaks. For example, poly(dimethylsiloxane) fluids, that is, silicone fluids, are very effective in breaking a wide variety of foams since their surface tension is low compared to almost any foam surface. Of course, as in the other uses of silicone fluids, there is the problem that a little may be fine, but a little extra can cause all kinds of problems.

Another effective way of breaking foams, that is not practical to use with coatings, but which illustrates the principle involved, is to spray a foam with a mist of ethyl ether. When an ethyl ether droplet lands on the surface of a bubble, it flash evaporates, dropping the temperature. The lower temperature locally increases the surface tension and neighboring liquid rushes to cover the area of high surface tension, thinning the wall of the bubble and causing breakage.

Several companies sell lines of proprietary antifoam products and offer test kits with small samples of their products. The paint formulator then evaluates the antifoam products with the paint that is giving foaming problems to find one that overcomes, or at least minimizes, the problem. While it is possible to predict what additive will break a foam in a relatively simple system, such predictions are difficult in latex paints because of the wide variety of components that could potentially be at the foam interface.

GENERAL REFERENCE

P. E. Pierce and C. F. Schoff, *Coating Film Defects*, Federation of Societies for Coatings Technology, Blue Bell, PA, 1988.

REFERENCES

1. G. P. Bierwagen, *Prog. Org. Coat.*, **3**, 101 (1975).
2. G. P. Bierwagen, *Prog. Org. Coat.*, **19**, 59 (1991).
3. J. Schwartz, *J. Coat. Technol.*, **64** (812), 65 (1992).
4. L. E. Scriven and C. V. Sternling, *Nature*, **187** (4733), 186 (1960).
5. S. E. Orchard, *Appl. Sci. Res.*, **A11**, 451 (1962).
6. T. C. Patton, *Paint Flow and Pigment Dispersion*, 2nd ed., Wiley-Interscience, New York, 1979, p. 554.
7. W. S. Overdiep in D. B. Spalding, Ed., *Physicochemical Hydrodynamics*, V. G. Levich, Festschrift, Vol. II, Advance Publications Ltd., London, 1978, p. 683.
8. W. S. Overdiep, *Prog. Org. Coat.*, **14**, 159 (1986).
9. P. E. Pierce and C. F. Schoff, *Coating Film Defects*, Federation of Societies for Coatings Technology, Blue Bell, PA, 1988.
10. F. J. Hahn, *J. Paint Technol.*, **43** (562), 58 (1971).
11. K. Tachi, C. Okuda, and K. Yamada, *J. Coat. Technol.*, **62** (791), 19 (1990).
12. Ref. [5], p. 572.
13. W. S. Overdiep, *Prog. Org. Coat.*, **14**, 1 (1986).
14. Ref. [5], pp. 578–579.
15. S. H. Wu, *J. Appl. Polym. Sci.*, **22**, 2769 (1978).
16. D. R. Bauer and L. H. Briggs, *J. Coat. Technol.*, **56** (716), 87 (1984).
17. L. W. Hill and Z. W. Wicks, Jr., *Prog. Org. Coat.*, **10**, 55 (1982).
18. W. H. Ellis, *J. Coat. Technol.*, **55** (696), 63 (1983).
19. R. M. Christenson, T. R. Sullivan, S. K. Das, R. Dowbenko, J. W. Du, and R. L. Pelegrinelli, U. S. Patent 4,055,607 (1977); J. M. Maklouf and S. Porter, U. S. Patents 4,147,688 and 4,180,619 (1979), M. S. Andrews and A. J. Backhouse, U. S. Patent 4,180,619 (1979); A. J. Backhouse, U. S. Patent 4,268,547 (1981); H. J. Wright, D. P. Leonard, and R. A. Etzell, U. S. Patent 4,290,932 (1981).
20. S. Ishikura, K. Ishii, and R. Midzuguchi, *Prog. Org. Coat.*, **15**, 373 (1988).
21. L. B. Brandenburger and L. W. Hill, *J. Coat. Technol.*, **51** (659), 57 (1979).
22. F. Fink, W. Heilen, R. Berger, and J. Adams, *J. Coat. Technol.*, **62** (791), 47 (1990).

23. R. Berndimaier, J. W. Du, D. R. Haff, J. C. Kaye, E. D. Kelley, R. LaGala, J. McGrath, J. M. McKeon, U. Schuster, M. Sileo, L. Waedle, S. Westerveld, and M. E. Wild, *J. Coat. Technol.*, **62** (790), 37 (1990).
24. M. Schnall, *J. Coat. Technol.*, **63** (792), 95 (1991).
25. M. Schnall, *J. Coat. Technol.*, **61** (773), 33 (1989).
26. Z. W. Wicks, Jr. and G. F. Chen, *J. Coat. Technol.*, **50** (638), 39 (1978).
27. B. C. Watson and Z. W. Wicks, Jr., *J. Coat. Technol.*, **55** (698), 59 (1983).

CHAPTER XXIV

Mechanical Properties

Critical properties of most coatings films relate to their ability to withstand use without damage. The range of requirements is very large. The coating on the outside of an automobile should withstand being hit by a piece of flying gravel without film rupture. The coating on the outside of a beer can must be able to withstand the abrasion involved when cans rub against each other during shipment in a railroad car. The coating on wood furniture should not crack when the wood expands and contracts as a result of changing temperatures in winter shipments or due to swelling and shrinkage resulting from changes in moisture content of the wood. The coating on aluminum siding must be flexible enough to withstand fabrication of the siding and hard enough to resist scratching of the surface during installation on the house. The list could go on and on.

In this and the following three chapters, various aspects of the durability of coatings are discussed—mechanical integrity in this chapter, exterior durability in Chapter 25, adhesion in Chapter 26, and corrosion protection by coatings in Chapter 27. The introductory part of this chapter discusses some of the broad problems of developing, evaluating, and testing long-lived coatings. While aimed primarily at mechanical integrity problems, much of the discussion is applicable to all these aspects of the durability of coatings.

Development of coatings with adequate durability is made complex by the very wide range of conditions to which coatings are exposed. It is a safe generalization to say that the only way to know how a coating will perform in actual use is to apply the coating to the final product, use the product over its lifetime, and see whether the coating performs satisfactorily. But in many cases, the lifetime of the product can be very long. The coating on the outside of an automobile should maintain its integrity and appearance for well over five years. The coating on furniture should perform satisfactorily for 20 or more years. No laboratory tests are available that will permit satisfactory product performance predictions in many applications. But the formulator must have some way of judging the merits of a new formulation. The most powerful tool, available in a few laboratories, is a data bank of actual field use performance of previous formulations.

Formulators have made judgments about the effects of formulation changes on durability based on their years of experience. Older formulators have tried to pass on their accumulated experience to starting formulators. Historically, the changes made in developing new formulations were relatively small modifications of formulations with known field performance. If the change was significant, initial field use might be limited. In the automobile field, for example, it used to be common, after a promising new formulation had been developed, to coat just a few cars. Then the next year, if no problems were encountered, the new formulation might be adopted for one color on one model of automobile. The following year, the use might be extended to three or four colors on two or three models. Finally, if all this history was satisfactory, the new formulation might be widely adopted. Obviously, some mistakes were made but these results broadened the background of the formulators.

This evolutionary approach to formulating worked quite successfully. However, in recent years there has been substantial pressure to accelerate the process. This has resulted from increasing performance requirements, increasing pressures to reduce costs, and particularly the need to meet regulation requirements. Reduction in VOC emissions has been a major driving force but also other factors such as increasing recognition of possible toxic hazards, especially from long-term exposure to relatively low levels of some chemicals, have required changes in relatively short time spans.

Increasingly, data bases resulting from actual field use are being accumulated. Computers make possible the analysis of masses of data correlating actual performance with composition and application variables. For example, for years teams of representatives of automobile manufacturers and suppliers have surveyed cars in parking lots in various parts of the country. The serial number, which can be seen through the windshield, permits identification of the coatings put on that car. Pipelines are regularly monitored. Exterior siding performance is followed. Abrasion resistance of exterior beer can coatings can be related to shipment variables. History of paint performance on offshore oil rigs is followed.

The role of the computer is critical—it makes available to all formulators in a company all records rather than just the memory of a single formulator. The importance of accumulating data on coatings that fail is as great as the data on the coatings that are satisfactory. More use of this approach in the future is critical to future progress in formulating superior coatings. A word of caution: some technical people think in terms of maximizing just performance; economic factors are also critical. It is foolish to put an expensive coating that will last 20 years on a product, if the product itself will last only 5 years. The probability is high that a 5-year life coating will cost less than a 20-year life coating.

Of course, accumulating this data base takes time. Meanwhile, the formulator must have tools to guide him/her. The formulator has three kinds of needs in considering the mechanical properties of films. In order to select the most appropriate components of the coating, the formulator needs to understand the relationships between composition and mechanical properties. Laboratory tests are needed to follow the effects of changes in formulation. Also, appropriate quality control tests are needed to check that production batches of the coating will perform as well as the approved formulation. Unfortunately, in many cases people working in the field do not realize how complex the problems are. It is far too common for coatings formulators and coatings users to assume that a quality control test can predict performance. That is almost never the case.

Dickie [1] has proposed a methodology for systematically considering the factors involved in service life prediction. He suggests that predictive models can provide the framework for assessing the importance and relevance of information that is available and the insight they may give on what may be missing from the evaluation of a given material or application.

24.1. BASIC MECHANICAL PROPERTIES

Clearly, the most powerful tool potentially available to the coatings formulator is an understanding of the relationship between composition and the basic mechanical properties of films. Most coatings formulators were educated as chemists, not as engineers, and few have had any education in mechanical properties. Terms like loss modulus, storage modulus, tan delta, and the like have little meaning to them. Because of the tremendous diversity of coatings, the relatively small volume of most types of coatings, and this lack of understanding of the physics of behavior of films, the coatings industry lagged in trying to apply these concepts to coatings. The plastics, rubber, and fiber industries have used such concepts for many years in developing products with superior performance.

In 1977 Loren Hill [2] published a review paper that discussed stress analysis as a tool for understanding coatings performance. He did an excellent job of presenting an introduction to stress analysis in terms that a coatings formulator could understand. Much of the information that he presented had to be discussed based on examples from plastics, rubber, and fiber work; there were few papers that presented examples of such analysis from studies of coatings. Ten years later, the application of stress analysis to coatings had mushroomed. Hill [3] published a monograph in 1987 that updated the growing understanding in the field. References [4] and [5] are recent review papers. Further rapid progress in broadening the understanding can be anticipated.

Basic to understanding the study of mechanical properties of coatings is recognition that coating films are viscoelastic materials. Chapter 19 deals with the flow aspects of rheology; the deformation aspects of rheology are covered here. The mode of deformation can be elastic and/or viscous. In ideal *elastic deformation* (*Hookean* deformation), a material elongates under a tensile stress in direct proportion to the stress applied in conformance with Hooke's law, as exemplified by a steel spring. When the stress is released, the material will return to its original dimensions essentially instantaneously. On the other hand, an ideal viscous material, a Newtonian fluid, will elongate, that is flow, when a stress is applied in direct proportion to the stress, but will not return to, or even toward, its original dimensions when the stress is released. The deformation is permanent.

Almost all coating films are *viscoelastic*, that is, they exhibit intermediate behavior. A thermoplastic film generally does not recover its original shape after deformation; the viscous flow part of the deformation is permanent. In cross-linked films, if there is no yield point, the recovery of the original dimensions may be complete even though there was viscous flow. The stress on the cross-links supplies the force to reverse the viscous flow. If there is a yield point, there will be partial but not complete recovery of the original dimensions.

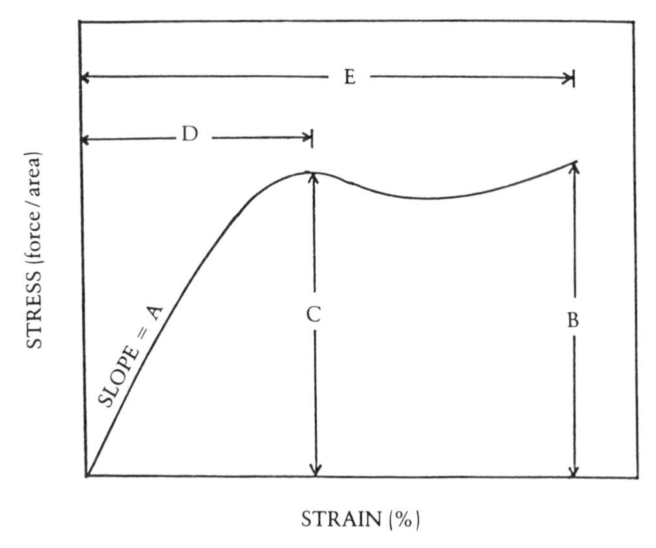

Figure 24.1. Stress-strain plot. (From Ref. [3], with permission.)

Figure 24.1 shows a schematic plot of the results of a stress–strain test, in which a sample of coating film is elongated at a constant rate (*strain*) and the resulting *stress* is recorded (methods for measurement of mechanical properties are discussed in Section 24.4). By convention, the stress (force per unit of cross-sectional area) is based on the original dimensions. Strain is expressed in terms of percent elongation of the sample. The slope (A) of the initial, essentially straight line, portion of the graph is the *modulus*, that is, the ratio of stress/strain. Thus, modulus is analogous to viscosity. One must be careful to know how the term modulus is used in a specific case. In the initial part of this plot, modulus is independent of strain. However, as strain increases, the ratio is no longer constant and the modulus depends on the strain. At the end of the curve, the sample broke. This point is defined in two ways: *elongation-at-break* is a measure of how much strain (E) can be withstood before breaking; the tensile strength or *tensile-at-break* is a measure of the stress (B) when the sample breaks. The area under the curve represents the *work-to-break* (energy vol^{-1}). Quite commonly, as shown in Figure 24.1, at an intermediate strain, the stress required for further elongation decreases. The maximum stress (C) at that point is called the *yield point*. Yield point can also be designated in two ways: *elongation-at-yield* (D) and *yield strength* (C).

An ideal elastic material deforms virtually instantaneously when a stress is applied and recovers its original shape virtually instantaneously when the stress is released. Ideal viscous flow is time dependent; the flow continues as long as a stress is applied. The rate of deformation depends on the viscosity of the material. Elastic deformation is, over a wide range, almost independent of temperature whereas viscous deformation is dependent on temperature. As a result, viscoelastic deformation is very dependent on the temperature and the time over which a stress is applied. If the rate of application of stress is rapid, the response can be primarily elastic response; if the rate of application of stress is low, the viscous component of the response will be proportionally higher and the elastic response relatively

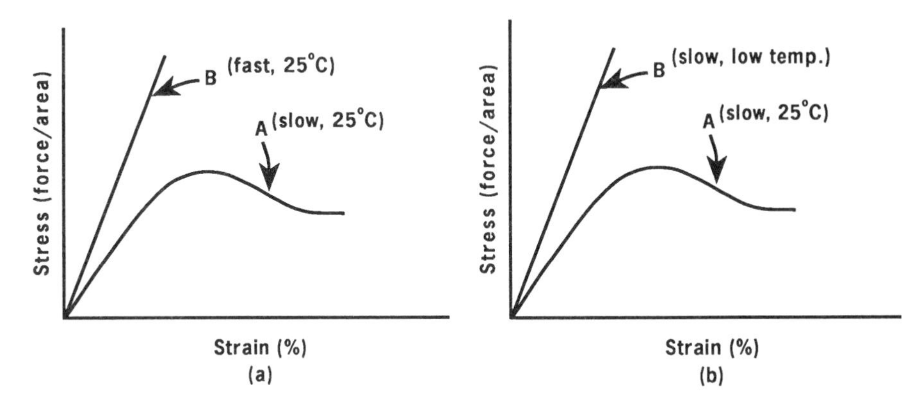

Figure 24.2. Schematic diagrams of the effects of rate of application of stress (*a*) and temperature (*b*) on stress–strain responses.

lower. Similarly, if the temperature is low, the response can be primarily elastic, whereas at a higher temperature, the viscous response will be proportionally greater.

These differences are illustrated in Figure 24.2*a* and *b*. Plot *a* shows schematically the results of pulling a coating film at two different rates at the same temperature. Curve A is the same curve shown in Figure 24.1; curve B results from a more rapid application of the stress. In curve A there is time for the sample to undergo some viscous flow along with the elastic deformation. In curve B the stress was applied at such a rapid rate that there was little time for viscous flow and the elastic response dominated. Note that in the figure, as commonly occurs in real samples, the elongation-at-break is much less and the tensile-at-break is greater when the rate of application of stress is high.

Plot *b* shows, schematically, the results of pulling the films at the same rates, equal to that for the slower rate in *a*, but at two different temperatures. Curve A is at the same temperature as in *a*; curve B is at a much lower temperature. At the lower temperature, the viscosity was much higher so that even at the slower rate of extension there was essentially no viscous flow and elastic deformation dominated. At the higher temperature, the viscosity was low enough to permit substantial viscous flow during the stretching. Note that in the figure, as commonly occurs is real samples, the elongation-at-break is less when the temperature is low; the tensile-at-break is higher.

Plots *a* and *b* are identical. The rates of application of stress and temperatures were chosen so that the change in the viscous response would be the same. In viscoelastic materials, the effects of higher rates of application of stress and lower temperatures are in the same directions. It is possible to do time–temperature superpositioning of curves mathematically. If one's instrument cannot operate at as high a rate of application of stress as one would want to evaluate stress–strain behavior, one can operate at a lower temperature and then calculate the data points at higher stress rates, as discussed in Ref. [3].

Transient experiments are also run in either creep or relaxation modes rather than a tensile mode. In creep experiments, a constant stress is applied and the resulting strain is determined. For a viscoelastic sample, strain is observed to

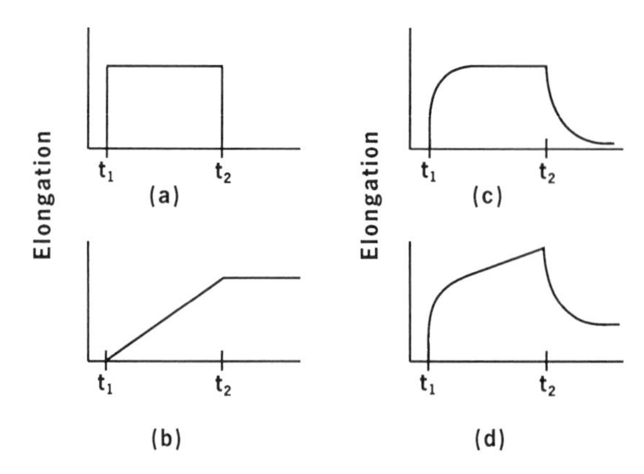

Figure 24.3. Schematic plots of creep test responses. A constant stress is applied at t_1 and removed at t_2: (*a*) ideal elastic solid, (*b*) Newtonian fluid, (*c*) viscoelastic sample with cross-links which result in almost complete recovery, and (*d*) viscoelastic sample which shows incomplete recovery. (Adapted from Ref. [3], with permission.)

increase over time in a nonlinear manner. An ideal elastic material undergoes an instantaneous strain with stress application at time t_1, that stays constant until the strain is removed at time t_2, as shown in Figure 24.3*a*. A Newtonian fluid subjected to a creep test undergoes a constant rate of strain increasing linearly with time, as shown in Figure 24.3*b*. Typical responses for cross-linked and thermoplastic coatings (viscoelastic) are shown in Figure 24.3*c* and *d*, respectively.

In a relaxation test, one applies an instantaneous strain elongating the sample, then follows the change in stress with time. The stress stays constant with time for an ideal elastic material and a Newtonian liquid exerts no stress. For viscoelastic samples the stress is initially high and drops to lower values (relaxes) with time.

Stress–strain analysis can also be done dynamically, that is, by utilizing instruments that apply an oscillating strain at a specific frequency. The stress and strain vary according to sine waves due to the alternation from the oscillations. The stress and the phase angle difference between applied strain and resultant measured stress are determined. For an ideal elastic material, the maximums and minimums occur at the same angles since there is an instantaneous stress response to an applied strain; the phase shift is 0°. For a Newtonian fluid, there would be a phase shift of 90°. Viscoelastic materials, on the other hand, show an intermediate response, as is illustrated in Figure 24.4. If the elastic component is high, the phase shift (δ) is small; if the elastic component is low compared to the viscous component, the phase shift is large. The phase shift along with the maximum applied strain, ε_0, and the maximum measured stress, σ_0, are used to calculate the dynamic properties.

The *storage modulus* (E'), sometimes called *elastic modulus*, which is a measure of the elastic response, equals $(\sigma_0 \cos \delta)/\varepsilon_0$. Its magnitude and physical significance are similar to moduli obtained from the initial straight line slope of a stress–strain curve, such as shown in Figure 24.1. The term storage reflects that E' measures the recoverable portion of the energy imparted by the applied strain. The *loss modulus* (E'') is a measure of the viscous response. E'' equals $(\sigma_0 \sin \delta)/\varepsilon_0$; the term loss reflects that viscous flow leads to the dissipation (as heat) of part of the energy

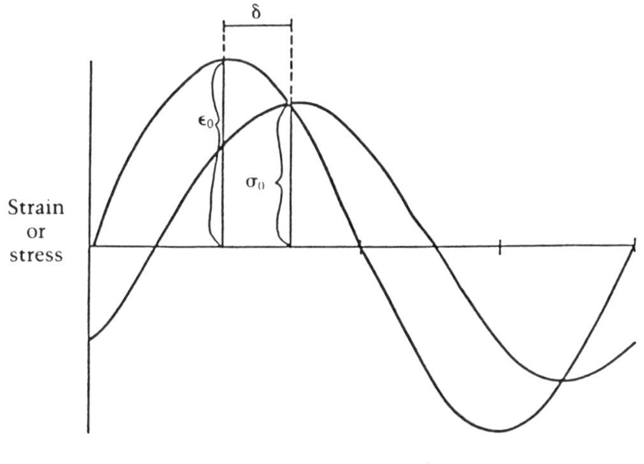

$$\text{Storage modulus} = E' = \frac{\sigma_0 \cos \delta}{\epsilon_0}$$

$$\text{Loss modulus} = E'' = \frac{\sigma_0 \sin \delta}{\epsilon_0}$$

$$\text{Loss tangent} = \frac{E''}{E'} = \tan \delta$$

Figure 24.4. Dynamic mechanical analysis plot as a sinusoidal strain is applied and a sinusoidal stress is determined. (From Ref. [3], with permission.)

imparted by the applied strain. The square of total modulus equals the sum of the squares of the storage and loss moduli. The ratio (E''/E') is called the *loss tangent* since all the terms cancel except the ratio sin δ/cos δ, corresponding to the tangent of the angle, commonly referred to as *tan delta.*

Dynamic mechanical analysis has the advantage over stress–strain curves, such as shown in Figure 24.1, that the elastic and viscous components of a modulus can be separated. The frequency of the oscillation is a variable related to the rate of application of strain. The higher the frequency, the greater the elastic response, that is, the smaller the phase angle; the lower the frequency, the greater the viscous response, that is, the larger the phase angle. The higher the frequency, the less time there is for viscous flow, hence elastic response will dominate and vice versa. Similarly, lowering temperature will also reduce viscous flow, decreasing the phase angle; and higher temperature will give greater viscous flow, increasing the phase angle. Generally, it is possible to run experiments over a wider range of frequencies in dynamic tests than the range of rates of applications of stress possible in linear stress–strain experiments. On the other hand, in dynamic tests, it is not possible to determine tensile-at-break (tensile strength), elongation-at-break, or work-to-break since the sample must stay intact to run the test.

The stress–strain analysis discussed above is based on elongation of samples, corresponding to application of tensile stress. Oscillating (dynamic) and linear (static) stress–strain analysis can also be carried out by application of shear forces. In shear tests, the stress is applied sideways—analogous to shear viscosity tests. The ratio of shear stress over shear strain is called shear modulus and is represented by the symbol G. Tensile modulus (E') equals three times shear modulus (G'). Recall (from Section 19.6.2) that the analogous extensional viscosity and shear viscosity have the same relationship.

24.2. FORMABILITY AND FLEXIBILITY

In many cases, a coated metal object is subjected to mechanical forces either to make the product, as in forming bottle caps or metal siding, or in use, as when a piece of gravel strikes the paint surface of a car with sufficient force to deform the steel substrate. To avoid film cracking during such distensions, the elongation-at-break must be greater than the extension of the film under the conditions of fabrication or distortion.

To illustrate some of the variables involved, let us consider the simpler case of a plastic, poly(methyl methacrylate) (PMMA), for which data on stress–strain relationships is available. Tensile stress–strain curves of PMMA (T_g = 378 K) as a function of temperature are shown in Figure 24.5. At low temperatures, there is no yield point, modulus is high, and elongation-at-break is low. Failures of this type are called brittle failures. The terminology confuses some people; the plastic at low temperature approaches being an ideal elastic material, but it is classified as showing a brittle failure because the elongation-at-break is so low. At higher temperatures, but still below T_g, greater elongations without breaking are possible, and there is a yield point.

Wu [6] has studied the modulus, elongation-at-yield, and elongation-at-break of PMMA and other plastics as a function of temperature. Figure 24.6 shows plots

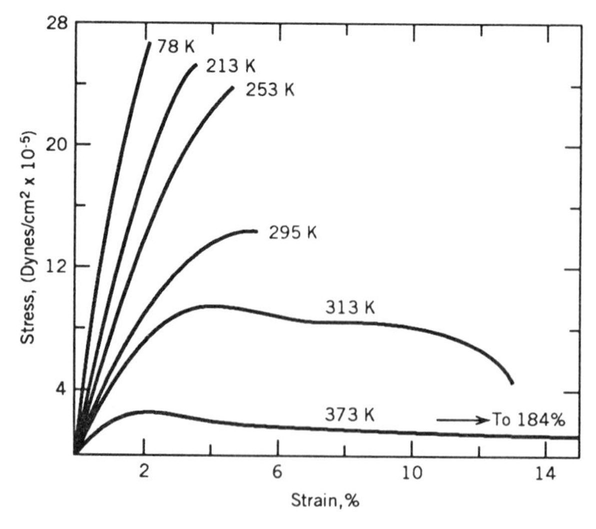

Figure 24.5. Stress–strain plots of PMMA as a function of temperature. (From Ref. [2], with permission.)

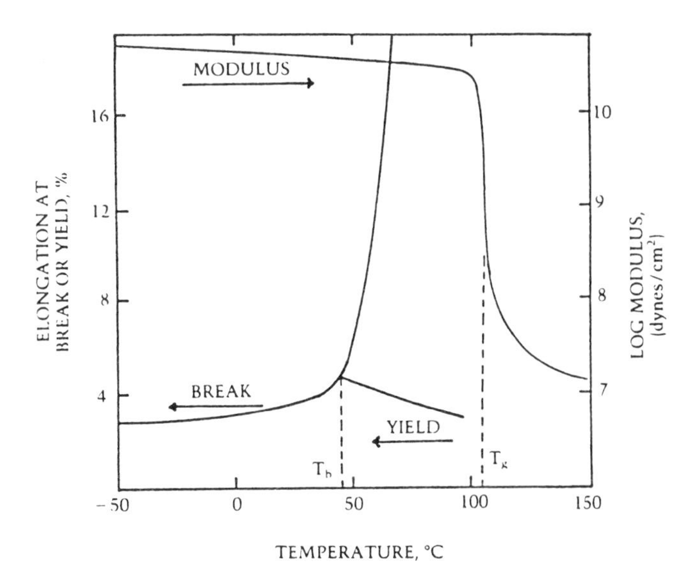

Figure 24.6. Elongation-at-break and -yield as functions of temperature, superimposed on a storage modulus, temperature plot. (From Ref. [3], with permission.)

of these three properties versus temperature. The abrupt drop in modulus at T_g is typical of all amorphous high molecular weight homopolymers. Similar data is in the literature for many such polymers. Fewer examples of the plots of the elongations appear. Note that the elongation-at-break starts to increase rapidly at a temperature below the T_g of the PMMA. Wu defines the temperature at the intercept of the elongation-at-break and the elongation-at-yield plots as the *brittle-ductile transition temperature* (T_b). (T_b can also be defined as the intercept of plots of tensile strength and yield strength versus temperature.) Below T_b, the polymer is brittle; between T_b and T_g it is hard and ductile; above T_g the polymer becomes increasingly soft. Plastics can be deep-drawn above T_b; it is not necessary to be above T_g, which is contrary to the common, but erroneous, definition of T_g as being the temperature below which an amorphous material is brittle.

Copolymers of MMA with methacrylates of longer chain alcohols, such as *n*-butyl methacrylate, have lower T_g and T_b values. There is substantial variation in the difference between T_g and T_b between thermoplastic polymers as illustrated by the data in Table 24.1. Since these polymers are viscoelastic, the deformation is dependent on rate of application of stress as well as temperature. Both T_g and

Table 24.1. Transition Temperatures of Homopolymers

Polymers	T_g (°C)	T_b (°C)	$T_g - T_b$
Polystyrene	100	90	10
Poly(methyl methacrylate)	105	45	60
Poly(vinyl chloride)	80	10	70
Bisphenol A polycarbonate	150	−200	350

T_b are dependent on the rate of application of stress: the higher the rate, the higher the T_g and T_b values.

The brittle-ductile transition temperature has been studied less with thermosetting systems. The limited data available indicate that polymers with a low degree of cross-linking still show a differential between T_b and T_g, but the differential decreases as cross-link density (XLD) increases. Thus, if fabrication involves significant elongation of a coating, thermoset coatings are fabricated at temperatures above T_g. The T_g of cross-linked polymers depends on several factors: structure of the segments between cross-links, XLD, amount of dangling chain ends, and the extent, if any, of cyclization of the backbone [7].

It has been known for many years that there is a proportional relationship between XLD and modulus of low XLD elastomers above T_g, but it had been uncertain whether the relationship could be extended to relatively higher XLD coatings. Hill [5,8] has now demonstrated that this relationship also holds for melamine–formaldehyde (MF) cross-linked acrylic and polyester coating films. The relationship is surprisingly simple, as shown in Eq. (24.1), where XLD, ν_e, is the number of moles of *elastically effective network chains* per cubic centimeter of film. An elastically effective network chain is one that is connected at both ends to the network at different junction points—short cyclical chains and dangling ends are not elastically effective. The variables E and G are tensile and shear storage moduli, respectively. Since E'' and G'' are low at temperatures well above T_g, $E \cong E'$ and $G \cong G'$.

$$E = 3\,\nu_e\,RT \qquad (\text{or } G = \nu_e\,RT)\ (T >> T_g) \tag{24.1}$$

Thus, at least for the classes of films studied, it is possible to calculate XLD (ν_e) from modulus/temperature plots [8]. Note that ν_e divided by the film density provides the moles of network chains per gram. The inverse, grams per mole of network chains, corresponds to the average molecular weight of network chains, frequently called average molecular weight between cross-links, $\overline{M}_c$. Commonly, $\overline{M}_c$ is erroneously defined and used as being the molecular weight per branch point; one must be careful in reading any paper dealing with molecular weight of network chains and XLD.

Cross-link density can also be calculated with an assumed interaction parameter, from the extent of swelling of a film by solvent [5]. While cross-linked films do not dissolve in solvent, solvent does dissolve in a cross-linked film. As cross-links get closer together, that is, as XLD increases, the extent of swelling decreases.

Equation (24.1) can be used to predict the storage modulus above T_g from the XLD. In a system with stoichiometric amounts of two reactants whose functional groups react completely, one can estimate the XLD from the equivalent weights and the average functionality. If the reactant mixture contains molecules of several different functionalities, calculation becomes more difficult. A more general approach is provided by the Scanlon equation (Eq. 24.2) [5].

$$\nu_e = \frac{3}{2}C_3 + \frac{4}{2}C_4 + \frac{5}{2}C_5 + \cdots \tag{24.2}$$

The C values are the concentrations of reactants with functionality 3–5 (or more) expressed in units of moles per cubic centimeter of final cured film. The volume of the final film depends on the density of the cured film and the loss of volatile byproducts of the reaction. Equation (24.2) does not include a term for difunctional reactants because these reactants do not create junction points in a network; they only extend the network chains.

Although the Scanlon equation is convenient for stoichiometric reactions with complete conversion, it does not apply to other cases. For nonstoichiometric mixtures and/or incomplete conversion, Miller–Macosko equations are useful general equations. Bauer [9] has selected the Miller–Macosko equations most useful in coatings, given examples of their applications, and provided a computer program.

Thus one can, at least in theory, design a cross-linked network to have a desired storage modulus above T_g by selecting an appropriate ratio of appropriate reactants. By proper selection of the structures between cross-links and cross-link density, one can design the T_g of the cross-linked network. Further research is needed to see how general the relationships are, but the implications for the future of coatings formulation are obviously enormous.

The properties are, of course, affected by the extent to which the cross-linking reaction has been carried to completion. Incomplete reaction leads to lower XLD and hence lower storage modulus above T_g. The extent of reaction can be followed by determining storage modulus as a function of time [10]. As cross-linking continues, storage modulus increases until a terminal value is reached.

Determination of dynamic mechanical properties has proved to be a valuable tool in studying cross-linking of hydroxy-functional resins with Class I MF resins [5,8]. It was found that as the stoichiometric ratio of methoxymethyl groups from the MF resin to hydroxyl groups increased from values less than 1, the storage modulus above T_g of fully cured films increased up to the point that the ratio became one. As discussed in Section 6.3.1.2, these results show that all of the functional groups on the Class I MF resin can react with hydroxyl groups and the reaction is not limited by steric hindrance as had been thought earlier. When the amount of MF resin was increased so that excess methoxymethyl groups were present, the storage modulus above T_g increased at higher temperatures during the dynamic mechanical testing. This behavior is explained on the basis that the excess methoxymethyl groups can undergo self-condensation reactions during the determination. The self-condensation reaction is relatively slow and was incomplete during the baking cycle used in preparing the film, hence the reaction continued at the higher temperatures used in dynamic mechanical analysis leading to the higher storage modulus. Self-condensation during baking of coatings also occurs when excess MF resin is used; then, the extent of self-condensation increases as baking time and temperature increase.

It has been found that an additional factor, which can affect the mechanical properties of polymeric materials, is the breadth of the T_g region [11]. In some materials the slope of the line connecting the relatively level modulus below T_g and the relatively level modulus above T_g is steep, as shown in Figure 24.6. In others this slope is shallower, as shown in Figure 24.7. The same effect can be seen in tan delta plots, which exhibit various breadths. The breadth of the T_g can also be estimated by differential scanning calorimetry (DSC), but dynamic mechanical analysis gives a clearer picture.

As done in Figure 24.7, it is common to label the peak of the tan delta curve as the T_g. Some authors prefer to label the peak of the loss modulus plot as the T_g. As can be seen, there is a substantial difference. The peak of the loss modulus

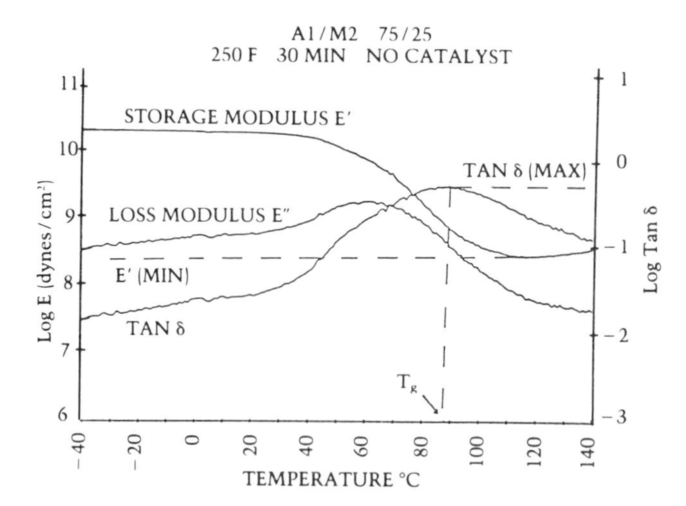

Figure 24.7. Dynamic properties of a highly cross-linked acrylic coating film (solid lines). Dashed lines indicate values, E'(Min), tanδ(Max), and T_g, that are measures of extent cure. The values of E'(Min) and T_g tend to increase, whereas tanδ(Max) tends to decrease as extent of cure increases. (From Ref. [3], with permission.)

plot is usually nearer to the T_g determined using differential scanning calorimetry (DSC). As noted before, T_g is dependent on the rate of heating in DSC and on the rate of application of stress—the frequency of oscillation as well as the rate of heating in a dynamic test. This dependence on experimental technique might lead one to wonder if T_g means anything. It does, but one must always be careful to compare T_g values determined in the same way. In using T_g with regard to fabrication or deformation of coatings, the most relevant T_g is that determined at rates most closely corresponding to the rates to be encountered in deformation of the coating.

Factors controlling the breadth of the T_g have only been partly elucidated, but broad peaks are frequently associated with heterogeneous polymeric materials. Blends of different thermoplastic resins often display two distinct T_gs, presumably because of phase separation. Other blends of thermoplastics have a single, often broad, T_g, presumably when phase separation is indistinct or when the phases are very small. With thermosetting polymers the T_g region is generally broader than with pure thermoplastics, and the breadth may vary considerably. Breadth of the distribution of chain lengths between cross-links is a factor, and blends of thermosetting resins such as acrylics and polyesters often display a single, broad T_g [8]. As a broad rule, to which there are probably many exceptions, materials with broad and multiple T_gs have better impact resistance than comparable polymers with sharp, single T_gs.

A different phenomenon that also relates to impact resistance is the presence in some homogeneous polymers of a second, usually small, tan delta peak at a much lower temperature than T_g. These peaks are called *low temperature loss peaks* or *β-transitions*. They are thought to result from the onset of some specific small-scale motion of part of the polymer molecules at low temperature. In plastics it is well established that tough, impact resistant materials generally have low temperature loss peaks; polymers made from bisphenol-A (epoxies and polycarbonates) are common examples. It is reasonable to speculate that coatings with low temperature loss peaks may also have good impact resistance (if adhesion is good), but the relationship is not well documented in the literature.

Understanding mechanical properties in coatings is much more complex than in most plastics. One reason is that coatings are used as thin films on a substrate. Interaction with the substrate affects the mechanical properties of thin films. The substrate can limit the extent of deformation that occurs. The substrate can act as an energy sink to dissipate the energy so that there can be less effect on the coating film. Adhesion can have a profound effect on ability to withstand fabrication. If the adhesion is good, fabrication and impact resistance of the films is almost always superior, perhaps because less stress is concentrated in the film. The extent of permanence of the deformation can have important effects. Thermoplastic films, as noted earlier, are more likely to be permanently deformed than cross-linked films. When a cross-linked film on a metal substrate is deformed by the fabrication, it is held in the deformed state by the metal substrate. As a result, there is a stress within the film acting to pull the film off the substrate. This recoverable extension has an associated restoring force that can overcome the adhesion. Coatings have been known to pop off post-formed metal bottle caps when a jar is sitting on a supermarket shelf.

Stress within films can also arise during the last stages of solvent loss and/or cross-linking of films. Both solvent loss and cross-linking result in shrinkage. If this shrinkage occurs when the temperature is near the T_g of the film, the resulting internal stresses may persist indefinitely (see Section 26.2 for further discussion of the effects of internal stresses on adhesion).

Film thickness is also an important variable in the ability of a coating film to withstand fabrication without cracking. Thin films can be used for deeper draws than thick films. In making coated exterior siding, the hardness of the film can be increased without encountering cracking by limiting the film thickness. Of course, thinner films of pigmented coatings give poorer hiding; a common compromise in this case is at 20–25 μm film thickness. The bottoms of two-piece fish cans are coated as flat sheets with a relatively highly cross-linked phenolic coating that is quite brittle so as to minimize swelling with the fish oil. Such cans can only be successfully formed without cracking if the film thickness is of the order of 5 μm or less.

Relatively little basic work has been published on the effect of pigmentation of films. As discussed in Section 21.1, in many cases, as the PVC of the films is increased up to CPVC, the tensile strength of the films increases. It is also possible that imperfections resulting from some types of pigmentation may lead to defects from which crack propagation is facile, as discussed in connection with fracture mechanical adhesion failure in Section 26.2. In some cases, the storage modulus above T_g is increased by the presence of pigment, and for pigmented coatings one cannot expect a direct proportional relationship between storage modulus above T_g and XLD [5]. However, if the pigment content is constant in a series of films varying in the polymeric portion, the relative values of E' should still indicate relative XLD for the pigmented films.

Another important variable can be the timing of the fabrication or flexing after curing of the film. It is very common for films to become less flexible as time goes on. It is common, particularly in air dry films, for some solvent to be retained in films. Since most coatings films have T_g values near or a little above room temperature, solvent loss may be very slow (see Section 15.1.4). Solvents generally act as plasticizers, so that as solvent is lost, T_g and storage modulus increase and films tend to become less flexible. In the case of cross-linkable coatings, if the cross-linking reaction was not complete, the reaction may continue slowly, increasing XLD and hence storage modulus, and decreasing flexibility. Again, this is particularly likely to occur in air dry coatings since the reaction rates are likely to become mobility rate controlled and hence the last part of the reaction is slow, as discussed in Section 3.2.3. Reactions occurring during the use life of the film, especially during exterior exposure (see Section 25.1), can result in embrittlement.

Hardening of baked cross-linked coatings over time is also fairly commonly observed. Although in some cases further volatile loss or continued cross-linking may be responsible, another possible factor is discussed briefly in Section 2.1.4. If a polymer is heated above its T_g and then cooled rapidly (quenched) to a temperature below its usual T_g, the T_g is commonly found to be lower than if the sample had been cooled slowly [12]. During rapid cooling more and/or larger free volume holes are frozen into the matrix than with slower cooling that provides greater chances for molecular motion. It has been found that on storage, the molecules in quenched films slowly move even though the temperature is somewhat below T_g and the free volume decreases. The process is commonly called *densification*; since it can result in a change in properties with aging with no chemical change, it is also called *physical aging*. With the decrease in free volume and the increase in density, the T_g increases and cracking during fabrication is more likely.

This phenomenon has been widely observed in plastics but is only beginning to be considered in coatings. It might be particularly likely to occur when coatings are baked on metal at high temperatures and then cooled rapidly after coming out of the oven. Hill [13] has suggested that densification may be a common cause of embrittlement during aging of baked coating films. Greidanus [14] has studied the physical aging at 30°C of polyester/MF cross-linked films that had been baked at 180°C and then quenched to 30°C. There was a small but reproducible increase in modulus with time at 30°C. The aging rate (i.e., the rate of increase of modulus) decreased with time. If the sample was heated again at 180°C and again quenched to 30°C, the modulus returned to its lower value and underwent physical aging again. Much further work is needed, but it is evident that physical aging can be an important phenomenon. Presumably, the extent of the effect is affected by the rate of cooling of a coated product after baking.

24.3. ABRASION RESISTANCE

Another important type of coating failure is by abrasion. At first blush, one is likely to suppose that harder materials are less likely to fail by abrasion. It certainly is true that hard materials are more difficult to scratch than softer materials. In some cases abrasion can occur by multiple scratching; in such cases it may well be true that harder (higher modulus) materials are more resistant. However, that is not generally the case. For example, rubber tires resist abrasion much better than steel tires even though steel is obviously much harder than rubber.

Table 24.2. Mechanical Properties of Floor Coatings

Floor Coating	Tensile Strength (psi)	Elongation (%)	Work-to-Break (in.-lb/in.^{3})	Taber[a] (rev/mil)
Hard epoxy	9000	8	380	48×10^3
Medium epoxy	4700	19	600	33×10^3
Soft epoxy	1100	95	800	23×10^3
Urethane elastomer	280	480	2000	36×10^3

[a]See Section 24.5.3.4 for a discussion of these results.

Evans [15a] studied the mechanical properties of a series of floor coatings with known actual wear life. He determined tensile-at-break, elongation-at-break, and work-to-break. His data are given in Table 24.2; the coatings are listed in order of increasing wear life. One might suppose that higher tensile strength would give higher abrasion resistance; the data show the reverse. (It should not be assumed from this limited data that abrasion resistance is always inversely related to tensile strength.) Elongation-at-break gave the proper rank order, but Evans concluded that work-to-break best represented the relative wear lives. Intuitively it seems reasonable that abrasion resistance would be related to work-to-break. However, work-to-break varies with rate of application of stress and should be determined at a rate comparable to that to be expected in use.

In studies on another series of coatings, Evans and Fogel [15b] determined that work-to-break (from stress–strain plots) did not always correlate with abrasion resistance, determined by loss of gloss in a ball mill abrasion tester, when the stress–strain tests were carried out at ambient temperatures. They reasoned that the strain rate settings on their instrument were too low relative to stress application in actual use. Using a time–temperature superposition relationship, they calculated that the tests at an accessible low strain rate setting should be carried out at -10°C in order to compensate for the (instrumentally inaccessible) higher rate of stress application in actual use at ambient temperature. The resulting work-to-break values did correlate with abrasion resistance for urethane films with T_g equal to or greater than -10°C.

Urethane coatings generally exhibit superior abrasion resistance combined with solvent resistance. This combination of properties may result from the presence of intersegment hydrogen bonds in addition to the covalent bonds. Without stress and at low levels of stress, the hydrogen bonds act like cross-links, reducing the swelling on exposure to solvent. At higher levels of stress, the hydrogen bonds can dissociate, permitting the molecules to extend without rupturing covalent bonds. When the stress is released, the molecules relax and new hydrogen bonds form. Urethanes are generally used as wear layers for flooring as well as top coats in aerospace applications, where this combination of properties is highly desirable. A shortcoming of urethanes is plasticization by absorbed moisture, which can disrupt the hydrogen bonds and thereby affect the properties of urethanes.

Factors in addition to work-to-break are involved in abrasion resistance. The coefficient of friction of the coating can be an important variable. For example, abrasion of the coating on the exterior of beer cans during shipment can be minimized by incorporation of a small amount of incompatible wax or fluorosurfactant in the coating. When the two coated surfaces rub against each other, the reduced surface tension, resulting from the additive, reduces the coefficient of friction so that transmission of the shear force from one surface to the other is minimized and abrasion is reduced.

Another variable is surface contact area. Incorporation of a small amount of a very small particle size SiO_2 pigment in the thin silicone coating applied to lens surfaces in plastic lens eyeglasses reduces abrasion, adding to the effect of the low surface tension of the silicone surface. The pigment particles reduce contact area, permitting the glasses to slide more easily over a surface. Another example of the same principle is the incorporation of a small amount of coarse SiO_2 inert pigment in wall paints to reduce *burnishing*. If a wall paint without such a pigment is frequently rubbed, as around a light switch, it will be abraded to a smoother surface, that is, it will burnish. The coarse inert pigment reduces burnishing apparently by reducing contact area.

Another approach that has been used for many years in resin-bonded pigment print colors on textiles is to incorporate rubber latex in the print paste. The latex particles are not soluble in the resin and end up as individual particles in the resin along with the pigment particles. The abrasion resistance is markedly improved by the latex addition. Similar work is now being done to improve the abrasion resistance of continuous coatings. Presumably, the relatively soft rubber particles act to dissipate stresses on the film, minimizing the chance of a stress concentration leading to film rupture. Lee [16] has reviewed abrasion resistance as one of several types of wear in a broad approach to fracture energetics and surface energetics of polymer wear.

Closely related to abrasion resistance are scratch and mar resistance. The term abrasion is generally used to describe wearing away of surfaces, scratching is used to describe more localized disruption of the surface. Mar resistance implies not only scratch resistance but also resistance to surface contamination such as *metal marking*. When one draws a coin across the surface of a coating, sometimes a dark streak is left, marring the surface.

If the surface of the coating is harder than the object scraping across the surface, scratching does not occur. If, as is generally the situation, the surface of the coating is softer, the two principal factors reducing scratch and mar resistance seem to be, as is the case with abrasion resistance, work-to-break and surface friction. Scratching and marring can be reduced by lubricating the surface so as to reduce friction. Additives can be incorporated in a coating formulation to increase slip. Modified polysiloxanes have been reported to be particularly effective [17]. Care must be exercised in selection of the particular grade of silicone additive and the amount of the additive used so as to minimize marring, scratching, and metal marking without causing other defects such as cratering.

24.4. MEASUREMENT OF MECHANICAL PROPERTIES

Most instruments require free films for the measurement of mechanical properties. Two major disadvantages of utilizing free films are: (1) the interaction of the film with the substrate can have major effects on some film properties and (2) free films are sometimes difficult to prepare and handle. Test results are generally less variable with thick films than thin films; however, for many reasons, thick films may not provide data applicable to thin films. Cutting the free films may result in nicks or cracks along the edge of the film. When subject to stress, such films commonly tear easily, starting at the imperfection, leading to meaningless results for that sample.

Preparation of thin unsupported films can be difficult. In some cases it is possible to make a film by drawing down the coating on a release paper with a wire wound bar. Release papers are coated with low surface tension materials to minimize adhesion; but if the surface tension is lower than the surface tension of the coating being applied, there is the possibility of crawling, that is, the coating tries to minimize surface free energy by drawing up into a ball (see Section 23.3 for a discussion of crawling). One tries to find a release paper with a low enough surface tension so that adhesion will be poor, but high enough so that crawling will not occur.

A generally more effective method is to apply the coatings to tin-plated steel panels. After cure, one end of the panel is placed in a shallow pool of mercury. Mercury creeps up the panel under the coating because it forms an amalgam with the tin, and the film comes free of the panel. Unfortunately, mercury vapor is toxic and great care must be taken to minimize the hazard of personnel being exposed to mercury vapor. The safety regulations of some laboratories do not permit such use of mercury. Even when one has the free film, there can frequently be problems in handling, especially for films with T_g values above room temperature. Such thin films tend to be brittle and easily broken.

With free films and, to an even greater extent, with supported films, one must be very careful about changes that may occur during testing, such as loss of residual solvent as well as continuing reaction. Obviously, the test results reflect such changes and are not representative of the initial film. Storage conditions can be critical. Most films absorb some water from the atmosphere. If the T_g is near room temperature and, especially if the film has groups such as urethanes that hydrogen bond strongly with water, the T_g and film properties can be strongly affected by the humidity conditions in storage, since water acts as a plasticizer. In actual use the films will encounter a variety of humidity conditions and hence a variation in properties. Comparisons should always be done with samples that have been stored at the same temperature and humidity. Commonly, 25°C and 50% relative humidity are used.

A wide range of instruments are available and new instruments come on the market frequently. The most commonly used instrument for tensile experiments (nondynamic) is the Instron tester. The sample is mounted between two jaws of the tester; great care must be taken to be sure that the film is completely in line with the direction of pull. The instrument can be run with a range of rates of jaw separation, but even the highest rates are slow compared to the rates of stress application found in many real life situations. As has been discussed, this problem can be partly overcome by running the tests at low temperatures. The method has the advantage that the stress can be increased until the film fails, making possible determination of elongation, tensile strength, tensile modulus, and work-to-break

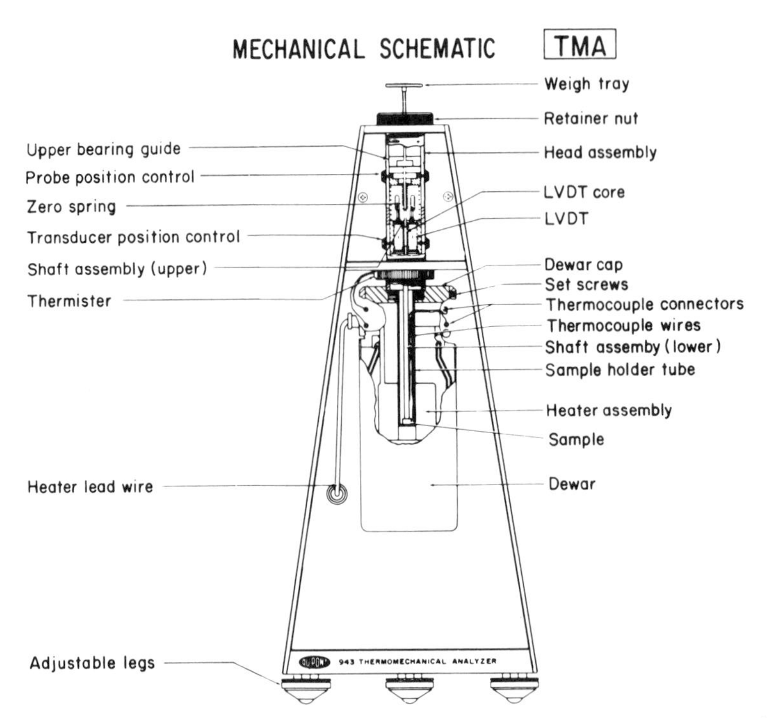

Figure 24.8. Schematic diagram of a thermal mechanical analyzer. (From Ref. [4], with permission.)

data. However, one can not separate the viscous and elastic components of the mechanical properties.

Another type of instrument, called a thermal mechanical analyzer (TMA), is a penetrometer that permits measurement of indentation versus time or temperature. A particular advantage over tensile instruments is that these instruments include a furnace and temperature programmer so that heating, cooling, and isothermal operations are possible. They can also be used with films on a substrate. A schematic diagram is shown in Figure 24.8.

An example of a use of the TMA is to measure the *softening point*, which is related to the extent of cure of cross-linking films. Figure 24.9 shows a plot of probe penetration as a function of temperature for an undercured and a well-cured, 25-μm thick acrylic coil coating. The softening points for the two samples are marked on the graph. The softening point is related to, but not identical with, T_g; it is frequently used as an index of flexibility [4].

There are various dynamic mechanical test instruments available. The most versatile are generally those where the sample is subjected to an oscillating strain by attachment under tension to a fixed clamp on one end and a vibrating clamp on the other. Oscillating stresses are thereby imparted to the sample. In most cases, a range of frequencies can be used and the change in properties occurring over a wide range of temperatures can be determined. The most sophisticated of the instruments are set up in line with a computer that analyzes the data and provides storage and loss modulus and tan delta figures and plots as functions of temperature. A schematic diagram is shown in Figure 24.10 [18].

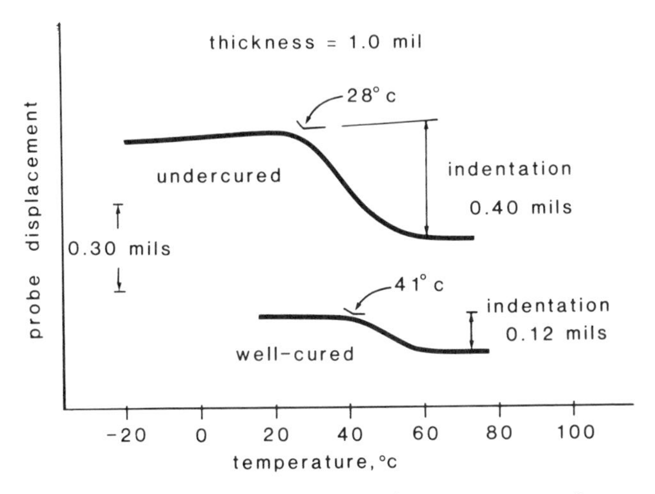

Figure 24.9. TMA plot of probe displacement against temperature for an undercured and a well-cured acrylic coil coating. (From Ref. [4], with permission.)

Another type of instrument for dynamic mechanical analysis is a torsional pendulum. In its simplest form, a film is fastened in jaws the lower end of which is attached to a disk to which weights can be added. The lower weight is twisted, setting up a pendulum motion whose decay can be analyzed to give the dynamic properties. Recently, it has been most widely used not with film but with a fiber braid that is saturated with a liquid cross-linking system. The instrument has the advantage that it can be used to follow changes in dynamic properties starting with liquid coatings as reactions occur on the braid. It has the disadvantages that the sample is not a film and that there are large surface areas of fiber/polymer interface that may affect properties. A schematic diagram is shown in Figure 24.11.

24.5. PAINT TESTS

A very wide variety of test methods has been established to characterize properties of paint films. In general, these do not permit calculation of the basic mechanical properties but rather test some combination of properties of the coating. There

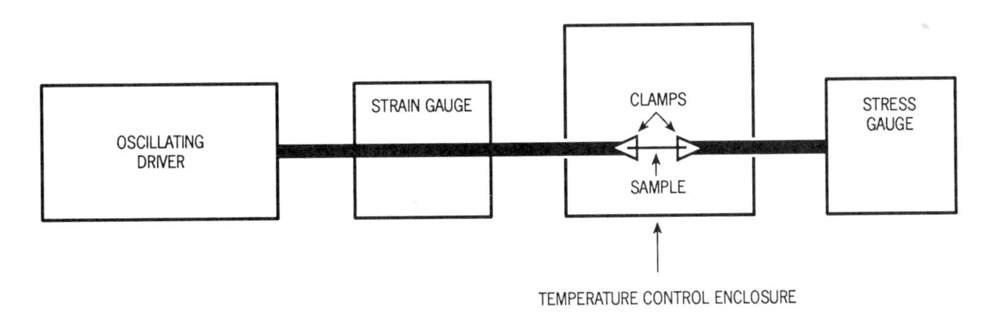

Figure 24.10. Schematic diagram of a dynamic mechanical analyzer.

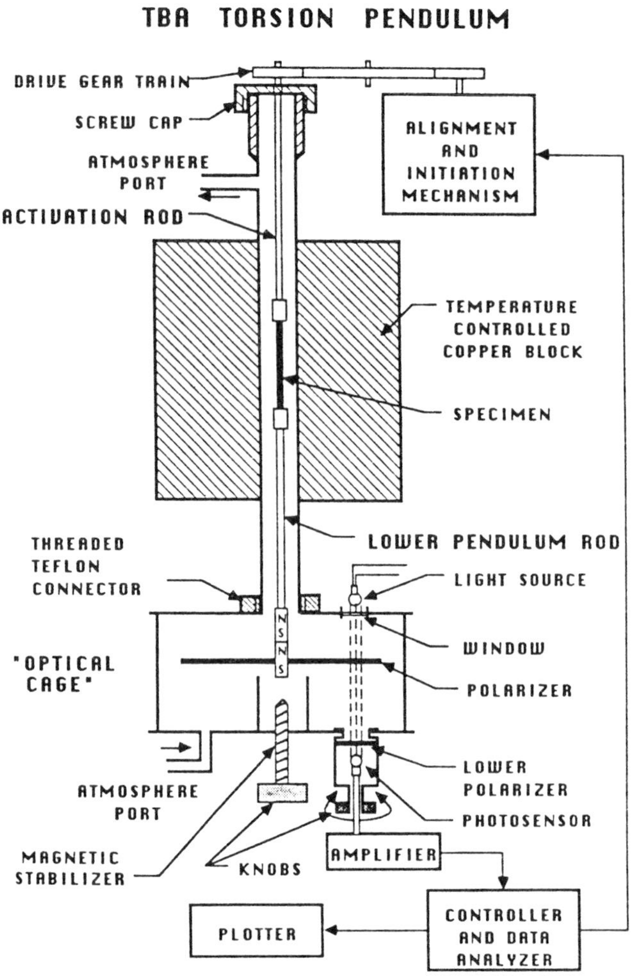

Figure 24.11. Schematic diagram of torsional braid analyzer. (From Ref. [4], with permission.)

are two very different categories of such tests: one type can be appropriate for prediction of actual use performance, the second type is suitable only for quality control tests. There are very real needs for both types of tests. Unfortunately, many people, including many who have worked in the field for many years, do not make the distinction. It is too common that paint tests which may be very appropriate for quality control are used to predict performance even though they are not capable of providing results that will permit performance predictions.

There are three broad classes of paint tests: field exposure tests, laboratory simulation tests, and a broad category of empirical tests. In all of these tests, one should always include standards of known performance along with the coatings to be tested. Such standards are often called controls.

24.5.1. Field Exposure Tests

As stated in the chapter introduction, the only reliable way of knowing how a coating will perform in use is to use it and see how it performs. But there is need for tests to predict performance before large scale use is made of the coating. The next best approach to actual use is to apply and use the coatings in field applications on a smaller scale and under especially stringent conditions that may accelerate possible failure. The more limited the test and the greater the degree of acceleration, the less reliable are the predictions; but, still carefully designed and analyzed tests of this type can be very useful. There are many examples; we cite a few to illustrate the principles.

New highway marking paints can be tested by painting stripes across the lanes of traffic instead of parallel to the traffic flow. In this way, the exposure is greater and many paints can be tested and compared on a limited length of highway. Controls with known use performance should be tested alongside the new materials. The test must be carefully designed to assure that all the samples receive the same amount of traffic wear. Tests should be done at different times of the year because such factors as the effect of snow plows, salt application to melt ice, and so forth must be taken into consideration where appropriate. Tests should be set up on the different types of highway materials such as concrete and asphalt to cite the simplest examples. New floor paints can be evaluated by a similar pattern.

In the case of coatings for bridges, ships, and chemical or petroleum refining factories, experimental materials along with standards of known performance can be applied to relatively small areas of such structures in actual use. If they fail, the cost of repainting is modest and the information obtained can be very useful. Great care should be exercised to record actual conditions during application and during the test period to assist in judging how generally applicable the information may be. If the potential advantage of a new coating is great enough, it may be possible to induce a highway authority to paint one or two small bridges. Generally, structures are selected where the conditions are especially severe. Of course, the time is long for such evaluations but at least the liability is limited in case of serious failure.

Automobiles painted with new coating systems are driven on *torture tracks* with gravel stretches, through water, under different climate conditions, and so on. Sample packs of canned goods are made; the linings are examined for failure and the contents are evaluated for flavor after various lengths of storage. The list of examples could go on and on.

24.5.2. Laboratory Simulation Tests

Many tests have been developed to simulate use conditions in the laboratory. The value of these tests depends on how well use conditions are simulated and, particularly, on how thorough a validation procedure has been used. An important key to the use of any test for performance prediction is the simultaneous evaluation of standards with known performances that cover the whole range from poor to excellent performance. It is not enough to use only the extremes of standards. While such information may be the first step in checking the possible applicability of the test, performance prediction requires more than two standard data points.

Enough replica tests must be run to provide a basis for setting the number of repeat tests that must be run to give results within the desired confidence limit range. Chemists commonly think of standard deviations but these are only 67% confidence limits; the odds are 33% that the proper value is outside the standard deviation range.

An example of a well-validated test is the laboratory shaker test devised to simulate the abrasion of can coatings when six-packs of beer are shipped by rail car [18]. Six-packs of beer are loaded on a shaker designed to simulate the pressures, rate of shaking, range of motion, and so on actually encountered in rail shipments. The test was validated using beer cans with known field performance. The times to reach different degrees of abrasion failure were established. In unpublished work, in the laboratories of several coatings suppliers and can manufacturers, the results were compared with known performance and with the results of standard abrasion tests. It was found that none of the standard laboratory abrasion tests gave satisfactory predictions, but this test did give results that could be used for performance prediction.

The automobile industry uses so-called *gravelometers* to evaluate resistance of coatings to chipping when being struck by flying gravel. Pieces of standard gravel or shot are propelled at the coated surface by compressed air under standard conditions. The tests have been standardized by comparison to a wide range of actual results, and have been found to give reasonably good predictions of actual performance.

Several laboratory devices are available that approximately reproduce stamping or other forming operations to test the ability of coated metal to withstand fabrication. Individual companies design such tests to match the conditions of their factory forming operations as closely as possible.

Commonly, such simulation tests only check one or a few of the performance requirements so they must be used along with other tests to predict overall performance. For example, the shaker test for beer cans obviously can give no information on the important ability of the coating to withstand the pasteurization procedure, but separate simulation tests are available for testing pasteurization performance.

In most cases, simulation tests are designed to be used for performance prediction rather than quality control. Generally, the sample preparation and testing time are too long to be useful in checking whether production batches of an established coating are equal to the standard.

24.5.3. Empirical Tests

A very wide range of paint tests have been developed and are widely used. In some cases, the results can be used as part of the data to predict performance. They are most likely to be useful in this regard in comparing formulations that are very similar to standards with known performance. In most cases, they are more appropriate for quality control tests. Fairly commonly, they are subject to a considerable range of error and many replicates should be run. They are frequently required in specifications, although, in many cases, they do not predict performance. Unfortunately, specifications are used not just for quality control but also as requirements for new coatings.

We can only mention a few of the most widely used tests. The American Society for Testing and Materials (ASTM) annually publishes books describing tests. Most of the tests of importance to the coatings field are in Volumes 06.01, 06.02, and 06.03, *Paint—Tests for Formulated Products and Applied Coatings*. While the books are published annually, most of the methods are not changed. Each method has a number such as "ASTM D 2933-74 (Reapproved 1981)." The 74 means that the test was approved in 1974. In this particular case, but not in most, the test was reapproved in 1981. If one finds a reference to a test as D 1876-71, and then looks it up in a 1992 ASTM book and finds D 1876-88, it means that the test was reviewed and revised in 1988. In general, one would use the new test method. Sometimes, a method will have a number like D 1459a. The "a" means that there was a minor rewriting of the method that did not change the basic method but perhaps clarified the writing or some other minor change. Sometimes people feel that because a test method is given an ASTM designation it can be used not just for quality control but also to predict performance. This is frequently not the case. However, following ASTM procedures for the various tests does provide the best chance for obtaining comparable test results.

Some ASTM methods include "precision statements," usually based on repeatability and reproducibility studies involving different laboratories. The precision statements should not be ignored, especially since many coatings professionals believe the tests are much more precise than they proved to be in actual ASTM round robin tests [19].

An excellent reference book in the field is the *Gardner–Sward Paint Testing Manual* [20]. The last edition was published in 1972 so it is not a source for new information, but it does provide descriptions of a wide range of test methods and summaries of each major class of properties, as well as background information and comparisons of the utility of various tests. Fortunately, a new edition is projected for 1993. Hill [3] provides an informative, brief discussion of the more important tests in his monograph on mechanical properties. Much of the discussion in the following sections is adapted from this monograph.

24.5.3.1. Hardness

Hardness is most closely related to modulus and many hardness tests can be thought of as one point modulus determinations. There are three types of tests used: indentation, scratch, and pendulum tests.

The most widely used indentation test is run with a Tukon indentation tester. An indenter with a diamond-shaped tip is pressed into the film with a fixed weight over a fixed time period. The indenter is raised and the indentation left in the film is measured with a calibrated microscope. The results are expressed in Knoop Hardness Numbers (KHN) that are related to the weight divided by the area of the indentation. Results are affected by film thickness: for the same coating, higher KHN values are obtained for thin films; in thin films, the hardness of the substrate affects the reading more than in thick films. Meaningful results are obtained with high T_g films, but intermediate materials may have partial recovery of the indentation in the time needed to move the sample under the microscope and make the measurement. Low T_g films give considerable response variation; the indenter may leave no indentation on a rubbery material. The careless tester may conclude that a rubbery material is very hard, even though it is obviously soft. In general, the method is most appropriate for baked coatings since these are more likely to have T_g values above the testing temperature.

The most widely used scratch test is the pencil hardness test. Pencils with hardnesses varying from 6B (softest) to 9H (hardest) are available. The "lead" (actually graphite and clay) in the pencil is not sharpened as for writing but is squared off by rubbing perpendicularly on abrasive paper. For the test, the pencil is held at a 45° angle to the panel and pushed forward with pressure just below that which will break the lead. The hardness is reported as the grade of pencil that does not cause any marring of the surface. The test is reproducible by an experienced tester to ± 1 hardness grade. Probably, the test reflects some combination of modulus and tensile strength.

The most widely used pendulum test in the United States is the Sward rocker. The rocker is a circular device made up of two rings joined with a glass level gauge, and weighted off center. The circumference of the rings rests on the panel. The rocker is rolled to a marked angle and released. The number of rocks (swings) required to dampen the motion down to a smaller fixed angle is determined. Polished plate glass is used for calibration so that it gives 50 rocks (a hardness reading of 100). Hard coatings give high readings (but less than 100). Soft coatings give low values. Dampening is caused by rolling friction as well as by mechanical loss. The results are dependent on film thickness and surface smoothness. It is probably most useful for following the increase of hardness of a coated panel during the drying of an ambient cure coating. While widely used to compare hardness of different coatings, such results have limited validity.

In Europe, Koenig and Persoz pendulums are most widely used. The pendulum makes contact with the coated panel through two steel balls. As the pendulum swings back and forth through a small angle, the weight is transferred between the two balls. Test results are reported as the time in seconds for the swing to be dampened from a higher to a lower angle (from perpendicular), from 6 to 3° in the case of the Koenig pendulum and from 12 to 4° in the Persoz test. Usually harder coatings give longer times. However, soft, rubbery coatings may also give longer times. Based on the reasonable assumption that the main contribution to dampening the pendulum is absorption of mechanical energy by the film, these apparently conflicting results can be explained in terms of the loss modulus. As shown in Figure 24.7, loss modulus values are low in both the regions below and well above T_g and are highest in the transition region. Low loss modulus could therefore account for longer dampening times for both soft, rubbery films with T_g values below ambient and hard films with T_g values well above ambient temperatures. This hypothesis predicts that dampening times for coatings in the transition region at ambient temperature may be very sensitive to temperature, since loss modulus goes through a maximum in this region.

24.5.3.2. Formability and Flexibility

One common type of flexibility test is a *mandrel bend test*, in which a coated panel is bent around a rod or cone (the mandrel) used as a support around which to bend a coated panel. The panel is bent with the coated side away from the mandrel. Any crack in the coating on the bend is reported as a failure. In the cylindrical mandrel test, a series of different mandrels is used and the smallest diameter that permits a bend without failure is reported as the mandrel flexibility.

In a *conical mandrel* test, one end of the panel is clamped and a lever is used to bend the panel over a cone. The distance from the small end of the cone to the end of a crack is reported. This distance, which is proportional to the radius of curvature of the mandrel at that point, can be used to calculate the percent elongation [20]. Thicker films crack more easily than thinner films because the percent elongation at the same distance along the mandrel is greater. The bent edge should be inspected with a magnifying glass to see hairline cracks, and the panels should be inspected the next day because cracks sometimes appear later. If adhesion is poor, the film comes loose from the panel. The severity of the test can be increased by testing at low temperature (by putting the panel and the tester in a freezer before testing). Obviously, the severity of the test is also affected by the rate of bending.

Another formability test, widely used in testing coil coatings, is the *T-bend test*. The coated panel is bent, coating side out, back on itself. If there is no crack at the edge, the result is reported as *zero T*. The zero means there is no additional layer of metal inside the bend. If the coating cracks, the panel is bent again back on itself. Repeated bends back over the original bend are made until the coating does not crack. The radius of curvature gets greater as the number of bends increases. The results are reported as 1T, 2T, 3T, and so forth, counting the number of thicknesses of metal inside the bend. Again the severity of the test is affected by temperature and rate of bending, and again the panels should be reexamined after some time has elapsed.

24.5.3.3. Impact Resistance

Impact tests evaluate the ability of the coating to withstand extension without cracking when the deformation is applied rapidly. A weight is dropped down a guide tube onto a hemispherical indenter that rests on the coated panel. An opening, opposite the indenter in the base support on which the panel rests, permits deformation of the panel. If the coated side is up, that is, it is directly hit by the indenter, the test is called a *direct impact test*. If the back of the panel is up, the test is called a *reverse impact test*. The weight is dropped from greater and greater heights until the coating on the panel cracks. In the United States, the results are reported in in.-lb, that is, the number of inches the weight falls times its weight. In the most common apparatus, the maximum is 160 in.-lb. Generally, reverse impact tests where the coating is extended are more severe than direct impact tests where the coating is compressed. The thickness, mechanical properties, and surface of the substrate substantially affect the results. If the metal panel is thick enough, it is not distorted by the impact and any coating will pass. While any one laboratory runs the test on the same metal substrate, another laboratory may use another substrate and comparison of the results would be meaningless.

24.5.3.4. Abrasion Resistance

A widely used test is the *Taber Abraser*. Two rotating abrasive wheels roll on a panel creating a circular wear path. The test is continued until the coating is worn through and the results are reported in numbers of rotations required to wear through one mil (25 μm) of coating. The results often do not correlate with field use tests. For example, with reference to the Taber abrasion results for the four floor coatings described in Table 24.2, the "hard epoxy" coating had the poorest abrasion resistance in actual field use but exhibited the highest rating in the Taber test. Such lack of correspondence is common. Another example was in the case of the beer can abrasion problem mentioned earlier. Generally, softer coatings tend to give poorer Taber Abraser results, probably because the abrasion disks rotate at a constant speed and therefore impart more energy to softer coatings. However, very soft coatings may clog the abrasive on the wheel resulting in high test results. The Taber Abraser may be of some value as a quality control test, particularly for coatings of similar hardness, but is commonly used inappropriately as a predictive test.

Another commonly used abrasion test is the *falling sand test*. Sand falls out of a hopper through a tube onto a coated panel that is held at a 45° angle to the stream of sand. The result is reported in liters of sand required to wear through a unit thickness of coating. Most people think this test is more reliable than the Taber test. The mechanisms involved in abrasion vary so widely and the rates of application of stress are so different that it is doubtful if any such single test can be devised that will be appropriate for predicting wear life of different coatings for different applications. Simulation tests, such as discussed earlier, are probably the only tests that can be useful for performance prediction.

24.5.3.5. Solvent Resistance

Solvent resistance is not a mechanical property, but it is included here because it is one of the properties that must be balanced with mechanical properties for many applications. It is also appropriate because resistance to swelling of cross-linked films is directly related to XLD which also affects many mechanical properties.

The most common test is the methyl ethyl ketone (MEK) double rub test. This can be done with a tissue soaked with MEK rubbed on the panel but is more conveniently done with a felt tip marker pen filled with solvent. The test can be mechanized so that there is one stroke back and forth (one double rub) on the film each second, and a timer can be used as a counter. A soft thermoplastic coating will rub off with very few rubs. In the case of thermosetting coatings, the number of rubs the coating withstands increases as the degree of reaction increases. The test is sensitive to the development of low cross-link density but insensitive to changes as the cross-link density gets higher. Usually, the test is stopped after 200 double rubs, and, therefore, a series of highly cross-linked coatings may all be reported to give 200+ double rub resistance even though there are differences in the extent of cross-linking.

Another type of solvent resistance test is to expose the coating to the solvent for a period of time, say 15 min, then carry out a pencil hardness test on the exposed area. For example, a test for aircraft top coats specifies that the coating shall not lose more than two pencil hardness units after exposure to hydraulic fluid for 15 min.

GENERAL REFERENCES

L. W. Hill, *Prog. Org. Coat.*, **5**, 277 (1977).

L. W. Hill, *Mechanical Properties of Coatings*, Federation of Societies for Coatings Technology, Blue Bell, PA, 1987.

D. J. Skrovanek and C. K. Schoff, *Prog. Org. Coat.*, **16**, 135 (1988).

REFERENCES

1. R. A. Dickie, *J. Coat. Technol.*, **64** (809), 61 (1992).
2. L. W. Hill, *Prog. Org. Coat.*, **5**, 277 (1977).
3. L. W. Hill, *Mechanical Properties of Coatings*, Federation of Societies for Coatings Technology, Blue Bell, PA, 1987.
4. D. J. Skrovanek and C. K. Schoff, *Prog. Org. Coat.*, **16**, 135 (1988).
5. L. W. Hill, *J. Coat. Technol.*, **64** (808), 29 (1992).
6. S. J. Wu, *J. Appl. Polym. Sci.*, **20**, 327 (1976).
7. H. Stutz, K.-H. Illers, and J. Mertes, *J. Polym. Sci., Part B: Polym. Phys.*, **28**, 1483 (1990).
8. L. W. Hill, *J. Coat. Technol.*, **59** (751), 63 (1987).
9. D. R. Bauer, *J. Coat. Technol.*, **60** (758), 53 (1988).
10. D. J. Skrovanek, *Prog. Org. Coat.*, **18**, 89 (1990).
11. M. B. Roller, *J. Coat. Technol.*, **54** (691), 33 (1982).
12. A. Eisenberg, "The Glassy State and the Glass Transition," in J. E. Mark, A. Eisenberg, W. W. Graessley, L. Mandelkern, and J. L. Koenig, Eds., *Physical Properties of Polymers*, American Chemical Society, Washington, DC, 1984, pp. 55–95.
13. L. W. Hill, personal communication, 1986.
14. P. J. Greidanus, *FATIPEC XIXth Congress Book*, Vol. I, 485 (1988).
15. (a) R. M. Evans, in R. R. Myers and J. S. Long, Eds., *Treatise on Coatings*, Vol. 2, Part I, Marcel Dekker, New York, 1969, pp. 13–190. (b) R. M. Evans and J. Fogel, *J. Coat. Technol.*, **47** (639), 50 (1977).
16. L. H. Lee, *Polym. Mater. Sci. Eng.*, **50**, 65 (1984).
17. F. Fink, W. Heilen, R. Berger, and J. Adams, *J. Coat. Technol.*, **62** (791), 47 (1990).
18. G. A. Vandermeersche, *Closed Loop*, 3, April, 1981.
19. R. D. Athey, Jr., *Amer. Paint and Coatings J.*, December 7, 38, 1992.
20. E. M. Corcoran, A. G. Roberts, and G. G. Schurr, Eds., *Gardner-Sward Paint Testing Manual*, 13th ed., American Society for Testing and Materials, Philadelphia, PA, 1972.

CHAPTER XXV

Exterior Durability

Exterior durability of coatings refers to the resistance to changes occurring during outdoor exposure; such changes are generally undesirable and include changes in mechanical properties, such as modulus and strength, loss of adhesion, discoloration, embrittlement, chalking, loss of gloss, and acid-etching. Thus, both aesthetic and functional properties are involved. The terms outdoor durability and weatherability are also used. Corrosion protection by coatings is discussed in Chapter 27.

The most common processes leading to degradation are photoinitiated oxidation and hydrolysis resulting from exposure to sunlight, air, and water. Interrelationships between these processes have been reported, including enhanced photoxidative degradation in high humidity as well as enhanced hydrolytic degradation during sunlight exposure. Hydrolytic degradation is also expected to be aggravated by exposure to acid (including acid rain). Other atmospheric degradants include ozone and oxides of nitrogen and sulfur. Changes in temperature and humidity may result in cracking arising from expansion and contraction of coatings or substrates.

The rates at which these processes occur vary widely depending on the exposure site(s), time of year, coating composition, and substrate. Consequently, exterior durability is best evaluated by actual field performance.

In general, coatings formulated for exterior durability should exclude or minimize resin components that absorb UV radiation at wavelengths longer than 280 nm, are readily oxidized, or are readily hydrolyzed. Absorption of UV below 280 nm is not important since these shorter wavelengths are screened from the earth by our protective ozone layer.

25.1. PHOTOINITIATED OXIDATIVE DEGRADATION

The general process of photoinitiated oxidation of polymers is a chain reaction as outlined in Scheme 1. Absorption of UV by a polymer or other coating component (P) produces highly energetic photoexcited states (P*), that may undergo bond cleavage to yield free radicals (P·). Free radicals undergo a chain reaction with O_2 (autoxidation) ultimately leading to polymer degradation. Hydroperoxides (POOH) and peroxides (POOP) are unstable products of photoinitiated oxidation, which readily

Scheme 1

Initiation

$$\text{Polymer (P)} \xrightarrow{\text{sunlight}} \text{P*} \tag{1}$$

$$\text{P*} \longrightarrow \text{Free radicals (P}\cdot) \tag{2}$$

$$\text{POOH} \xrightarrow{\text{sunlight}} \text{PO}\cdot + \cdot\text{OH}$$

Propagation

$$\text{P}\cdot + O_2 \longrightarrow \text{POO}\cdot \tag{3}$$

$$\text{POO}\cdot + \text{Polymer (P—H)} \longrightarrow \text{POOH} + \text{P}\cdot \tag{4}$$

$$\text{PO}\cdot\ (\cdot\text{OH}) + \text{Polymer (P—H)} \longrightarrow \text{POH } (H_2O) + \text{P}\cdot$$

Chain Termination

$$2\ \text{POO}\cdot \longrightarrow \text{POOP} + O_2$$

$$2\ \text{P}\cdot \longrightarrow \text{P—P} + \text{Disproportionation products}$$

$$\text{POO}\cdot + \text{P}\cdot \longrightarrow \text{POOP} + \text{Disproportionation products}$$

$$2\ \text{POO}\cdot \longrightarrow \text{Ketones (aldehydes)} + \text{Alcohols}$$

Chain Scission

$$\text{PO}\cdot \longrightarrow \text{Ketones} + \text{P}'\cdot$$

dissociate with sunlight and moderate heat to yield alkoxy (PO·) and hydroxy (HO·) radicals. These radicals are highly reactive toward hydrogen abstraction and yield polymer radicals (P·) which enter into the propagation stage of polymer degradation. Tertiary alkoxy radicals also tend to dissociate into ketones and a lower molecular weight polymer radical (P′·), thereby resulting in chain scission.

As shown in Scheme 1, chain propagation leading to oxidative degradation proceeds by hydrogen abstraction from the polymer. Therefore, functional groups that promote hydrogen abstraction should be minimized. A general ordering of functional groups that promote the abstraction of hydrogen atoms from adjacent carbon atoms is

Amines: $-CH_2-\text{NR}_2$ >

Hydrocarbons with methylene groups between two double bonds:

$-\text{CH}=\text{CH}-CH_2-\text{CH}=\text{CH}-$ >

Ethers, alcohols: $-CH_2-\text{O}-\text{R(H)}$ >

Hydrocarbons: $\text{Ph}-CH_2-$, $-\text{CH}=\text{CH}-CH_2-$, C_3CH,

Urethanes: $-CH_2-\text{NH}-\text{CO}-\text{OR}$ >

Esters: $-CH_2-\text{CO}-\text{O}-CH_2-$ >

Hydrocarbons: $-CH_2- > CH_3$,

Methylsiloxanes: $-\text{Si}(CH_3)_2-\text{O}-$

None of these functional groups absorb sunlight directly. In the absence of absorbing aromatic groups, absorption of sunlight by aliphatic resins occurs primarily by inadvertently present peroxide and ketone groups, both of which absorb ambient UV (above 280 nm). Essentially all resins used in coatings contain some hydroperoxides, as do most organic substances. Peroxides and ketones (as well as aldehydes) are also formed in photoinitiated oxidation (Scheme 1). Photolysis of peroxides and ketones is shown in Scheme 1 and Scheme 2, respectively. Photoxidation of aldehydes and ketones yields peracids, which are strong organic oxidants. Peracids may play a significant role in oxidative degradation.

Methyl-substituted silicones and silicone-modified resins exhibit high photoxidation stability, generally in proportion to the silicone content. The excellent exterior durability of fluorinated resins may be attributed, at least in part, to the absence (or reduced level) of C-H groups.

Aromatic groups with directly attached heteroatoms, as found in aromatic urethanes (Ar-NH-CO-OR) and bisphenol A expoxies (Ar-O-R), absorb UV above 280 nm and undergo direct photocleavage to yield free radicals that can participate in oxidative degradation. Coatings made using aromatic isocyanates yellow badly after only short exposures to UV radiation. Coatings based on BPA epoxides have

Scheme 2

Oxidation of Aldehydes and Ketones (Peracid Formation)

$$\mathrm{P{-}\overset{\overset{\displaystyle O}{\|}}{C}{-}P} \xrightarrow{\text{sunlight}} \mathrm{P{-}\overset{\overset{\displaystyle O}{\|}}{C}\cdot + P\cdot}$$

$$\mathrm{P{-}\overset{\overset{\displaystyle O}{\|}}{C}\cdot + O_2} \longrightarrow \mathrm{P{-}\overset{\overset{\displaystyle O}{\|}}{C}OO\cdot}$$

$$\mathrm{P{-}\overset{\overset{\displaystyle O}{\|}}{C}OO\cdot + PH} \longrightarrow \mathrm{P{-}\overset{\overset{\displaystyle O}{\|}}{C}OOH + P\cdot}$$

$$\mathrm{P\cdot + O_2} \longrightarrow \mathrm{POO\cdot}$$

$$\mathrm{P{-}\overset{\overset{\displaystyle O}{\|}}{C}H + POO\cdot} \longrightarrow \mathrm{P{-}\overset{\overset{\displaystyle O}{\|}}{C}\cdot + POOH}$$

$$\mathrm{P{-}\overset{\overset{\displaystyle O}{\|}}{C}\cdot + O_2} \longrightarrow \mathrm{P{-}\overset{\overset{\displaystyle O}{\|}}{C}OO\cdot}$$

$$\mathrm{P{-}\overset{\overset{\displaystyle O}{\|}}{C}OO\cdot + P\overset{\overset{\displaystyle O}{\|}}{C}H} \longrightarrow \mathrm{P{-}\overset{\overset{\displaystyle O}{\|}}{C}OOH + P{-}\overset{\overset{\displaystyle O}{\|}}{C}\cdot}$$

generally chalked rapidly on exposure outdoors. There has been a report that acrylic resins cross-linked with BPA epoxy containing hindered amine light stabilizer (see Section 25.2.3) stood up well when exposed in Florida [1]. Further work is required to see how general this result may be.

Highly chlorinated resins such as vinyl chloride copolymers, vinylidene chloride copolymers, and chlorinated rubber degrade by autocatalytic dehydrochlorination on exposure to either heat or UV. They must be formulated with stabilizers, as discussed in Section 25.5.

25.2. PHOTOSTABILIZATION

The lines of defense against photoinitiated oxidative degradation include the use of UV absorbers to reduce UV absorption by the polymer (Scheme 1, Eq. 1), use of excited state quenchers to compete with bond cleavage of P* (Scheme 1, Eq. 2), and antioxidants to reduce oxidative degradation (Scheme 1, Eqs. 3 and 4). Review articles on photo- (and thermal) stabilization of coatings, including degradative pathways are available [2].

25.2.1. UV Absorbers and Excited State Quenchers

Important characteristics for UV stabilizers, both absorbers and quenchers, are photostability together with chemical and physical permanence. Photostability requires that the photoexcited stabilizer (A* or Q*) can return to the ground state by converting the UV energy into thermal energy, as shown in Scheme 3. This process generally occurs by reversible intramolecular hydrogen transfer or *E*-*Z* (cis-trans) isomerization of double bonds.

Scheme 3

UV Absorber (A):

$$A \xrightarrow{\text{sunlight}} A^*$$

$$A^* \longrightarrow A + \text{Heat}$$

$$\text{Overall: Radiation} \longrightarrow \text{Heat}$$

Excited State Quencher (Q):

$$P \xrightarrow{\text{sunlight}} P^*$$

$$P^* + Q \longrightarrow P + Q^*$$

$$Q^* \longrightarrow Q + \text{Heat}$$

$$\text{Overall: Radiation} \longrightarrow \text{Heat}$$

An additional important feature of excited state quenchers is their effective quenching volume within which quenching of photoexcited polymers occurs efficiently. The effective quenching volume depends on the mechanism of energy transfer. This volume is generally characterized by a critical quenching radius that varies from less than 1 nm for exchange energy transfer (EET) (requiring overlap of molecular orbitals of photoexcited polymer and quencher) to as large as 10 nm for resonance energy transfer (RET) (which occurs by a dipole–dipole interaction). Since volume increases with the cube of the radius and required concentration of quencher is inversely proportional to quenching volume, it follows that increasing quenching radius by 10-fold allows a reduction in quencher concentration of 1000-fold.

These considerations, including the requirements for quenching by RET, have been discussed and related to the prospects for stabilizing aromatic polymers [3]. An important requirement for RET quenching is strong UV absorption by the stabilizer in the wavelength region in which the photoexcited polymer fluoresces. On the other hand, stabilization by UV absorbers requires strong absorption in the wavelength region in which the photoexcited polymer also absorbs. For this reason, stabilizers, which function as UV absorbers in some coatings, may perform as excited state quenchers in other coatings. The most effective stabilizers probably function by both roles in many coatings.

One cannot eliminate UV absorption by the resin components by adding a UV absorber; one can merely reduce absorption by the binder to reduce the rate of photodegradation reactions. Since absorption increases as the path length increases, UV absorbers are most effective in protecting the lower parts of a film, for example, the base coat under a clear top coat, and least effective in protecting the layer at the air interface.

A critical consideration in the design and selection of UV absorbers is their absorption spectra. In general terms, one would like to have very high absorption of UV radiation from 280 through 380 nm. To avoid color effects of adding absorber, ideally there would be no absorption above 380 nm. A wide range of UV absorbers are available commercially, exhibiting different UV absorption spectra.

Substituted 2-hydroxybenzophenones, 2-(2-hydroxyphenyl)-2*H*-benztriazoles, and 2-(2-hydroxyphenyl)-4,6-phenyl-1,3,5-triazines are classes of UV absorber/quencher stabilizers.

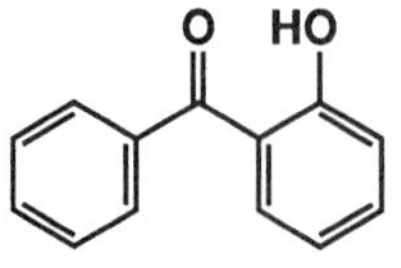

O HO

Hydroxybenzophenones

HO

N N N

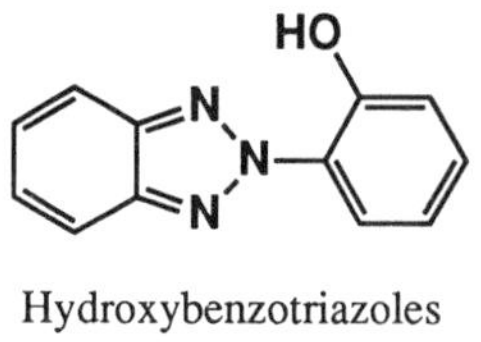

Hydroxybenzotriazoles

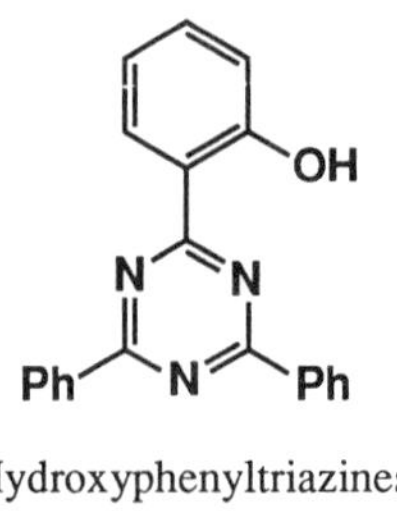

OH

N N Ph N Ph

Hydroxyphenyltriazines

These UV stabilizers function by intramolecular hydrogen transfer. Ultraviolet absorption or excited state quenching yields photoexcited states that convert their excess electronic energy into chemical energy by undergoing intramolecular hydrogen transfer to yield thermodynamically unstable intermediates. The unstable intermediates spontaneously undergo reverse hydrogen transfer to regenerate the UV absorber/quencher with conversion of the chemical (potential) energy into heat. The process is illustrated with 2-hydroxybenzophenone.

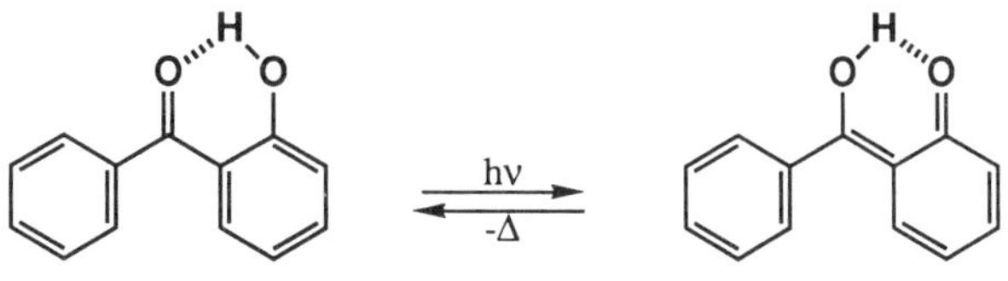

$$\xrightleftharpoons[-\Delta]{h\nu}$$

It is important that the UV absorber be soluble in the coating film. Several grades of the various absorbers are available commercially with different substituents on the aromatic rings that provide for solubility in different polymer systems. Commonly, the absorber is added to the top coat of a multicoat system. However, it must be remembered that, especially in baking systems, migration may result in the absorber being distributed throughout the whole coating, hence reducing the concentration in the top coat. This effect has been demonstrated by analyzing sections through a film of clear coat/base coat automotive finishes [4]. In one combination of coatings, where UV absorber was added only to the clear coat, the absorber content throughout the whole film, both clear coat and base coat, was essentially uniform. In a second case, with a different type of base coat, a major fraction of the absorber stayed in the clear coat part of the combined film.

Another critical requirement of a UV absorber is permanence. If a UV absorber has even a small vapor pressure, it can slowly volatilize from the surface over the long term periods for which durability is desired. Analysis for UV absorber as a function of depth into the film initially and after one year exterior exposure (on a Florida black box) can show significant losses from the exposed films, particularly near the surface of the film [4]. Hydroxyphenyltriazines have very low vapor pressures and are quite permanent. Longer term permanence may be achieved by utilizing polymer-bound or oligomeric photostabilizers. Generally, the latter approach is taken.

25.2.2. Antioxidants

Antioxidants are classified into two groups: preventive and chain-breaking antioxidants. Preventive antioxidants include peroxide decomposers, which reduce hydroperoxides to alcohols and become oxidized into harmless products. Examples of peroxide decomposers are sulfides and phosphites that are initially oxidized to sulfoxides and phosphates, respectively, as shown in Eqs. 5 and 6 with dilauryl thiodipropionate (LTDP) and triphenylphosphite. Further reactions may occur.

$$POOH + \underset{\text{LTDP}}{S(CH_2CH_2CO-OC_{12}H_{25})_2} \longrightarrow POH + O=S(CH_2CH_2CO-OC_{12}H_{25})_2 \quad (5)$$

$$POOH + (PhO)_3P \longrightarrow POH + (PhO)_3P=O \quad (6)$$

Metal complexing agents are a second type of preventive antioxidants. They are used to tie up transition metals such as cobalt, which, otherwise, can catalyze conversion of hydroperoxides into peroxy and alkoxy radicals by redox reactions, as outlined in Scheme 4. The overall reaction occurs catalytically by interconversion of Co(II) and Co(III). The resulting peroxy and alkoxy radicals promote oxidative degradation, as shown above in Scheme 1.

Scheme 4

$$POOH + Co^{+2} \longrightarrow POO\cdot + H^+ + Co^{+3}$$

$$POOH + Co^{+3} \longrightarrow PO\cdot + HO^- + Co^{+2}$$

$$\text{Overall: } 2\ POOH \longrightarrow POO\cdot + PO\cdot + H_2O$$

Note that the same type of transition metals are used as catalysts (so-called driers) for oxidative cure of drying oils and alkyds (see Section 9.2.2). Thus, the reactions in Scheme 4 are involved in both oxidative cross-linking and oxidative degradation. It follows that drier concentrations should be minimized to reduce subsequent degradation during exterior exposure. Antioxidants that function as metal complexing agents reduce the ability of transition metals, such as cobalt, to participate in redox conversion of hydroperoxides into degradative peroxy and alkoxy radicals.

The tetrafunctional bidentate imine, derived from *o*-hydroxybenzaldehyde (salicylaldehyde) and tetraaminomethylmethane, is reported to complex a large number of transition metals effectively, including Co, Cu, Fe, Mn, and Ni (represented by M^{+n} in complex 1) [5].

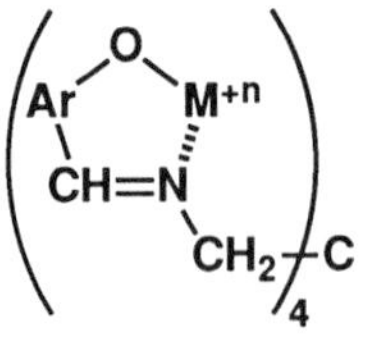

$$\left(\begin{array}{c} Ar - O - M^{+n} \\ | \quad\quad \vdots \\ CH=N - CH_2 \end{array} \right)_4 C$$

(Ar = 1,2-disubstituted benzene)

1

Chain-breaking antioxidants function by interfering directly with the chain propagation steps of autoxidation, shown in Scheme 1. An example is a hindered phenol, 2,2′-methylenebis(4-methyl-6-tert-butylphenol), which reacts with peroxy radicals in competition with hydrogen abstraction from a polymer (PH), to yield the resonance-stabilized, less reactive, phenoxy radical shown in Eq. 7.

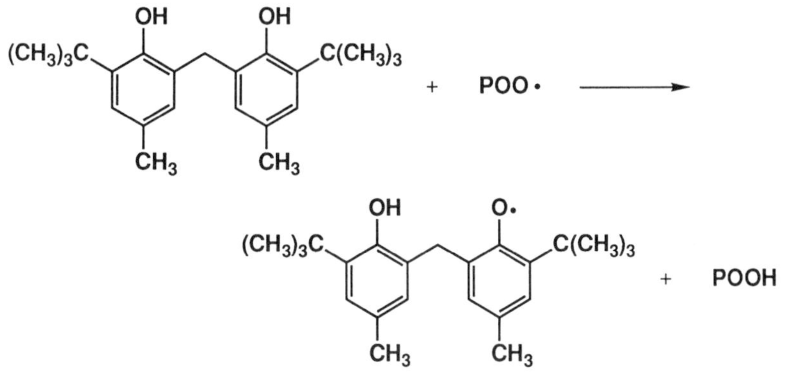

$$+ \; POO\cdot \longrightarrow \; + \; POOH \qquad (7)$$

Note in Eq. 7 that hydroperoxides are generated; this is a basis for synergistic stabilization of polymers by combination of peroxide decomposers and chain-breaking phenolic antioxidants. Synergistic stabilization signifies that the combination of stabilizers is more effective than the additive effect of each stabilizer by itself.

25.2.3. Hindered Amine Light Stabilizers

Unquestionably, the most important advance in the stabilization of coatings in recent decades has been the development of hindered amine light stabilizers (HALS), first introduced in the early 1970s. These HALS are derivatives of 2,2,6,6-tetramethylpiperidine **2**. They are reported to function both as chain-breaking antioxidants [6] and transition metal complexing agents [7]. Note that the 2,2,6,6-methyl groups prevent oxidation of the ring carbons attached to the nitrogen.

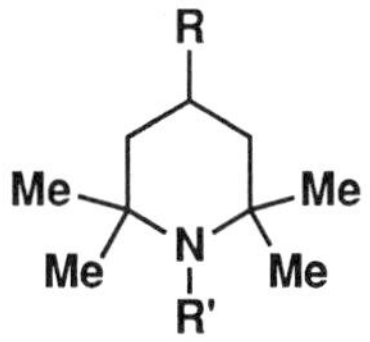

(R'= H, alkyl, alkanoyl, alkoxy)

2

A variety of HALS compounds are available. Note that R in the general formula **2** is usually a diester group that joins two piperidine rings; this increases the molecular weight and hence decreases the volatility of the compounds. The first commercial HALS compounds, still used to a degree, had R′ = H. Later versions with R′ = alkyl exhibit better long-term stability. Alkanoyl HALS compounds with R′ = COR″ are not basic and are suitable for the stabilization of acid-catalyzed cross-linking systems. Most recently, hydroxylamine ethers (R′ = OR″) have been introduced.

A key feature of HALS derivatives is their photoxidative conversion into nitroxyl radicals ($R_2NO\cdot$) that react with carbon-centered radicals by disproportionation

Scheme 5

$$R_2N{-}R' + O_2 \xrightarrow{UV} R_2NO\cdot$$

$$R_2NO\cdot + P\cdot \longrightarrow R_2NOH + R_2NOP \quad (8)$$

$$R_2NOH(P) + POO\cdot \longrightarrow R_2NO\cdot + POOH(P) \quad (9)$$

and combination to yield corresponding hydroxyl amines and ethers, respectively, as shown in Scheme 5 (Eq. 8). The hydroxylamines and ethers, in turn, react with peroxy radicals to regenerate nitroxyl radicals, also shown in Scheme 5 (Eq. 9). In this manner, HALS derivatives interfere with propagation steps involving both carbon-centered and peroxy radicals in autoxidation (see Scheme 1). In contrast to nitroxyl radicals, hindered phenols do not react with carbon-centered radicals. Consequently, the chain-breaking antioxidant activity of hindered phenols is limited to reaction with peroxy radicals and oxygen is required as a costabilizer.

Apparently, HALS derivatives undergo rapid photoxidation to form nitroxyl free radicals on exterior exposure of coatings. In an exposed film only a small fraction (about 1%) is present as the nitroxyl radical, the major storage components being the corresponding hydroxylamine (R_2NOH) and ethers (R_2NOP). Continued stabilization requires the presence of the nitroxyl radical and its disappearance is followed shortly by rapid polymer degradation. Probably the ultimate demise of HALS occurs, at least in part, by oxidation of the nitroxyl radical accompanied by opening of the piperidine ring. Transition metals and peracids are potential oxidants for this process.

Combinations of UV absorbers and HALS compounds act synergistically [8]. The presence of the UV absorber reduces the rate of generation of radicals, whereas the HALS compound reduces the propagation rate, that is, oxidative degradation by the radicals. A further factor may well be that UV absorbers are relatively inefficient at protecting the upper surface of a film; in contrast, HALS compounds can effectively scavenge free radicals even at the surface. Results of analysis of coating films after exterior exposure show that a significant amount of HALS derivatives remain after two years of black box Florida exterior exposure [4].

25.2.4. Pigmentation Effects

Many pigments can absorb UV radiation. In fact the strongest UV absorber known is fine particle size carbon black. Furthermore, many carbon blacks have structures with multiple aromatic rings and, in some cases, with phenol groups on the pigment surface. Such black pigments have both UV absorber and antioxidant activity. Enhanced exterior durability is commonly obtained with carbon black pigmented coatings. Other pigments absorb UV radiation to varying degrees. For example, 50-μm coatings pigmented with fine particle size, transparent iron oxide pigments virtually absorb all radiation below about 420 nm [9]. This strong absorption is particularly useful in stains for use over wood, since the pigmented transparent coating protects the wood from photodegradation.

Scheme 6

$$TiO_2 + \text{Light} \longrightarrow TiO_2^*(e/p)$$

$$TiO_2^*(e/p) + O_2 \longrightarrow TiO_2(p) + O_2\cdot^-$$

$$TiO_2(p) + H_2O \longrightarrow TiO_2 + H^+ + HO\cdot$$

$$H^+ + O_2\cdot^- \longrightarrow HOO\cdot$$

$$2\ HOO\cdot \longrightarrow H_2O_2 + O_2$$

$$TiO_2^*(e/p) + H_2O_2 \longrightarrow TiO_2 + 2\ HO\cdot$$

Rutile TiO_2 absorbs most UV radiation strongly. Absorption is a function not only of wavelength and concentration but also of particle size of the pigment [10]. Optimum particle size for absorption of UV by rutile increases from 0.05 μm for 300-nm to 0.12 μm for 400-nm radiation. This is smaller than the optimum particle size of 0.19 μm for hiding of visible light (see Section 16.2.3). Nevertheless, TiO_2 white pigment with an average particle size of 0.23 μm still absorbs UV strongly. Anatase TiO_2 also absorbs most UV radiation strongly, although not as strongly as rutile in the near UV. Thus, TiO_2, especially rutile TiO_2, functions as a UV absorber in coating films.

However, TiO_2 also acts to accelerate photodegradation of films on exterior exposure. Titanium dioxide pigments can accelerate *chalking* of paint films, that is, degradation of the organic binder and exposure of unbound pigment particles on the film surface that rub off easily like chalk off a blackboard. Degradation of the binder is enhanced by interaction of photoexcited TiO_2 with oxygen and water to yield oxidants, as shown in Scheme 6 [11].

Photoexcitation of TiO_2 results in promotion of a low-energy valence band electron into the higher-energy conduction band, thereby creating a separated electron(e)/hole(p) pair, signified by $TiO_2^*(e/p)$ in Scheme 6. Electron capture by O_2 (reduction) and hole capture by H_2O (oxidation) result in regeneration of ground-state TiO_2 and ultimately lead to hydroperoxy and hydroxy radicals that can participate in oxidative degradation of the binder, as shown in Scheme 1.

Anatase TiO_2 is substantially more active in promoting oxidative degradation than the rutile crystal form. The photoactivity of TiO_2 pigments is generally reduced by coating the pigment particles with silica and/or alumina that presumably function as a barrier layer against the electron transfer redox reactions. Treated rutile pigments are commercially available that accelerate chalking to only very minor extents. A laboratory test has been developed to compare the photoreactivity of various grades of TiO_2 [12]. Various other stabilizing additives, including HALS, have also been reported [13].

Formulating coatings for exterior uses is further complicated by the existence of pigment/binder interactions. It is not unusual to find that a pigment will show excellent color retention after exterior exposure when formulated with one class of resins as the binder, but poor durability when formulated with another class.

For example, thioindigo maroon had excellent color retention when used in nitrocellulose lacquers, but poor stability when used in acrylic lacquers. While one can use experience in other systems as an initial basis for selecting pigments for use with a new class of resins, actual field tests are required to assure that each combination of pigments and resins is indeed suitable.

25.3. DEGRADATION OF CHLORINATED RESINS

All highly chlorinated resins undergo dehydrochlorination on exposure to either heat or UV. The ultimate products are conjugated polyenes. As the number of conjugated double bonds increases, the polymer progressively discolors, finally absorbing all wavelengths of radiation completely, that is, becoming black. The resulting highly unsaturated polymer undergoes autoxidation, resulting in a high degree of cross-linking and embrittlement.

The mechanism of degradation of poly(vinyl chloride) has been very extensively studied. It is generally agreed that dehydrochlorination is promoted by the presence of activated chlorine on a tertiary carbon or on a carbon allylic to a double bond. Chlorines on tertiary carbons result from chain transfer to polymer, resulting in a branch on a carbon bearing a chlorine (see Section 4.2.1). As described in Section 4.2.1, it has been proposed that at least one major weak point results from the

$$\text{Ⓟ}\!\sim\!CH_2-\overset{H}{\underset{Cl}{C}}-CH_2-\overset{H}{\underset{Cl}{C}}\cdot + CH_2{=}CH-Cl \longrightarrow$$

$$\text{Ⓟ}\!\sim\!CH_2-\overset{H}{\underset{Cl}{C}}-CH_2-\overset{H}{\underset{Cl}{C}}-\overset{H}{\underset{Cl}{C}}-CH_2\cdot$$

$$\text{Ⓟ}\!\sim\!CH_2-\overset{H}{\underset{Cl}{C}}-CH_2-\overset{H}{\underset{Cl}{C}}-\overset{H}{\underset{Cl}{C}}-CH_2\cdot + CH_2{=}CH-Cl \longrightarrow$$

$$\text{Ⓟ}\!\sim\!CH_2-\overset{H}{\underset{Cl}{C}}-CH_2-\overset{H}{\underset{Cl}{C}}-CH{=}CH_2 + H-\overset{H}{\underset{Cl}{C}}-\overset{H}{\underset{Cl}{C}}\cdot$$

$$\text{Ⓟ}\!\sim\!CH_2-\overset{H}{\underset{Cl}{C}}-CH_2-\overset{H}{\underset{Cl}{C}}-CH{=}CH_2 \longrightarrow$$

$$\text{Ⓟ}\!\sim\!CH_2-\overset{H}{\underset{Cl}{C}}-CH{=}CH-CH{=}CH_2 + HCl$$

addition of a vinyl chloride monomer in a head-to-head fashion to the growing polymer chain followed by chain transfer to monomer [14]. The resulting allylic chloride is highly susceptible to dehydrochlorination that generates a new allylic chloride with two conjugated double bonds. Progressive dehydrochlorination is favored because the growing number of conjugated double bonds further increases the lability of allylic chlorides down the chain.

A wide variety of stabilizing agents has been found. Since the dehydrochlorination reaction is catalyzed by hydrogen chloride, HCl traps, such as epoxy compounds and basic pigments, are useful. Diels–Alder dienophiles can act as stabilizers; the Diels–Alder addition breaks up the chain of conjugated double bonds. Dibutyltin diesters are effective stabilizers. It is proposed that the activated chlorines are interchanged with ester groups of the tin compounds to form the more stable ester-substituted polymer molecules. Dibutyltin maleate is a particularly effective stabilizer since it acts both as a tin compound and a dienophile. Barium, cadmium, and strontium soaps act as stabilizers. Choice of stabilizer combinations can be very system specific, especially depending on whether stabilization is needed against heat, UV, or both. In the case of UV stabilization, use of UV absorbers can further enhance stability.

25.4. HYDROLYTIC DEGRADATION

A general ordering of functional groups that are subject to hydrolysis is esters > ureas > urethanes. The inherent tendency of functional groups to hydrolyze can be reduced by steric interference, for example, by placement of alkyl groups in the vicinity of the susceptible groups, such as esters (see Section 8.1.1). The alkyl groups may also reduce hydrolysis by increasing hydrophobicity in the vicinity of the susceptible groups. Rates of hydrolysis are also influenced by neighboring groups. For example, phthalate half esters, in which the groups are ortho, are more readily hydrolyzed under acidic conditions, such as are generally encountered in exterior exposure, than isophthalate half esters, in which the groups are meta (see Section 8.1.2).

Hydrolysis of polyesters results in backbone degradation. On the other hand, the backbones of (meth)acrylic resins are completely resistant to hydrolysis since the linkages are all carbon–carbon bonds. Acrylate (and particularly methacrylate) ester side groups are very resistant to hydrolysis owing to steric effects of the acrylic backbone.

When polyester or acrylic resins are cross-linked with MF resins the activated ether linkages formed are somewhat susceptible to hydrolysis. Their hydrolysis is enhanced by residual acid catalysts (generally sulfonic acids) used to catalyze cross-linking (see Section 6.3.1.1). Curing temperatures can be reduced by increasing the concentration of sulfonic acid in the coating, but the sulfonic acid remaining in the cured film enhances susceptibility to hydrolysis. An apparently ideal solution would be the utilization of transient (or fugitive) acid catalysts that either leave the film or become neutralized after cure. It has been reported that *p*-toluenesulfonic acid is slowly leached out of films by water; thus, hydrolytic stability may improve as a coating is exposed in areas with high rainfall.

Since the introduction of base coat/clear coat finishes for automobiles, a new problem, called *acid etching* or *acid spotting*, has developed. Small, unsightly spots appear in the clear coat surface, sometimes within days in a climate like Saint Augustine, Florida. The spots are uneven, shallow depressions in the clear coat surface. Presumably, they result from hydrolytic erosion of resin in the area of a droplet of water containing a significant acid concentration. Urethane/polyol clear coats seem less susceptible to acid etching than many MF/polyol clear coats. A variety of approaches have been undertaken to minimize the problem; see Section 33.1.3.2 for a discussion of the problem.

Acrylic/urethane coatings are reported to be more effectively stabilized by HALS than are acrylic/melamine coatings [14]. This result might be attributed to the presence of acid catalysts in the melamine coatings that reduce the activity of HALS derivatives. It might also reflect the tendency of acrylic/melamines to degrade primarily by hydrolysis; HALS derivatives do not stabilize against hydrolytic degradation. On the other hand, hydrogens on carbons alpha to the nitrogen of urethanes are relatively easily abstracted by free radicals, resulting in oxidative degradation. Since HALS derivatives are potent stabilizers against photoxidation, their effective stabilization of urethanes can be rationalized.

Hydrolytic degradation of acrylic/urethane [15] as well as acrylic/melamine [16] coatings is reported to be accelerated by UV exposure. These results can be rationalized, in general terms, by an increase of hydrophilic groups, such as hydroperoxides, alcohols, ketones, and carboxylic acids, resulting from photoxidation. Photoxidation may also occur at specific sites to generate groups that are more susceptible to hydrolysis. Photodegradation of acrylic/melamine coatings is also reported to be accelerated in high humidity [16]. This result has been attributed to conversion of formaldehyde (from hydrolysis) into performic acid, a strong oxidant.

The rate of hydrolysis is affected by factors other than the ease of hydrolysis of bonds in the film. For example, if the T_g of the film is sufficiently greater than the temperature the coating encounters during exterior exposure, hydrolysis is very slow even for relatively easily hydrolyzed ester groups. Some films can be plasticized by water, that is, their T_g is lowered by water, perhaps by interrupting intermolecular hydrogen bonds. In considering the possible effect of T_g on hydrolytic stability, one should bear in mind that the critical T_g is the T_g in the presence of water.

Silicone coatings, which are highly resistant to photodegradation, are subject to hydrolysis at cross-linked sites, where silicon is attached to three oxygens [17,18]. Apparently, the electronegative oxygens facilitate nucleophilic attack at Si by water (see Section 13.3). The reaction is reversible, so the cross-links can hydrolyze and reform. If a silicone-modified acrylic or polyester coating is exposed to water over long periods or is used in a climate with very high humidity, the coating can get softer. It is common to include some MF resin as a supplemental cross-linker in the formulation. Apparently the MF cross-links with the acrylic or polyester are more hydrolytically stable than the bonds between the silicone resin and the polyester or acrylic.

25.5. OTHER MODES OF FAILURE ON EXTERIOR EXPOSURE

While exposure to the UV radiation of sunlight and the hydrolytic effects of rainfall and humidity are probably the major causes of exterior failure of coatings, many, many other phenomena can occur. Thermoplastic coatings exposed in hot sun and then subjected to a thunderstorm can get dimples where rain drops hit. To avoid this failure, a high-T_g polymer is used.

Automobile finishes can undergo microcracking in environments where there is a rapid change of temperature, such as when the sun goes down or a car is driven out of a warm garage on a very cold winter day. The stress built up by the differential of coefficients of expansion of the coating and the steel as the temperature drops rapidly can lead to shrinkage in excess of the elongation-at-break of the coating at that rate of shrinkage. Low molecular weight thermoplastic coatings are particularly susceptible to such failure.

When paint is applied to wood, it must be able to withstand the elongation that results from the expansion of the wood when moisture is absorbed into the wood. Otherwise, grain cracking will occur. This failure mode can occur with interior coatings; however, it is more likely to happen with exterior coatings, particularly with alkyds or drying oils that embrittle with exterior exposure. Grain cracking of alkyd-based paints applied to exterior wood siding and trim has been common. As a result of their greater exterior durability and extensibility, acrylic latex paint films seldom fail this way.

Another common problem of exterior, oil-based house paints on wood is blistering. The blistering results from accumulation of water in wood siding beneath the paint layer. The vapor pressure of the water increases with heating by the sun and blisters form to relieve the pressure. Since latex paints have higher moisture vapor permeability than oil-based paints, the water vapor can pass through the latex paint film, relieving the pressure before blisters grow.

However, the high moisture vapor permeability of latex paint films can lead to failures of other types. For example, if calcium carbonate fillers are used in an exterior latex paint, *frosting* can occur. Water and carbon dioxide permeate into the film, dissolving the calcium carbonate by forming soluble calcium bicarbonate, a solution of which can diffuse out of the film. At the surface of the film, the equilibrium changes direction and the calcium bicarbonate is converted back to a deposit of calcium carbonate. Stalactites may look nice in a cave but home owners don't usually appreciate having little stalactites growing on the surface of the paint underneath the eaves of their house.

Contaminants in the atmosphere can lead to exterior paint failure. Hydrogen sulfide leads to black staining of white lead pigmented paints. Lead pigments are no longer used in consumer paints in the United States, but the organomercury derivatives used as biocides can yield black stains of mercuric sulfide when exposed to hydrogen sulfide. Organomercury biocides are no longer permitted for interior use but are permissible and are used in exterior paints.

Dirt retention can be a difficult problem with latex house paints. Latex paints must be designed to coalesce at relatively low application temperatures. At warmer temperatures, soot and dirt particles that land on the paint surface may stick tenaciously and not be washed off by rain. This problem is minimized with low gloss paints because the high pigment content increases the viscosity of the paint film surface. The problem is, of course, particularly severe in areas where large amounts of soft coal are burned.

Often one needs to be a detective to determine the origin of a paint failure. For example, there were suddenly numerous reports of ugly dirt spots on houses in Bismarck, North Dakota. It turned out that the flight pattern of airplanes taking off from Bismarck airport had been changed. Droplets of oil from the incomplete combustion of jet fuel were landing on the surface of the paint, softening it, and making a good adhesive for the black dust blowing off North Dakota fields. The tremendous range of things that can happen to paint outdoors makes the problem of predicting performance from the results of any simplified test conditions very difficult.

25.6. TESTING FOR EXTERIOR DURABILITY

Unfortunately, no test is available that reliably predicts the exterior durability of a coating. The only way to determine whether a coating will be durable for some length of time in a particular environment is to apply the coating to the product and use the coated product in that environment to determine its lifetime. This is not a desirable state of affairs, especially for coatings that are expected to be durable after many years of exterior exposure. What can one do?

As stressed in the introduction to Chapter 24, the most important potential tool for the coatings formulator is to accumulate a data bank of durability data based on actual field experience with real coated products. For example, representatives of automobile, coatings, steel, and steel treatment material manufacturers have, for many years, gone to parking lots in various cities and accumulated data on the condition of coatings. From the car's serial number, the formulations of the top coat and primer, baking conditions, as well as the type of steel and surface treatment can be retrieved. By analyzing such data over many years, the coatings and automobile manufacturers accumulate an increasingly useful data bank for the prediction of performance of new coatings. Furthermore, data are available to test hypotheses for relating structure to performance. The establishment of structure/performance relationships is useful in the design of new systems with improved durability. The data can also be used to evaluate laboratory tests. Similar data bases are accumulated by some manufacturers of maintenance paints, marine coatings, and coil coatings for exterior siding.

25.6.1. Accelerated Outdoor Testing

Obviously, accelerated test methods are desirable to permit prediction of performance in shorter times than possible by actual field use. The most reliable accelerated tests are outdoor fence exposures of coated panels, especially if they are carried out in several locations with quite different environments. Even at one location, there are sometimes substantial variations in conditions that lead to variability of test results [19]. The most commonly used exposure is in south Florida with the panels facing south at an angle 5° from the perpendicular.

Southern Florida has a subtropical climate with relatively high humidity and relatively high average sunshine level. On the other hand, Arizona has more hours of sunshine per year; cities in the northeastern states have higher levels of acid in the atmosphere; the Denver area has higher UV intensity, because of the high altitude, and the change of temperature from day to night is greater. It is important to maintain records of the weather over the exposure period. The panels are examined periodically to compare their appearance before and after exposure. Usually at least part of the panel surface is cleaned for the comparison. Ease of cleaning, change in gloss, change in color, degree of chalking, and any gross film failures are reported. Changes in color are particularly difficult to assess since any change in gloss or any chalking will change the color even if there has been no change at all in the color of the components of the coating. The effects of changes in gloss and chalking on the color of the components can be minimized by cleaning the panels, and then coating part of each panel with a thin layer of clear gloss lacquer. The lacquer layer minimizes the effect of differences in surface reflection on the color.

Martin [20] presents a good basic review of the differences between exposure conditions in Florida and Arizona. Test fence exposures can eliminate some formulations as inadequate after a few months of exposure; however, several years may be required to permit one to conclude that resistance to exposure at that location under those conditions is adequate.

Probably the next most useful test method that can give results in substantially shorter times is EMMAQUA (Equatorial Mount with Mirrors for Acceleration plus water) exposure testing [20]. DSET Laboratories, Inc. has exposure facilities outside Phoenix, Arizona that permit enhancing the intensity of sunlight on the panel surfaces by a factor of 8 times the level from direct exposure. The high intensity is achieved by reflecting additional sunlight from rotating mirrors that follow the sun to maintain a position perpendicular to the sun's direct beam radiation at all times. A wind tunnel distributes air both over and under the samples to prevent excessive sample temperatures. Since rainfall and humidity in the desert are low, the test facilities permit periodic spraying of water on the panel surfaces.

Comparisons based on equal EMMAQUA exposure energies per unit area, expressed in MJ m^{-2} (i.e., langleys), and especially on equal exposure in the UV spectral region, are the most reproducible and can be approximately translated into Arizona or Florida exposure times based on relative solar intensity, that is, energy per unit time. The UV comparison gives results that correlate most closely with real time exposures. Narrow-band UV exposure data, for example, a 10-nm band centered around 313 nm, may prove even more reliable as a basis for comparing exposures. In addition, spraying with water at night has been found to give results that more closely correspond to actual field exposure than spraying during the daylight hours as was done in earlier procedures. A few weeks of EMMAQUA exposure is said to give results approximately comparable to three years fence exposure in Arizona.

Another means of accelerating film degradation that has been fairly widely used is called *black box exposure*. The panels are mounted at 5° to the horizontal on black boxes rather than on a fence with the backs of the panels open to the air. There is a substantial increase in the temperature of the coating when exposed to sunshine, which accelerates degradation reactions. Unfortunately, the temperature increase and, therefore, the extent of acceleration can vary substantially from coating to coating. An important variable that could affect the extent of change due to black box exposure is, obviously, the relationship between the temperatures and the T_g values of the films.

25.6.2. Early Detection Methods

Early detection methods are based on highly sensitive techniques for detecting the onset of degradation. They are best carried out on coated panels (or real objects) subjected to short-term ambient weathering in the field.

Studies have been made by various surface analysis techniques, such as scanning electron microscopy (SEM) to detect physical changes, as well as X-ray photoelectron spectroscopy (XPS) and Fourier transform infrared spectroscopy (FTIR) to detect chemical changes. The idea is that physical and/or chemical changes at the surface, which are detectable by sensitive instrumentation, may serve as indicators of impending failures, such as loss of gloss before the changes are great enough to be seen visually. Specific information on chemical changes may also be obtained.

A promising early detection technique utilizes electron spin resonance (ESR) spectrometry to monitor changes in free radical concentrations in exposed films. The rate of disappearance of stable nitroxyl radical incorporated into TiO_2 pigmented acrylic/melamine coatings and subjected to ambient or accelerated short-term UV exposure has been correlated with loss of gloss in long-term Florida exposure [8]. Similarly, the rate of buildup of phenoxy radicals from bisphenol A epoxy resins incorporated into acrylic/melamine coatings exposed to UV radiation in an ESR spectrometer cavity has been correlated with cracking of coating films exposed in a QUV weathering device [21]. The latter technique was used to evaluate UV stabilizers.

Utilization of ESR spectrometry to monitor the rate of disappearance of nitroxyl radicals in acrylic/melamine coatings allows the calculation of photoinitiation rates (PR) of free radical formation, which were found to correlate with rates of loss of gloss (GLR): $GLR \propto (PR)^{1/2}$. This proportional relationship of GLR with $(PR)^{1/2}$ is consistent with a free radical process and results from termination by second-order radical–radical reactions [8]. Photoinitiation rates determined by this method have also been utilized to evaluate experimental conditions for the synthesis of acrylic polyols by free radical polymerization, including the effect of initiator, temperature, and solvent on the exterior durability of the resulting acrylic/melamine coating [22].

The nitroxyl early detection method has also been utilized to investigate the synergistic stabilizing effect of a UV-absorber/quencher, 2-(2-hydroxy-3,5-bis (1,1-dimethylbenzyl)-phenyl)-2*H*-benzo-triazole, and a HALS derivative, bis(2,2,6,6-tetramethyl-4-piperidinyl) sebacate, in acrylic/urethane coatings. The UV-absorber/quencher reduces the photoinitiation rate (PR) of free-radical formation; whereas the HALS derivative reduces the propagation rate by lowering the concentration of free radicals.

A somewhat more direct application of ESR to studying the photostability of coating films is the determination of free radical concentration after UV irradiation of films at a temperature of 140 K, well below their T_g values [23]. Under these conditions, free radicals are stable. The method is particularly appropriate for evaluating stabilizers such as HALS compounds by comparing the radical concentrations with and without stabilizer. It is said that useful comparisons can be made in three hours [8].

Recent studies have demonstrated quantitative relationships between photoxidation rates and hydroperoxide concentrations in acrylic/urethane and acrylic/MF coatings [24]. The rate of photoxidation in both types of coatings was linear in hydroperoxide concentration, as determined by titration of cryogenically ground samples of coatings.

An interesting variation on accelerated weathering for evaluating the binder in TiO_2 pigmented films is to accelerate the pigmentation effect on exterior durability by substituting the surface-treated rutile TiO_2 that would be used in a commercial coating with highly photoactive, untreated anatase TiO_2.

25.6.3. Accelerated Weathering Devices

A variety of laboratory devices for accelerating exterior durability studies is available. See ASTM methods G23-90 and G53-88 for descriptions of various devices and operating procedures. The various devices expose panels to UV sources with different wavelength distributions. The panels are also subjected to cycles of water spray (or high humidity). Although these tests are widely used, results frequently do not correlate with actual exposure results. A general problem with accelerated weathering methods is the difficulty in accelerating the effects of radiation, heat, and moisture uniformly, not to mention the effects of other atmospheric degradants. The predictive value of accelerated weathering with artificial light sources is particularly questionable when a light source includes wavelengths less than 280 nm. It has been shown that variability of performance of the test instruments can be a major problem, especially when comparing results from laboratory to laboratory [25].

A recent evaluation of accelerated weathering devices for a polyester/urethane coating, using photoacoustic-FTIR spectroscopy, concluded that none of the conventional devices were suitable, including an Atlas Weather-O-Meter, housing a carbon arc with Corex D filters, an Atlas xenon arc with borosilicate inner and outer filters, and a Q-Panel QUV weathering device with FS-40 fluorescent bulbs [26]. All artificial light sources resulted in the loss of isophthalate groups in the coating, which was not observed during accelerated weathering in Florida (5° south) and Arizona (EMMAQUA). It is interesting that these results correlate with the general wisdom that polyesters, presumably with phthalate or isophthalate groups, perform worse than acrylics in accelerated weathering devices relative to their performance in Florida or Arizona exposure.

The unnatural weathering, which resulted in the loss of isophthalate groups, was attributed to excessive amounts of short wavelength light (lower than 280 nm) from the artificial light sources. A comparison of spectra of sunlight (Miami average optimum on 3/20/84) and artificial weathering devices that illustrates this disparity is provided in Figure 25.1. The FS-40 fluorescent bulbs, utilized in Atlas UVCON, and the Q-Panel QUV devices exhibit strong unnatural emission below 300 nm. The carbon and xenon arcs appear to contain UV with only slightly shorter wavelength than sunlight. Apparently, even small amounts of higher-energy UV radiation are critical and should be filtered out.

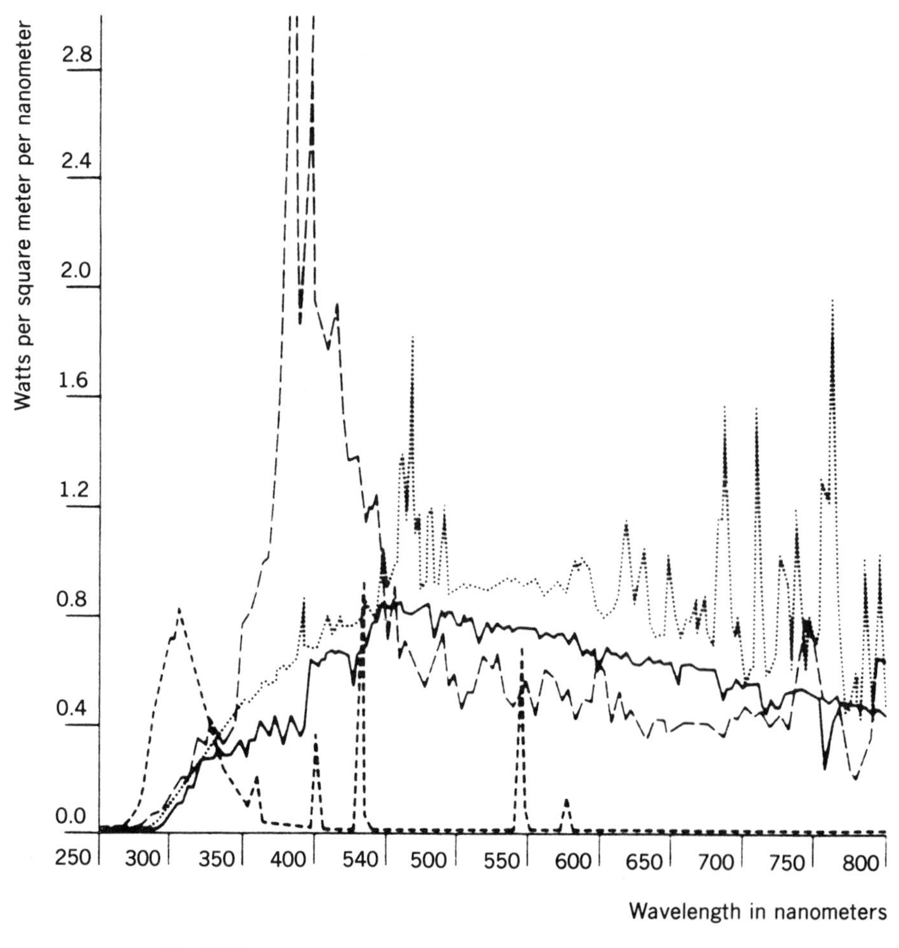

—— **Sunlight**
Miami Average Optimum Direct Global Radiation Measured 45 S 3 20 84

– – – **Sunshine Carbon Arc**
As used in Atlas Weather-Ometer 6500 Watt Xenon Lamp with Borosilicate inner and outer filters 340 nm control (35 W m)

········· **Xenon Arc Lamp**
As used in Atlas Weather-Omerter 6500 Watt Xenon Lamp with Borosilicate inner and outer filters 340 nm control (35 W m)

- - - - - **FS-40 Fluorescent Sun Lamp**
commonly used in the Atlas UVCON and the Q-Panel Q-U-V Accelerated Weathering Tester as per A S T M-G 53

Figure 25.1. Wavelength distribution of radiation from various sources. (Courtesy Atlas Electric Devices Company, Chicago, IL.)

Is the use of accelerated weathering devices better than nothing? Probably not. As an example of the shortcomings of laboratory testing, early evaluations of HALS by QUV testing were not promising; this might have led to their abandonment. Fortunately, remarkable outdoor durability results were also obtained which prompted continued development. Many examples of reversals of results comparing coatings with known exterior durability have been found. One example of poor prediction by a lab test is given in a technical bulletin [1] comparing durability of acrylic urethane coatings and acrylic epoxy-based coatings containing HALS stabilizers; QUV exposure predicted substantially better durability for the urethane type coating, but Florida exposure results were roughly comparable.

Not only are there examples of poorer coatings showing better laboratory test results and vice versa, but there are also disparities and reversals when comparing different laboratory test methods. Thus, one may optimize a coating formulation for performance in a particular weathering device with a significant possibility that the coating will perform poorly outdoors or in another weathering device. More serious, as might have occurred with HALS, is the possibility of rejecting a superior material on the basis of accelerated laboratory testing. The argument that laboratory tests are at least useful for eliminating coatings that exhibit clearly inadequate performance is specious, since accumulated knowledge of mechanisms of failure is adequate to eliminate such formulations without the need for testing.

Although accelerated weathering devices can be improved by closer correspondence of UV emission with sunlight, it appears unlikely that a completely satisfactory device will ever be developed because the variations in exterior environments to which coatings are exposed are so great. This situation serves to emphasize the need to accumulate data banks of real field use results and to correlate these results with structure to develop a basic understanding of the mechanisms of failure.

REFERENCES

1. Anon., *Technical Bulletin, Experimental Resin OR-1285*, Rohm & Haas Co., Philadelphia, PA, 1990.
2. P. J. Schirmann and M. Dexter, "Light and Heat Stabilizers for Coatings," in *Handbook of Coatings Additives*, L. J. Calbo, Ed., Marcel Dekker, New York, 1987, pp. 225–269. See also M. Dexter, "Antioxidants," in *Encyclopedia of Polymer Science and Engineering*, 2nd ed., Vol. 2, Wiley, New York, pp. 73–91, and M. Dexter, "UV Stabilizers," in *Kirk–Othmer Encyclopedia of Chemical Technology*, 3rd ed., Vol. 23, Wiley, New York, 1983, pp. 615–627.
3. (a) E. L. Breskman and S. P. Pappas, *J. Coat. Technol.*, **48** (622), 34 (1976); (b) S. P. Pappas, L. R. Gatechair, E. L. Breskman, and R. M. Fischer in *Photodegradation and Photostabilization of Coatings*, ACS Symposium Series, No. 151, S. P. Pappas and F. H. Winslow, Eds., American Chemical Society, Washington, DC, 1981, pp. 109–116.
4. H. Bohnke, L. Avar, and E. Hess, *J. Coat. Technol.*, **63** (799), 53 (1991).
5. J. D. Shelton, "Stabilization Against Thermal Oxidation," in *Polymer Stabilization*, W. L. Hawkins, Ed., Wiley-Interscience, New York, 1971, pp. 80–84.
6. D. J. Carlsson, J. P. T. Jensen, and D. M. Wiles, *Makromol. Chem.*, Suppl. 8, 79 (1984).
7. S. P. Fairgrieve and J. R. MacCallum, *Polym. Degrad. Stabil.*, **8**, 107 (1984).
8. J. L. Gerlock, D. R. Bauer, L. M. Briggs, and R. A. Dickie, *J. Coat. Technol.*, **57** (722), 37 (1985).
9. R. F. Sharrock, *J. Coat. Technol.*, **62** (789), 125 (1990).
10. P. Stamatakis, B. R. Palmer, C. F. Boren, G. C. Salzman, and T. B. Allen, *J. Coat. Technol.*, **62** (789), 95 (1990).

11. H. G. Voelz, G. Kaempf, H. G. Fitzky, and A. Klaeren, "The Chemical Nature of Chalking in the Presence of Titanium Dioxide Pigments," in Ref. 2(b), pp. 163–182.
12. G. Irick, Jr., G. C. Newland, and R. H. S. Wang, "Effect of Metal Salts on the Photoactivity of Titanium Dioxide: Stabilization and Sensitization Processes," in Ref. 2(b), pp. 147–162.
13. J. H. Braun, *J. Coat. Technol.*, **62** (785), 37 (1990).
14. W. H. Starnes, Jr., *Pure Appl. Chem.*, **57**, 1001 (1985). See also G. Georgiev, L. Christiv, and T. Gancheva, *J. Marcomol. Sci.–Chem.*, **A27**, 987 (1990).
15. D. R. Bauer, M. J. Dean, and J. L. Gerlock, *Ind. Eng. Chem. Res.*, **27**, 65 (1988).
16. D. R. Bauer, *Prog. Org. Coat.*, **14**, 193 (1986).
17. L. H. Brown, "Silicones in Protective Coatings," in *Treatise on Coatings*, Vol. I, Part III, R. R. Myers and J. S. Long, Eds., Marcel Dekker, New York, 1972, pp. 536–563.
18. Y.-C. Hsaio, L. W. Hill, and S. P. Pappas, *J. Appl. Polym. Sci.*, **19**, 2817 (1975). See also S. P. Pappas and R. L. Just, *J. Polym. Sci., Polym. Chem. Ed.*, **18**, 527 (1980).
19. R. M. Fischer, W. P. Murray, and W. D. Ketola, *Prog. Org. Coat.*, **19**, 151 (1991).
20. J. L. Martin, *Proc. Adv. Coat. Technol. Conf. Eng. Soc.*, Detroit, MI, 1991, p. 219.
21. S. Okamoto, K. Hikita, and H. Ohya-Nishiguchi, in *Proc. XVII Fatipec Congress*, Venice, 1986, pp. 239–255.
22. J. L. Gerlock, D. R. Bauer, L. M. Briggs, and J. K. Hudgens, *Prog. Org. Coat.*, **15**, 197 (1987).
23. A. Sommer, E. Zirngiebl, L. Kahl, and M. Schonfelder, *Prog. Org. Coat.*, **19**, 79 (1991).
24. D. R. Bauer, D. F. Mielewski, and J. L. Gerlock, *Polym. Degrad. Stab.* **38**, 57 (1992).
25. R. M. Fischer, W. D. Ketola, and W. P. Murray, *Prog. Org. Coat.*, **19**, 165 (1991).
26. D. R. Bauer, M. C. P. Peck, and R. O. Carter, III, *J. Coat. Technol.*, **59** (755), 103 (1987).

CHAPTER XXVI

Adhesion

Adhesion is one of the most important characteristics of most coatings. Unfortunately, there is inadequate basic scientific understanding of the wide range of variables that affect adhesion. Although this chapter deals with experimentally determined principles, some comments are based only on reasonable deductions that fit in with accumulated experience. One difficulty in dealing with the subject is defining what adhesion means. In most cases, a coatings formulator means by adhesion—How hard is it to remove the coating? A physical chemist would think in terms of the work that would be required to separate the two interfaces that are adhering. These can be very different considerations. The latter is only one aspect of the former. Commonly, removal of a coating requires breaking or cutting through the coating and pushing the coating out of the way as well as separating the coating from the substrate.

Consider as an extreme example the adhesion of the plastic covering on a piece of wire used in electrical connections. The covering must have good "adhesion" since it must stay on the wire to protect against short circuits, shocks, and so on. However, in order to attach the wire to a fixture, one wants to be able to make a cut through the wire covering and then easily slip the coating off the metal. In this case, it is desirable to have minimum interactive forces between the plastic and the copper, but considerable toughness so that it is difficult to remove the coating accidentally.

26.1. SURFACE MECHANICAL EFFECTS ON ADHESION

The resistance to separation of coating and substrate is complicated by the potential importance of a mechanical interlocking effect on separation. Consider the schematic representations in Figure 26.1. In the case of a very smooth interface between coating and substrate, as shown in sketch A, the only forces holding the substrate and the coating together are the interfacial attractive forces per unit of geometric area. In the case of a very rough surface on a microscopic scale, as represented in sketch B, two other factors are of great importance. In some places there are undercuts in the substrate. In attempting to pull the coating off the substrate, one would either have to break the substrate or break the coating in order to separate them. The situation is analogous to using a dovetail joint to hold two pieces of wood together. Another

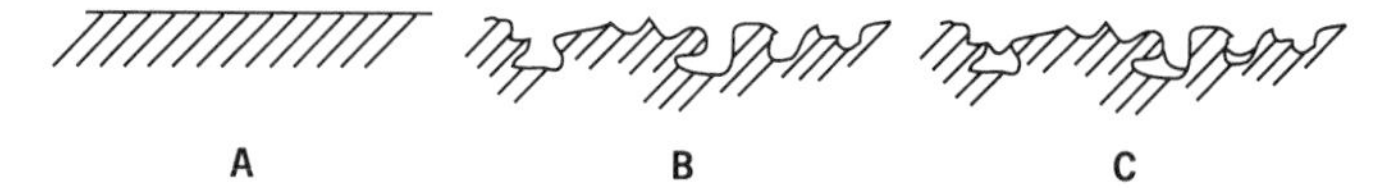

Figure 26.1. Schematic representations of geometries of surface interactions between coating and substrate. A, smooth interface between coating and substrate. B, rough surface on a microscopic scale. C, rough surface with incomplete penetration of coating.

factor is that the actual contact area between the coating and the rough substrate is substantially larger than the geometric area. In most cases force per unit of geometric area is the factor involved in determining how great a force is required to remove the coating. For these two reasons, better adhesion can generally be obtained if the surface of a substrate is roughened before coating.

However, as can be seen in sketch C, surface roughness can be a disadvantage. If the coating does not completely penetrate into the microscopic pores and crevices in the surface, dovetail effects will not be realized and the actual interfacial contact area can be smaller than the geometric area. Furthermore, when water permeates through the film to the substrate, there will be areas of contact of water with coating-free substrate which can be a major problem, especially when a function of the coating is to protect a steel substrate against corrosion.

The importance of surface roughness is, of course, widely recognized, but the effects have not been subjected to a wide range of scientific studies. Much of the discussion that follows, therefore, is based on interpretations of practical experience rather than rigorous scientific investigation. Such theories can be useful; however, one must always remember that they are speculative. One must be ready to abandon or revise such theories if they are found to be incompatible with results of more rigorous studies.

The scale of the roughness of the surfaces can vary from macroscopic to microscopic to submicroscopic. Perhaps it is most important to consider the situation on a microscopic and submicroscopic scale. What factors control the rate of penetration of a liquid into such pores and crevices? The question can be considered as being analogous to the penetration of a liquid into a capillary. Equation 26.1 shows the variables that affect the distance of penetration L (cm) into a capillary of radius r (cm) in time t (s) where γ is surface tension in mN m^{-1}, θ is contact angle, and η is viscosity in Pa·s.

$$L = 2.24 \left[\left(\frac{\gamma}{\eta} \right) (r \cos \theta) t \right]^{1/2} \tag{26.1}$$

The rate of penetration (Lt^{-1}) is greatest if surface tension of the coating is high. However, there is an upper limit to this surface tension effect because the rate is strongly affected by contact angle. The rate is fastest when cosine of the contact angle is 1, that is, when the contact angle is zero. The cosine can only be 1 if the surface tension of the liquid is less than that of the solid substrate. The radius of the capillary is a variable of the substrate, not of the coating.

The variable over which the coating formulator has the greatest control is the viscosity. It is important to recognize that on the scale of microscopic and submicroscopic crevices and pores, the pigment particles in the coating are large compared to at least some of the surface irregularities and, therefore, the critical viscosity is that of the continuous, that is, external phase of the coating, not the bulk viscosity of the coating. The lower the viscosity of the external phase, the more rapid will be the penetration. Since in most cases viscosity of the vehicle increases after application, it can also become important to keep that viscosity low for a long enough time for penetration to approach completion.

Since viscosity of resin solutions increases with increasing molecular weight, one would expect that lower molecular weight resins would provide superior adhesion after cross-linking if everything else were equal. Viscosity is most dependent on weight average molecular weight, therefore, one would expect that $\overline{M}_w$ would be particularly important. These hypotheses have been confirmed in the case of epoxy resin coatings on steel [1].

Coatings with low viscosity external phases, slow evaporating solvents, and relatively slow cross-linking rates have been found, in general, to give better adhesion. It is also true that, in general, baking coatings give better adhesion than do air dry coatings. (The term "air dry" is widely used but is potentially confusing. It generally means just that the film is formed at ambient temperature without added heat and does not imply that oxygen is required for cross-linking.) When the coated article goes into the oven, temperature increases, viscosity of the external phase decreases, and penetration into the surface irregularities can increase. This is only one of several possible explanations for the advantages of baking coatings when adhesion is critical.

Volume changes during cure may also have a mechanical effect on adhesion. For example, when water freezes in a nonstick fluorocarbon-coated pan, the adhesion is remarkably good. Expansion of liquid water in depressions on the coating surface during solidification promotes intimate contact with the surface, enhancing adhesion by mechanical dovetail effects.

Bailey and co-workers [2] investigated the phenomenon of volume changes during polymerization. Prior to their studies, all known polymerization reactions occurred with shrinkage because the nonbonding distance between the monomers is longer than the resulting covalent bonds in the polymer. It was known that shrinkage is relatively low for monomers that polymerize by ring-opening, since, in these cases, some covalent bonds are also converted into longer nonbonding distances. Bailey's group extended this concept further by synthesizing bicyclo and spiro monomers, in which two rings open during polymerization, and demonstrated that expansion can occur. For example, ring-opening polymerization of spiro orthocarbonate **1** occurs with about 4% expansion. The polymerization can be initiated with either free radicals or cations and is shown is Scheme 1 for radical initiation.

Scheme 1

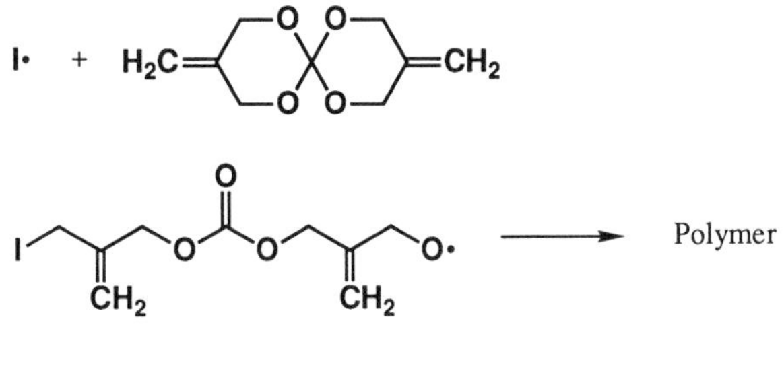

Polymer

Bailey's group has reported that expanding monomers promote adhesion in adhesive and composite applications.

26.2. EFFECTS OF INTERNAL STRESS, SHRINKAGE, AND FRACTURE MECHANICS ON ADHESION

Internal stresses in coatings amount to forces that counteract adhesion. In other words, less external force is required to disrupt the adhesive bond. When solvent evaporates from a thermoplastic coating (lacquer), in the early stages the polymers can accommodate the resulting voids by bond-relaxation and shrinkage occurs. However, as film formation proceeds, T_g rises and free volume is reduced, it becomes more difficult for the polymer to accommodate the voids from solvent evaporation, bonds become fixed in thermodynamically unstable conformations, and internal energy (stress) increases. This phenomenon is particularly likely to occur with coatings in which T_g approaches the film-forming temperature [3].

In the case of thermosetting coatings, the cross-linking reactions lead to formation of covalent bonds that are shorter than the distance between two molecules before they react. This volume reduction, like that resulting from solvent evaporation, also results in void volume. When such reactions occur at temperatures near the T_g of the cross-linking film, stresses will result from the inability of the coating to undergo shrinkage. As the rate of cross-linking increases, stresses also tend to increase, since less time is available for polymer relaxation to occur. An extreme example is UV curing of acrylated resins by free radical polymerization that can proceed in a fraction of a second at ambient temperatures (see Section 32.2). Shrinkage, measured by thermomechanical analysis (TMA), has been shown to lag significantly behind polymerization [4]. The high rates of polymerization together with the relatively large shrinkage, which accompanies polymerization of double bonds, are major contributors to the generally observed poor adhesion of UV cure acrylated resins to smooth metal surfaces. Heating after UV cure relaxes the cross-link network and often improves adhesion.

Adhesion to tin plate and aluminum is more satisfactory with UV cured epoxy resins (by cationic polymerization) that are used to coat can ends for two-piece cans. An important factor is a heating step, following the UV curing, which promotes relaxation of the coating. Lower shrinkage of ring-opening epoxy polymerization relative to acrylate chain addition may also contribute to better adhesion (see Section 32.3).

The most common statement that internal stresses result from shrinkage is misleading; rather, internal stresses result from the inability of coatings to shrink. It is true that cross-linking reactions that are accompanied by relatively high degrees of shrinkage are more likely to develop internal stresses if the shrinkage cannot be accommodated by polymer relaxation.

When thermosetting epoxies are cured above ultimate T_g values, internal stresses are introduced primarily during the cooling process between T_g and ambient temperature, since polymer relaxation is restricted in this temperature range. Incorporation of an expanding spiro orthoester monomer into such a thermosetting bisphenol A epoxy coating results in reduced internal stresses. However, the stress reduction could be attributed to corresponding lower T_g values and not to an expansion effect [5]. Internal stress reduction and promotion of adhesion by expanding monomers without the undesirable effect of lowering T_g probably requires that a significant extent of the ring-opening polymerization occurs during the later stages of cure for systems in which T_g approaches the cure temperature.

Nonuniform curing, particularly in the later stages, as well as film defects or imperfections in the film can lead to localized stresses that generally exert adverse effects on adhesion [6]. Localization of stresses at imperfections falls within the discipline of *fracture mechanics*. The effects of fracture on abrasion resistance are considered in Section 24.3. The phenomena have been most extensively studied in adhesive bonding of substrates [7], but clearly must affect the adhesion of coatings as well. The principle is obvious; if there is a local imperfection in a film, any stress applied to that part of the film could be concentrated at that location. The more localized the imperfection, that is, the smaller the area of the imperfection, the greater the resultant stress, which is force per unit area, and the greater the probability of forming a crack. Once the crack starts, the stress concentrates at the point of the crack leading to crack propagation. If the crack propagates to the coating/substrate interface, the stress concentration could well cause the film to delaminate.

While the principle has been recognized, identifying the causes of imperfections is more difficult. Pigment particles with sharp crystal corners and air bubbles are examples of potential sites for concentration of stresses. On the other hand, as noted in Section 24.3, it has been proposed that incorporation of particles of rubber may lead to dissipation of stresses. Presumably such stress dissipation would reduce the probability of fracture mechanical adhesive failure.

26.3. RELATIONSHIPS BETWEEN WETTING AND ADHESION

Wetting is a major and perhaps limiting factor in adhesion. In other words, if the coating does not spread spontaneously over the substrate surface so that there is intermolecular contact between the substrate surface and the coating there cannot be interactions and hence no contribution to adhesion. The relationships between wetting and adhesion were extensively studied by Zisman [8]. A liquid will spread spontaneously on a substrate if the surface tension relationships (interfacial free energy) are appropriate. If the surface tension of the liquid is too high, a drop of the liquid will stay as a drop on the surface—it is said to have a contact angle of 180°. If a liquid has a sufficiently low surface tension, it will spread spontaneously on the substrate—it is said to have a contact angle of 0°. At intermediate surface tensions, there will be intermediate contact angles. A schematic drawing of a drop with intermediate surface tension is shown in Figure 26.2. The relationship between contact angle θ and the surface tensions of the substrate, γ_S, the liquid, γ_L, and the interfacial tension between the solid and the liquid, γ_{SL} for a completely planar surface is given in Eq. 26.2.

$$\cos\theta = \frac{(\gamma_S - \gamma_{SL})}{\gamma_L} \tag{26.2}$$

It is evident that maximum adhesion requires a contact angle of 0°. In general,

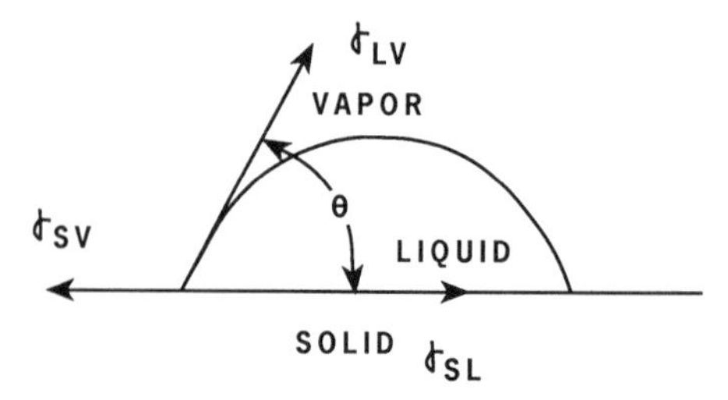

Figure 26.2. Schematic representation of contact angle.

it is sufficient to conceptualize the relationships and say that adhesion requires that the liquid have a lower surface tension than the substrate to be coated. Experimental determination of contact angles of complex systems like coatings, especially on rough substrates with heterogeneous composition, is difficult [9]. From a practical standpoint, it is useful to do cruder but easier experiments. One can apply a drop of a coating on the substrate, put the sample in an atmosphere saturated with the solvents in the coating, and watch spreading. If the droplet of coating stays as a small ball, spreading is poor and adhesion problems should be expected. If, however, the drop spreads out to a thin wide circle, the coating meets at least one criterion for good adhesion.

It is also important to carry out a second type of experiment; namely, mechanically spread the coating on the substrate, under a solvent-saturated atmosphere, and watch it. A liquid that does not spread when a drop is put on the surface may spread under mechanical forces, but may then ball up again when the pressure is removed. Generally, a liquid that spreads spontaneously in the first experiment will remain as a spread out film in the second experiment. Sometimes, however, a liquid that can be spread on the surface, will draw up into droplets or at least pull away in sections from the substrate when standing without solvent evaporation.

Consider, for example, applying *n*-octyl alcohol to a clean steel surface. The surface tension of *n*-octyl alcohol is much lower than the surface tension of steel, and it spontaneously spreads on steel. However, if one spreads a film of *n*-octyl alcohol out on steel, it draws up to form droplets on the surface of the steel. The low surface tension of the *n*-octyl alcohol results from the linear hydrocarbon chain; however, after spreading on the polar surface of the steel, the hydroxyl groups on the *n*-octyl alcohol molecules interact with the surface so that a monolayer of oriented *n*-octyl alcohol molecules forms on the surface. In effect, this has made a new surface, aliphatic hydrocarbon, which has a low surface tension, lower in fact than the surface tension of *n*-octyl alcohol. The *n*-octyl alcohol above the monolayer *dewets*.

The behavior of *n*-octyl alcohol illustrates a principle important in formulating coatings; one must be careful about using additives with single polar groups and long hydrocarbon chains in coatings that are to be used directly on metals. A classic example is the poor adhesion that can result from use of dodecylbenzenesulfonic acid as a catalyst (see Section 6.3.1.1).

Reconsidering the example of freezing water in a fluorocarbon-coated pan (see Section 26.1) in the light of surface tension effects, it is even more remarkable that good adhesion results, since the surface tension of water is substantially higher than that of fluorocarbons. This behavior serves to illustrate the importance of mechanical effects on adhesion.

26.4. ADHESION TO METAL SURFACES

The surface tension of a clean metal surface (usually metal oxide) is higher than that of any potential coating. However, metal surfaces are frequently contaminated with oil, and such surfaces can have very low surface tensions. Whenever possible, it is desirable to clean the surface of the metal before applying a coating. Sometimes this has been done by wiping with rags wet with solvent; a more efficient method is *vapor degreasing*. The object is hung from a conveyor which carries it into a tank just above boiling chlorinated solvent. The cold steel surface acts as a condenser, condensing solvent onto the surface of the steel where it dissolves oils. The solution drips off the object, removing the oil. The solvent is purified by distillation so it can be recycled.

Alternatively, surfactant solutions can be used for cleaning oil off the surface of the metal. Considerable care must be exercised in selecting the surfactants and in rinsing the surface after cleaning. It is possible that surfactant could strongly adsorb on the surface, creating a hydrocarbon layer on the surface of the metal instead of clean metal.

A further alternative for cleaning steel is sandblasting. The surface of the steel, including oil, rust, and other contaminants, is removed, leaving a rough surface. This method of cleaning is widely used for steel structures like bridges and tanks. Sandblasting leaves the steel surface too rough for products such as automobiles and appliances.

Although sandblasting is very effective, there can be significant health hazards to workers from the inhalation of silica dust. A wide variety of alternative blasting techniques are being used or evaluated [10]. These include use of steel grit instead of sand, use of water-soluble abrasives including sodium bicarbonate and salt, cryogenic cleaning by blasting with dry ice pellets, ultra-high pressure water-jetting, steam cleaning, vacuum blasting using aluminum oxide, and, for cleaning softer metals such as aluminum, plastic pellet blasting.

Formulations are tested on laboratory panels, but it is important to remember that the surfaces of these panels are not the same as those of the product to which the coating will be applied commercially. Laboratory tests should also be carried out on actual ultimate substrate or at least sample pieces of metal that will be used in production, simulating, to the extent possible, factory cleaning and treating procedures. It has even been shown that commercial test panels vary—for example, the side of the panel next to the wrapping was shown to have a different surface analysis than inner panels in the packages [11]. Washing the panels with warm water and rinsing with acetone before coating generally improved adhesion significantly.

26.4.1. Surface Preparation

When good adhesion, corrosion protection, and a relatively smooth surface are required, it is most common to chemically treat the surface of the metal. In the case of

Scheme 2

$$HPO_4^{2-} \rightleftharpoons H^+ + PO_4^{3-}$$

$$2H^+ + Fe \longrightarrow Fe^{2+} + H_2$$

$$Fe^{2+} + [Ox] \longrightarrow Fe^{3+}$$

$$Fe^{3+} + PO_4^{3-} \longrightarrow FePO_4 \downarrow$$

$$3Zn^{2+} + 2PO_4^{3-} \longrightarrow Zn_3(PO_4)_2 \downarrow$$

steel, the treatments are generally called *conversion coatings* or *phosphate coatings*. A wide variety of phosphate coatings is used. In the simplest cases, the steel object is immersed in a bath of acid zinc phosphate solution. A coprecipitate of zinc and ferric phosphates is formed on the steel surface as shown in Scheme 2.

Depending on the zinc concentration in the treatment bath, different crystals can be deposited. At relatively high zinc concentrations, the coating is dominantly hydrated zinc phosphate, $Zn_3(PO_4)_2{\cdot}4H_2O$, called hopeite. Under relatively zinc-starved conditions, the crystals have been identified as $Zn_2Fe(PO_4)_2{\cdot}4H_2O$, phosphophyllite [12]. The performance of the conversion coating is strongly affected by the uniformity and degree of surface treatment. Zinc phosphate coatings are generally applied in the range of 1.5–4.5 g m^{-2}. A wide variety of other phosphate coatings are used.

The reactions shown are straightforward. Achieving the high rates of reaction required to permit minimum dwell times for treatment is more difficult. Proprietary formulations have been developed that can reduce times to the order of seconds. The treatment quality is dependent on the time, temperature, and pH; these and other variables must be closely controlled. The treated surface must be thoroughly rinsed so as to remove any soluble salts since these could lead to blister formation when water vapor permeates through a coating film applied over the soluble salts. It is common for the last rinse to contain a low concentration of chromic acid to protect against corrosion. (See Sections 27.3.1.1, 30.2, and 33.1.1 for further discussion of conversion coating.)

The mechanism of action of the phosphate crystal layer is not fully understood. One factor may be that a dense mesh of crystals is deposited and firmly attached to the surface of the steel. The coating penetrates into the crystal mesh giving a mechanical attachment. Also the interfacial area for interaction is greater than for a relatively smooth steel surface. It is also possible that hydrogen-bond interactions between these crystals and the resin molecules may be stronger (less readily displaced by water) than those between the steel surface and the resin molecules.

The situation with aluminum is quite different than with steel. The surface of aluminum is a thin, dense, coherent layer of aluminum oxide. For many applications, no treatment other than cleaning is required. However, for applications where there might be exposure to salt, surface treatment is necessary. Most aluminum treatments are chromate treatments. A common example is an acid bath containing chromate, fluorides, and a ferricyanide salt as an accelerator. The resultant coating is said to have the following composition: $6Cr(OH)_3{\cdot}H_2CrO_4{\cdot}4Al_2O_3{\cdot}8H_2O$.

To provide greater protection against corrosion, steel coated with zinc is widely used in construction and automobiles. Several types of zinc-coated steel are used; the best known is galvanized steel. There can be large variations in adhesion depending on the condition of the zinc layer of the galvanized steel. If zinc-coated steel has been exposed outdoors before coating, there will have been some degree of surface oxidation leading to the presence of a combination of ZnO, $Zn(OH)_2$, and $ZnCO_3$; all are basic and somewhat soluble in water. Therefore, it is even more important than in primers for steel to use saponification resistant resins in primers for galvanized steel. Resins such as alkyds, which are relatively readily saponified, are likely to give poor adhesion. Zinc surfaces on automobile bodies are treated with zinc, manganese, nickel phosphate conversion coatings before electrodeposition of primer [13].

26.4.2. Coating–Substrate Interaction

The surface of clean steel is not iron; rather hydrated iron oxides are present as a monolayer on the iron—not a layer of rust particles, but a monolayer of hydrated oxide [14]. Adhesion of coatings to this surface is promoted by developing hydrogen-bond interactions between groups on the resin molecules and the oxide and hydroxide groups on the surface of the steel. (In the case of phosphate conversion coated steel, there is the further possible hydrogen-bond interaction with the phosphate groups.) Some authors [15] prefer to view the interactions in terms of association between soft acids and soft bases.

It follows that adhesion is promoted by using resins that have multiple hydrogen-bond donating and accepting groups. Adhesion is promoted by such groups as carboxylic acid (strongly hydrogen donating), amine (strongly hydrogen accepting), hydroxyl, urethane, and amide (the latter three being both hydrogen donating and accepting). One might assume that having a very large number of such substituents on a molecule would be desirable. However, it is known from adsorption studies that if there are large numbers of polar groups, at equilibrium the adsorbed layer can be very thin. The principle can be illustrated by considering a polymer molecule with an aliphatic backbone chain with polar groups on every other carbon atom. At equilibrium, adsorption of adjacent polar groups is favored sterically, resulting in a thin adsorbed layer with the polar groups on the steel surface and only methylene groups exposed to the rest of the coating. The interactions between the rest of the coating and these methylene groups would be expected to be weak, resulting in a weak boundary layer and poor cohesion. In contrast, if a smaller number of hydrogen-bond donating (accepting) groups are scattered along a resin chain, adsorption of resin molecules may occur with loops and tails sticking up from the surface, with some of the polar groups adsorbed on the surface and some on the loops and tails where they can interact with the rest of the coating. (The adsorption situation is analogous to that on pigment surfaces discussed in Section 20.1.3.)

On those parts of the resin molecule in the loops and tails, there can be groups that can have hydrogen-bond interactions with other molecules in the coating and functional groups which can cross-link with the cross-linker in the coating. Both BPA epoxy resins and their derivatives very commonly provide excellent adhesion to steel. These resins have hydroxyl groups and ether groups along the chain which can provide for interactions with both the steel surface and other molecules in the coating. It may also be important that the backbone consists of alternating flexible 1,3-glyceryl ether and rigid bisphenol A groups. It seems logical that such a combination could provide the flexibility necessary to permit multiple adsorption of hydroxyl groups onto the surface of the steel, together with the rigidity to prevent adsorption of all the polar groups. The remaining polar groups could then participate in cross-linking reactions or hydrogen bonding with the rest of the coating. References [1] and [16] discuss the effects of some variations in epoxy resin compositions on adhesion.

There is need for further research on interactions and orientation at the interface of coatings and substrates. There have been many studies of adsorption of polymer molecules on metal surfaces and, in general terms, the results are consistent with the picture given above. However, these studies involve adsorption from dilute solutions. In some cases, they have been observed over relatively long time intervals, permitting equilibrium conditions to develop. In our example of a polymer molecule with polar groups on alternate carbon atoms, initially, many molecules could be adsorbed, forming a thick adsorbed layer, but the equilibrium condition would favor adsorption of all, or substantially all, the polar groups of individual molecules leading to the thin adsorbed layers described. With polydisperse molecular weight adsorbents, low molecular weight species are commonly adsorbed first; but, at equilibrium, they are displaced by higher molecular weight molecules with larger numbers of polar groups.

But what really happens when a coating is applied? The resin is in a relatively concentrated (not a dilute) solution and the solvent evaporates in a relatively short time. There may not be time for equilibrium to be established. Depending on the coating, those groups that happened to be near the surface when the film was applied might remain there and could lead to poor adhesion, even if the same resins could provide good adhesion given the opportunity for appropriate orientation and equilibration to occur. Such a scenario is compatible with improved adhesion experienced with slow evaporating solvents that allow more time for orientation and also permit more complete penetration into surface crevices. Perhaps another reason that baking systems commonly lead to improved adhesion is that there is greater opportunity for appropriate orientation of molecules at the steel/coating interface at the higher temperature.

An important approach to studying adhesion has resulted from the development of sensitive surface analysis techniques. The surface of a steel substrate can be studied by Auger analysis that provides information on chemical composition. The presence of organic compounds on the surface of some cold-rolled steels has been detected. These organic compounds apparently become embedded in the surface of the steel during coil annealing of steel. If this happens, it becomes very difficult to obtain high-quality phosphate conversion treatments on the steel [17]. Such steels lead more commonly to adhesion failures and to inferior corrosion protection by coatings.

X-Ray photoelectron spectroscopy (XPS) can be used to study the surface of steel from which a coating has been removed and to study the underside of the coating that was in contact was the steel. This technique can be particularly powerful in showing where failure occurred; that is, whether failure was between the steel and the coating or between the main body of the coating and a monolayer (or very thin layer) of material on the surface of the steel. Still another valuable analytical procedure for very thin surface layers is attenuated total reflectance (ATR) FT-IR. Such techniques have thus far been most useful in diagnosing problems; further use of these and other analytical techniques for understanding the mechanism of adhesion can be anticipated in the future.

While achieving strong interaction between coating and steel is critical for achieving good adhesion, it is, perhaps, even more important to develop interactions that cannot be easily displaced by water. One reason that the presence of multiple groups on resin molecules such as hydroxyl groups on epoxy resins may be desirable is that some may remain bonded to the steel while others are reversibly displaced by water. This phenomenon has been termed *cooperative adhesion* [18].

It has been found empirically that amine groups on the cross-linked resin molecules promote corrosion protection. Explanations for the effect are controversial; one hypothesis is that the amine groups interact strongly with the steel surface and are not as easily displaced by water from the surface as are hydroxyl groups. Phosphate groups are another substituent group that has been found to impart improved adhesion, especially in the presence of water (i.e., wet adhesion). For example, the use of epoxy phosphates (see Section 11.6) in epoxy coatings improves both adhesion and wet adhesion [16].

26.4.3. Covalent Bonding to Glass and Steel Substrates

Stronger interactions with the substrate surface should be possible by forming covalent bonds rather than the more readily displaced hydrogen bonds. One approach along these lines is the use of so-called reactive silanes. They are very effective in the enhancement of the adhesion of coatings to glass [19]. A variety of reactive silanes are available. They all have a trialkoxysilyl group attached to a short hydrocarbon chain, the other end of which has a functional group such as amine, mercaptan, epoxy, vinyl, and so forth. The alkoxysilyl group can react with hydroxyl groups on the surface of glass and with other alkoxysilyl groups after hydrolysis, so the surface of the glass becomes covalently bonded to a series of hydrocarbon tails substituted with reactive groups that can cross-link with the coating being applied.

For example, in formulating an epoxy-amine coating for glass, one could add 3-aminopropyltrimethoxysilane to the amine package of the two-package coating. After application, the trimethoxysilyl group can react with silanol groups on the surface of the glass to generate siloxane bonds, as shown in the first step of Scheme 3. The silyl methoxy groups can also react with water to produce silanol groups that can, in turn, react with remaining silyl methoxy groups to generate polysiloxane groups at the glass surface—second step in Scheme 3. The terminal amine groups can, of course, react with epoxy groups in the resin so that the coating is multiply bonded to the surface of the glass—third step in the scheme.

When water vapor penetrates through the coating to the glass interface, hydrolysis of some of the interfacial Si–O bonds should be expected. However, if there are multiple interfacial bonds, at least some of the bonds are expected to remain intact and to prevent the coating from delaminating. Furthermore, hydrolysis is reversible, so that the hydrolyzed bonds can reform. Before the advent of reactive silanes, it was very difficult to formulate coatings that maintained adhe-

Scheme 3

```
   OH    OH
-----|-----|------        +   (MeO)3Si(CH2)3NH2      ------->
—O–Si–O–Si–O—
Glass Surface

H2N(CH2)2CH2      CH2(CH2)2NH2                     H2N(CH2)2CH2      CH2(CH2)2NH2
    (MeO)2Si      Si(OMe)2                               —O-Si—O—Si-O—
            O      O                H2O                         O      O
-------|------|-------          -------->             -------|------|-------
   —O–Si–O–Si–O—                                         —O–Si–O–Si–O—

              BPA-CH2CH(OH)CH2NH(CH2)2CH2      CH2(CH2)2NHCH2CH(OH)CH2-BPA
                                   —O-Si—O—Si-O—
BPA Epoxy                                 O      O
-------->                        -------|------|-------
                                   —O–Si–O–Si–O—
```

sion to glass after exposure to a humid atmosphere; now such coatings are readily available; all contain reactive silanes.

Reactive silanes have also been added to coatings with the objective of improving adhesion to steel surfaces [19]. There has been some evidence of improvement of adhesion. Perhaps the trialkoxysilyl group can react with hydroxyl groups attached to iron, but this reaction is unproven. Reactive silanes have not been widely adopted as additives to improve the adhesion of coatings to steel. It thus appears that the improvement is not significant.

Another approach to achieve chemical bonding to steel is the use of resins containing groups that can form coordination complexes with ferrous and ferric compounds. For example, one can make resins with acetoacetic ester substituents (see Section 13.5). Such esters are highly enolized and can coordinate with metal ions, including ferrous and ferric salts. Preliminary reports indicate the possibility of some improvement in adhesion and corrosion protection [20]. Because of the potential hydrolysis of acetoacetic esters, evaluation over relatively long time intervals will be required to assess their commercial utility.

26.5. ADHESION TO PLASTICS AND COATINGS

In contrast to the situation with clean steel and other metals, it is very possible that there will be a problem wetting the surface of plastic substrates with a coating. Wetting and adhesion can be affected by the presence of mold release agents on the surface of a molded plastic part. Mold release agents should be avoided if at all possible. If essential, release agents should be selected that are relatively easily removed from the molded part and great care should be exercised to remove all traces of them from the plastic surface.

Even after cleaning, the surface tensions of some plastics are lower than those of many coatings. The contact angle between the coating and the substrate should be 0° to permit spreading. Determination of contact angle can be experimentally difficult owing to surface roughness and inhomogeneity [9].

Polyolefins in general have low surface tension and it can, therefore, be difficult to achieve good adhesion of coatings. Attainment of satisfactory adhesion to polyolefins generally requires treatment of the surface to increase its surface tension. This can be done by oxidation of the surface to generate polar groups such as hydroxyl, carboxylic acid, and ketone groups. The presence of these groups not only increases surface tension so that wetting is possible with a wider range of coating materials, but also provides hydrogen-bond acceptor and donor groups for interaction with complementary groups on the coating resin molecules. A variety of processes can be used to oxidize the surface [21]. The surface of films, flat sheets, and cylindrical objects can be oxidized by flame treatment with gas burners using air/gas ratios such that the flames are oxidizing flames. Oxidation can also be accomplished by subjecting the surface to a corona discharge atmosphere; the ions and free radicals generated in the air by the electron emission serve to oxidize the surface of the plastic. Various chemical oxidizing treatments can be very effective; the most widely used has been an aqueous potassium dichromate/sulfuric acid solution. Adhesion to untreated polyolefins can be assisted by applying a thin *tie coat* of a low solids solution of a chlorinated polyolefin or chlorinated rubber. Ryntz [21] has reviewed the various approaches and provided the results of various types of surface analysis. See Section 34.3 on coating plastics for further discussion of surface treatments.

While adhesion between coatings and plastic substrates can be enhanced by hydrogen-bond interactions, still further enhancement can be obtained if the temperature is above the T_g of the plastic substrate. At temperatures above T_g there is adequate free volume to permit resin molecules from the coating to move into the surface of the plastic. When the last of the solvent evaporates, the intermingled molecules increase adhesion. This type of interaction can be enhanced by having the structure of the coatings resin sufficiently similar to the structure of the plastic for the resin molecules to be somewhat soluble in the plastic substrate. In some cases, promotion of adhesion by heating the plastic substrate above its T_g is not feasible because the plastic substrate may undergo heat distortion.

A related approach to achieve adhesion to a plastic substrate, which can be effective at lower temperatures, is by selecting solvents that are soluble in the plastic. The solvent will swell the plastic, lowering its T_g, and facilitating the penetration of coating resin molecules into the surface of the plastic. The solvents should evaporate slowly to permit time for penetration to occur. Fast evaporating solvents like acetone can cause *crazing* of the surface of high T_g thermoplastics like polystyrene and poly(methyl methacrylate). Crazing is the development of large numbers of minute surface cracks; see Section 34.3 for further discussion of crazing and coatings for plastics. Many plastics have crystalline domains, such crystalline domains are less likely to be swollen by solvents.

Adhesion to other coatings, commonly called *intercoat adhesion*, is another example of adhesion to plastics. The same principles apply. The surface tension of the coating being applied must be lower than the surface tension of the substrate coating to permit wetting. The presence of polar groups in both coatings will permit hydrogen bonding; in the case of thermosetting coatings covalent bonding is possible, which will enhance intercoat adhesion. It has been found empirically that the presence of relatively small amounts of amine groups on resins commonly gives coatings with superior intercoat adhesion. Such comonomers as 2-(*N*,*N*-dimethylamino)ethyl methacrylate and 2-aziridinylethyl methacrylate are used in making acrylic resins to enhance intercoat adhesion.

Temperatures above T_g increase the probability of satisfactory adhesion. Use of compatible resins in the substrate coating and top coat also increases the probability of adhesion. Use of solvents in the coating that can swell the substrate coating is a commonly used technique for enhancing intercoat adhesion. Coatings with lower cross-link density are more swollen by solvents and, in general, are easier to adhere to than are coatings with high cross-link density.

Adhesion to high-gloss coatings is most difficult to achieve because of their surface smoothness. Gloss enamel coatings that have undergone excessive cross-linking on aging are particularly difficult surfaces to which to apply an adherent coating. Sanding to increase surface roughness may well be necessary to achieve intercoat adhesion.

One reason for formulating primers with low gloss is because low gloss paints have rougher surfaces and hence are easier to adhere to. When possible, increasing the pigment loading of a primer above CPVC facilitates adhesion of the top coat. Above CPVC the dry film contains pores. When a top coat is applied, vehicle from the top coat can penetrate into the pores in the primer providing a mechanical anchor to promote intercoat adhesion. Care must be exercised not to have PVC too much higher than CPVC or so much vehicle will be drained away from the top coat that the PVC of the top coat will increase, leading to a loss of gloss.

An essential requirement of many industrial coatings is *recoat adhesion*, that is, the ability of a coating to adhere to itself well enough that flawed or damaged objects can be repainted without extensive preparation. This can be a difficult requirement to satisfy, especially with highly cross-linked gloss enamels. Frequently, use of additives, particularly polysiloxanes, to overcome film defects during application (see Section 23.3) must be minimized to assure recoat adhesion.

26.6. TESTING FOR ADHESION

In view of the complexity of adhesion phenomena, it is not surprising that there is difficulty in devising suitable tests for adhesion. As is so often the case in coatings, the only really effective way of telling whether the adhesion of a coating on an object is satisfactory is to use the object and see whether the coating adheres over the useful life of the object.

The most common method formulators use to evaluate adhesion is to see how easily a penknife can scrape a coating from a substrate. By comparing the resistance of a new coating–substrate combination to combinations with known field performance, the formulator has some basis for performance prediction. While a penknife in the hand of an experienced person can be a valuable tool, it has major disadvantages as a test method. The experience is not easily transferred from one person to another. Even the technique for the test is not easily transferred. There is no real way of assigning a numerical value to the results. Thus, it does not provide a basis for following possible small changes in adhesion as a result of changes in composition that could aid in developing hypotheses to relate composition and adhesion.

Relatively satisfactory test methods for evaluating adhesives have been developed, but few, if any, of these methods are applicable to coatings. Many investigators have worked on a variety of methods in attempts to devise meaningful tests for evaluating the adhesion of coatings [22,23]. None of these tests are very satisfactory. For research purposes, the most useful technique is a *direct pull test.* A rod is fastened with an adhesive perpendicular to the upper surface of the coated sample. A razor blade is used to make a cut through the coating down to the substrate around the base of the rod. The panel is fastened to a support that has a perpendicular rod on its back surface so that the two perpendicular rods are lined up exactly opposite each other. The assembly is put into the jaws of an Instron Tester and the tensile force required to pull the coating off the substrate is recorded.

Since the procedure is subject to considerable experimental error, multiple determinations must be made. Experienced operators can achieve precisions of $\pm 15\%$. Obviously, it is necessary to find an adhesive that bonds the rod to the coating surface more strongly than the coating is bonded to the substrate. It is also essential that the adhesive not penetrate significantly into the coating to avoid any possible change of coating properties as a result of the adhesive. Cyanoacrylate adhesives are generally satisfactory for the purpose. The most difficult experimental problem is to have the rods aligned exactly with each other and perpendicular to the coating. If the rod is at even a slight angle to the surface, stress will be concentrated on only part of the substrate–coating interface and less force will be required to break the bond. Sometimes, the weakest component is the substrate—this may be nice for advertising purposes but does not provide a measure of the adhesive strength.

Another potential complication is cohesive failure of the coating; again no information on adhesion is obtained. In reading the literature, one often finds data from tests in which cohesive, adhesive, and mixed cohesive/adhesive failures have occurred. The authors may then discuss the improvement in adhesion from some change that resulted in a greater force to get adhesive failure as compared to another sample that failed cohesively. Clearly such comparisons are invalid. When there is cohesive failure, all that is known is that the adhesive strength is above the measured value (within experimental error).

One must use caution in interpreting the results even when the sample appears to have failed adhesively, that is, that the separation occurred at the substrate–coating interface. Sometimes when no coating can be seen on the substrate surface after the test, there is actually a monolayer (or very thin layer) of material from the coating left on the substrate surface. In this event, failure was not at the substrate surface, but between the material adsorbed on the surface and the rest of the coating. Surface analysis is useful in determining the locale of failure and the identity of the adsorbed material.

Fairly often there will be a combination of adhesive and cohesive failure. At least one possible explanation of such failures is that there was a fractural failure starting at some imperfection within the film with the initial crack propagating down to the interface. Again, the tensile values from samples that fail in this way cannot be compared to the tensile values of samples that failed adhesively.

It must be recognized that the direct pull test does not evaluate the potentially important factors related to differences in the difficulty of breaking through a coating film and of shoving it out of the way that were mentioned in the beginning of this chapter. In spite of all the difficulties, direct pull tests are the most useful available. Instruments have also been devised for direct pull tests under field conditions. The method is quite widely used for quality control in high-performance maintenance and marine coatings. A serious disadvantage is that the test is destructive and the tested area must be repainted.

Probably, the most widely used specification test is called the *cross-hatch* adhesion test. Using a device with 11 sharp blades, a scratch mark pattern is made across the sample followed by a second set cut perpendicular to the first. A strip of pressure sensitive adhesive tape is then pressed over the pattern of squares and yanked off. The number of squares of coating left on the substrate is taken as a measure of the degree of adhesion. The test is subject to many sources of error, an important one being the rate at which the cuts are made. If the cuts are made slowly, they are likely to be even. However, if the cuts are made rapidly, it is possible that there will be cracks proceeding out from the sides of the cuts due to more brittle behavior at higher rates of application of stress. Obviously, the adhesive tape, the pressure with which it is applied, and the angle and rate at which the tape is pulled off the surface are other important variables. The degree of bending, if any, of the substrate during the test is important. It can also be argued that if all of these variables are controlled, either all the squares or none of them should be removed. In other words, the test may be useful in distinguishing between samples having very poor adhesion and those having fairly good adhesion; but it certainly has no quantitative meaning. Yet cross-hatch adhesion is the most widely specified test.

See Sections 27.3.1.1 and 27.5 dealing with corrosion for further discussion of factors affecting adhesion and testing for adhesion, especially in the presence of water.

GENERAL REFERENCES

K. L. Mittal, *Adhesion Aspects of Polymeric Coatings*, Plenum, New York, 1983.

S. R. Hartshorn, *Structural Adhesives: Chemistry and Technology*, Plenum, New York, 1986.

REFERENCES

1. P. S. Sheih and J. L. Massingill, *J. Coat. Technol.*, **62** (781), 25 (1990).
2. W. J. Bailey, R. L. Sun, H. Katsuki, T. Endo, H. Iwama, R. Tsushima, K. Saigo, and M. M. Bitritto, "Ring-Opening Polymerization," in *ACS Symposium No. 59*, T. Saegusa and E. Goethals, Eds., American Chemical Society, Washington, DC, 1977, p. 38.
3. D. Y. Perrera and D. Van den Eynden, *J. Coat. Technol.*, **59** (748), 55 (1987).
4. J. G. Kloosterboer, *Adv. Polym. Sci.*, **84**, 1 (1988).
5. M. Shimbo, M. Ochi, T. Inamura, and M. Inoue, *J. Mater. Sci.*, **20**, 2965 (1985).
6. V. E. Basin, *Prog. Org. Coat.*, **12**, 213 (1984).
7. A. V. Pocius, "Fundamentals of Structural Adhesive Bonding," pp. 23–68 and G. B. Portelli, "Testing, Analysis, and Design," pp. 407–449 in *Structural Adhesives: Chemistry and Technology*, S. R. Hartshorn, Ed., Plenum, New York, 1986.
8. W. A. Zisman, *J. Coat. Technol.*, **44** (564), 42 (1972).
9. M. Yekta-Fard and A. B. Ponter, *J. Adhesion Sci. Technol.*, **6**, 253 (1992).

10. J. Rex, *J. Protective Coatings and Linings*, October, 50 (1990).
11. B. S. Skerry, W. J. Culhane, D. T. Smith, and A. Alavi, *J. Coat. Technol.*, **62** (788), 55 (1991).
12. M. J. Dyett, *J. Oil Colour Chem. Assoc.*, **72**, 132 (1989).
13. C. K. Schoff, *J. Coat. Technol.*, **62**, (789), 115 (1990).
14. G. Reinhard, *Prog. Org. Coat.*, **15**, 125 (1987).
15. H. J. Jacobasch, *Prog. Org. Coat.*, **17**, 115 (1989).
16. J. L. Massingill, P. S. Sheih, R. C. Whiteside, D. E. Benton, and D. K. Morisse-Arnold, *J. Coat. Technol.*, **62** (781), 31 (1990).
17. S. Maeda, *J. Coat. Technol.*, **55** (707), 43 (1983).
18. W. Funke, *J. Coat. Technol.*, **55** (705), 31 (1983).
19. E. P. Pluddemann, *Prog. Org. Coat.*, **11**, 297 (1983).
20. P. Del Rector, W. W. Blount, and D. R. Leonard, *J. Coat. Technol.*, **61**, (771), 31 (1989).
21. R. A. Ryntz, *Polym. Mater. Sci. Eng.*, **67**, 119 (1992).
22. E. M. Corcoran, "Adhesion," in *Gardner–Sward Paint Testing Manual*, 13th ed., A. G. Roberts and G. G. Schurr, Eds., American Society for Testing and Materials, Philadelphia, PA, 1972, pp. 314–322. (A new edition of this book is in preparation.)
23. T. R. Bullett and J. L. Prosser, *Prog. Org. Coat.*, **1**, 45 (1972).

CHAPTER XXVII

Corrosion Protection by Coatings

Corrosion is a process by which a substance is worn away by chemical action. Economic losses from corrosion in the United States alone are estimated to exceed $100 billion per year and may be as much as 4% of the gross national product [1]. In this chapter we discuss the principles of corrosion and of the protective role of organic coatings; specific types of coatings for corrosion control are covered in Chapters 28, 29, 30, 33, and 36.

27.1. CORROSION OF UNCOATED STEEL

The major economic losses are from the electrochemical corrosion of steel and most of the discussion in this chapter deals with protecting steel by using organic coatings. Before considering the role of coatings, it is important to understand the factors affecting the corrosion of uncoated steel.

An electrochemical cell can be made by connecting plates of two different metals with a conductive wire and partly immersing them in an electrolyte (water containing some dissolved salt). An electrochemical reaction begins spontaneously. Electric current flows through the plates and wire, and ions migrate through the electrolyte. If one of the plates is zinc and the other copper, the zinc will be the anode of the cell and dissolve (corrode) and the copper will be the cathode and remain unchanged. The initial chemical reactions are:

$$\text{Anode:} \quad Zn \longrightarrow Zn^{2+} + 2e^-$$

$$\text{Cathode:} \quad 2H^+ + 2e^- \longrightarrow H_2$$

By studying the reactions that occur with many combinations of metal plates, tables have been devised in which the metals are arranged in an *electromotive series*. When any two metals are connected in a cell, the higher metal on the list is the anode and the lower one the cathode. Those metals involved in the discussions in this chapter are:

Magnesium (Mg)—most easily oxidized
Aluminum (Al)
Zinc (Zn)
Iron (Fe)
Tin (Sn)
Copper (Cu)—least easily oxidized

There are many kinds of steel; all are alloys of iron and carbon with other metals. Various steels corrode at greatly different rates depending on their composition and on the presence of mechanical stresses in the steel object. In a piece of steel, the composition varies from location to location; as a result, some areas of the surface are anodic relative to other areas that are cathodic. Stresses can also be a factor in setting up anode–cathode pairs. For example, cold-rolled steel has more internal stresses than hot-rolled steel and is generally more susceptible to corrosion. However, cold-rolled steel is widely used since it is stronger than hot-rolled. Internal stresses can be created in steel after rolling, for example, during fabrication or by the impact of a piece of gravel on an auto body. Stresses formed in this way may create anode and cathode sites.

Steel adsorbs a microscopic layer of water on its surface from water vapor in the air. If there are traces of soluble salt on the surface of steel, all the components of an electrochemical cell are present. Electrochemical reactions begin; in the absence of oxygen, the initial reactions are:

$$\text{Anode:}\quad Fe \longrightarrow Fe^{2+} + 2e^-$$

$$\text{Cathode:}\quad 2H^+ + 2e^- \longrightarrow H_2$$

If these were the only reactions occurring, corrosion would start, proceed briefly, and then virtually stop, except when the pH is low. It would stop because the hydrogen ions near the cathodic areas would be used up. This is an example of *polarization*—in this case, cathodic polarization. However, if oxygen is present dissolved in the water, a different reaction occurs at the cathode:

$$O_2 + H_2O + 4e^- \longrightarrow 4OH^-$$

In the presence of adequate oxygen, the cathodic reaction no longer depends on the concentration of H^+ ions and cathodic polarization is eliminated, that is, *depolarization* has occurred; corrosion can then continue:

$$2Fe + O_2 + 2H_2O \longrightarrow 2Fe^{2+} + 4OH^-$$

The rate of corrosion of steel depends on the concentration of oxygen dissolved in the water at the steel surface, as shown in Figure 27.1. At low concentrations, the rate increases with increasing dissolved oxygen concentration. At high concentrations, the rate declines because of *passivation*, discussed in Section 27.2.1. The equilibrium concentration of oxygen in water exposed to the atmosphere at 25°C is about 6 mL L^{-1}.

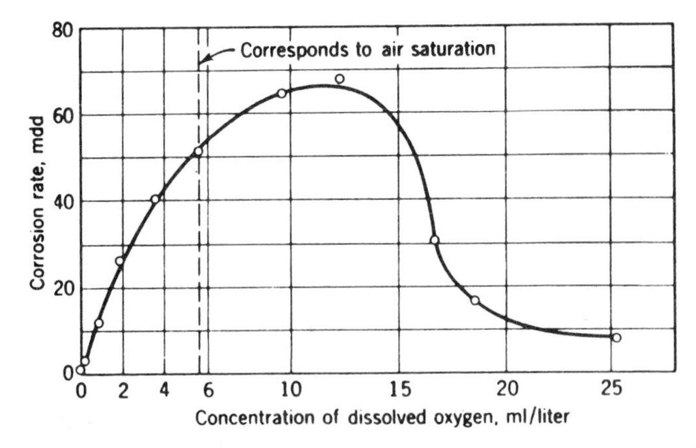

Figure 27.1. Effect of oxygen concentration on corrosion of mild steel in slowly moving distilled water, 48-hr test, 25°C. (From Ref. [2], with permission.)

Corrosion can occur at a significant rate only if there is a complete electrical circuit. Therefore, the rate of corrosion depends on the conductivity of the water at the steel surface. Dissolved salts increase conductivity, which is one reason that the presence of salts increases the rate of corrosion of steel. The effects of salts on corrosion rates are complex; the reader is referred to Ref. [2], or other general texts on corrosion, for detailed discussions.

An example of the effects observed is found in the relationship between NaCl concentration and corrosion rate shown in Figure 27.2. The dashed vertical line indicates the salt concentration in sea water. At higher salt contents, the rate of corrosion decreases since the solubility of oxygen decreases as the NaCl concentration increases.

The rate of corrosion also depends on pH, as shown in Figure 27.3. Since iron dissolves in strong acid even without electrochemical action, it is not surprising that corrosion is most rapid at low pH. Corrosion rate is nearly independent of pH between about 4 and 10. In this pH region, the initial corrosion causes a layer of ferrous hydroxide to precipitate near the anode. Subsequently, the rate is controlled by the rate of oxygen diffusion through the layer. Underneath, the surface

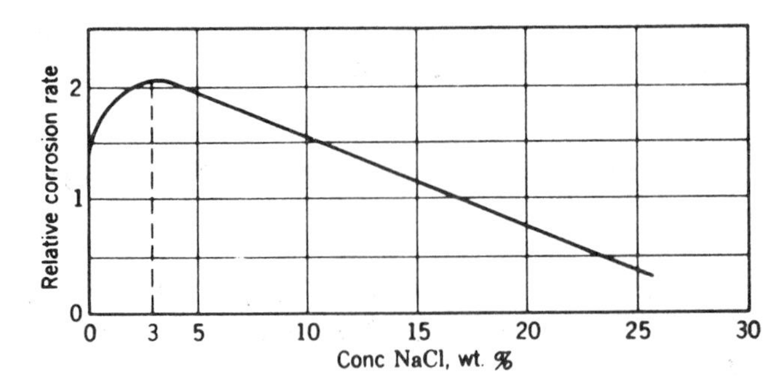

Figure 27.2. Effect of sodium chloride on corrosion of iron in aerated solutions at room temperature (composite of data from several investigators). (From Ref. [2], with permission.)

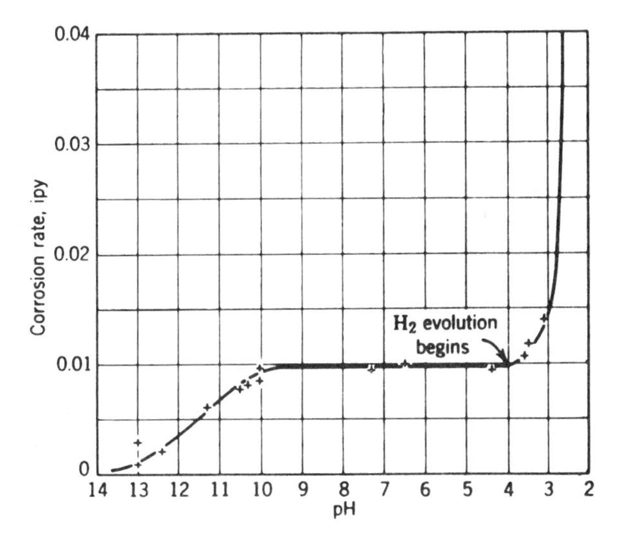

Figure 27.3. Effect of pH on corrosion of iron in aerated soft water at room temperature. (From Ref. [2], with permission.)

of the iron is in contact with an alkaline solution whose pH is about 9.5. When the environmental pH is above 10, the increasing alkalinity raises the pH at the iron surface. The corrosion rate then decreases because of passivation, as discussed in Section 27.2.1.

Corrosion rate also depends on temperature, as shown in Figure 27.4. The reactions proceed more rapidly at higher temperatures, as indicated by the increase in corrosion rate in a closed system. However, the solubility of oxygen in water decreases as temperature increases so that in an open system, where the oxygen can escape, the rate of corrosion goes through a maximum at some intermediate temperature. The temperature at which corrosion rate maximizes is quite system dependent.

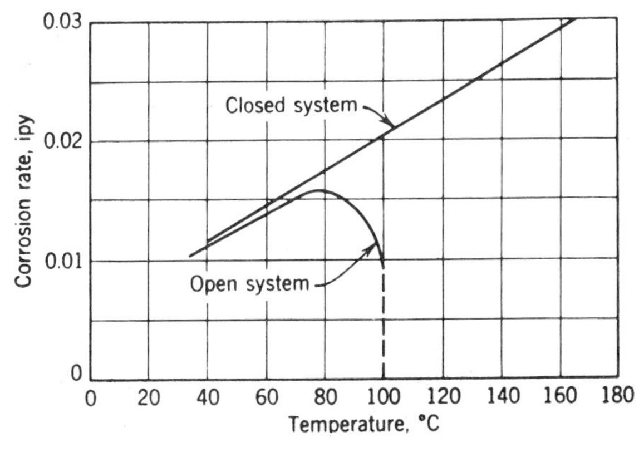

Figure 27.4. Effect of temperature on corrosion of iron in water containing dissolved oxygen. (From Ref. [2], with permission.)

27.2. CORROSION PROTECTION OF METALS

Three distinct strategies can be employed to control electrochemical corrosion without the use of organic coatings. First, one can employ chemical means to suppress the anodic reaction, as discussed in Section 27.2.1. Second, one can suppress the cathodic reaction, as discussed in Section 27.2.2. Third, one can cover steel with a barrier coat to prevent water and oxygen from contacting the surface, as discussed in Section 27.2.3.

27.2.1. Inhibition and Passivation

There are inhibitors that suppress corrosion by any of several mechanisms. One important class acts by retarding the anodic reaction, they are called *passivators*. The mechanisms of passivation are complex. A given passivator suppresses corrosion above some critical concentration, but it may accelerate corrosion at lower concentrations by cathodic depolarization. This phenomenon is illustrated by the effect of oxygen concentration on corrosion rate shown in Figure 27.1. Increasing oxygen concentration up to about 12 mL L^{-1} increases corrosion rate because it contributes to depolarization at the cathode. At higher concentrations, more oxygen reaches the surface than can be reduced by the cathodic reaction; beyond that concentration, oxygen is a passivating agent.

The mechanism of passivation has not been fully elucidated. According to one theory, if the oxygen concentration near the anode is high enough, ferrous ions are oxidized to ferric ions soon after they are formed at the anodic surfaces. Since ferric hydroxide is less soluble in water than ferrous hydroxide, a barrier of hydrated ferric oxide forms over the anodic areas. The iron is said to be passivated.

The critical oxygen concentration for passivation depends on conditions. It increases with dissolved salt concentration and with temperature, and it decreases with increase in velocity of water flow over the surface and pH. At about pH 10, the critical oxygen concentration reaches the value for air-saturated water (6 mL L^{-1}) and it is still lower at higher pH values. As a result, iron is passivated against corrosion by the oxygen in the air at sufficiently high pH values.

It is impractical to control corrosion by oxygen passivation below about pH 10 since the concentrations needed are in excess of those dissolved in water in equilibrium with air. However, a wide variety of oxidizing agents can act as passivators. Chromate, nitrite, molybate, plumbate, and tungstate salts are examples. As with oxygen, a critical concentration of these oxidizing agents is needed to achieve passivation. The reactions with chromate salts have been most extensively studied. It has been shown that partially hydrated mixed ferric and chromic oxides are deposited on the surface where they presumably act as a barrier to halt the anodic reaction.

Certain nonoxidizing salts such as alkali metal salts of boric, carbonic, phosphoric, and benzoic acids also act as passivating agents. Their passivating action may result from their basicity. By increasing pH, they may reduce the critical oxygen concentration for passivation below the level reached in equilibrium with air. Alternatively, it has been suggested that the anions of these salts may combine with ferrous or ferric ions to precipitate complex salts that form a barrier coating at the anode. Possibly, both mechanisms operate to some extent.

Many organic compounds are corrosion inhibitors for steel. Most are polar substances that would tend to adsorb on high-energy surfaces [3]. Amines are particularly widely used. Clean steel wrapped in paper impregnated with a volatile amine or the amine salt of a weak acid is protected against corrosion. Amines are also used in boiler water to minimize corrosion. The reason for their effectiveness is not clear. They may act as inhibitors because they are bases and neutralize acids. It may be that amines are strongly adsorbed on the surface of the steel by hydrogen bonding or salt formation with acidic sites on the surface of the steel. This adsorbed layer then may act as a barrier to prevent oxygen and water from reaching the surface of the steel.

Aluminum is higher in the electromotive series than iron and is more easily oxidized. Yet aluminum, generally, corrodes more slowly than steel. A freshly exposed surface of aluminum oxidizes quickly to form a dense coherent layer of aluminum oxide. In other words, aluminum is passivated by oxygen at concentrations in water that are reached in equilibrium with air. On the other hand, aluminum corrodes more rapidly than iron under either highly acidic or highly basic conditions. Also, salt affects the corrosion of aluminum even more than it affects the corrosion of iron; aluminum corrodes rapidly in the presence of sea water.

27.2.2. Cathodic Protection

If steel is connected to the positive pole of a battery or of a direct current source while the negative pole is connected to a carbon electrode, and both are immersed in salt water, the steel does not corrode. The impressed electrical potential makes the whole steel surface cathodic relative to the carbon anode. The result is the electrolysis of water, rather than corrosion of steel. This is an example of cathodic protection.

A related method is to connect the steel electrically to a piece of metal higher in the electromotive series than iron, Magnesium, aluminum, and zinc are examples. When a block of these metals is connected to steel and immersed in an electrolyte, the metal is the only anode in the circuit and all corrosion takes place at that anode. The metal is called a *sacrificial anode*. This method is often used to protect pipelines and the steel hulls of ships. Of course, the sacrificial anode is gradually used up and must be replaced periodically. Zinc and magnesium are generally the preferred sacrificial metals. Aluminum is often ineffective because a barrier layer of aluminum oxide forms on its surface (see Section 27.2.1). However, aluminum is appropriate for marine applications since it corrodes relatively easily in salt water.

Another important example of cathodic protection is the coating of steel with zinc to make galvanized steel. The steel is protected in two ways. Zinc functions as a sacrificial anode, and also acts as a barrier preventing water and oxygen from reaching the steel surface. Since zinc is easily oxidized, it is passivated by oxygen at concentrations below 6 $mL\ L^{-1}$. If the surface of the galvanized sheet is damaged and bare steel and zinc are exposed, the zinc corrodes but not the steel. After exposure to the atmosphere, the surface of the zinc becomes coated with a mixture of zinc hydroxide and zinc carbonate. Both are somewhat soluble in water and strongly basic.

27.2.3. Barrier Protection

In order for steel to corrode, oxygen and water must be in direct molecular contact with the surface of the steel. Barriers that can prevent oxygen and water from reaching the surface protect steel against corrosion. As noted above, the zinc layer in galvanized steel acts as a barrier. It may even be considered that a layer of irreversibly adsorbed small molecules can act as a barrier.

It is often incorrectly assumed that the tin coating on steel in tin cans acts electrochemically similarly to zinc in galvanized steel. Actually, tin is lower in the electromotive series than iron, so that the iron is the anode and the tin is the cathode. Before the can is opened, the tin coating is intact and acts as a barrier so that no water or oxygen can reach the steel. After the can has been opened, the cut bare edges expose both steel and tin to water and oxygen and the steel corrodes relatively rapidly.

27.3. CORROSION PROTECTION BY INTACT COATINGS

Two broad strategies can be used to design coatings to protect steel. In this section we discuss how organic barrier coatings can be designed to prevent, insofar as possible, oxygen and water from reaching the steel surface to cause corrosion. Barrier coatings can be very effective, and the current trend is to use the barrier strategy when it is anticipated that the coating can be applied so as to cover essentially all of the substrate surface and when the film will remain intact in service. However, when it is anticipated that there will not be complete coverage of the substrate or that the film will be ruptured in service, alternative strategies using coatings that can suppress electrochemical reactions involved in corrosion may be preferable; they are discussed in Section 27.4. It is seldom effective to try to use both strategies at the same time.

27.3.1. Critical Factors

Until about 1950, coatings were generally believed to protect steel by acting as a barrier to keep water and oxygen away from the steel surface. Then it was found by Mayne [4] that the permeability of paint films was so high that the concentration of water and oxygen coming through the films would be higher than the rate of consumption of water and oxygen in the corrosion of uncoated steel. Mayne concluded, therefore, that barrier action could not explain the effectiveness of coatings and proposed that electrical conductivity of coating films must be the variable that controls the degree of corrosion protection. Presumably, coatings with high conductivity would give poor protection as compared to coatings of lower conductivity. It was confirmed experimentally that coatings having very high conductivity afforded poor corrosion protection. However, in comparisons of films with relatively low conductivity, little correlation between conductivity and protection was found.

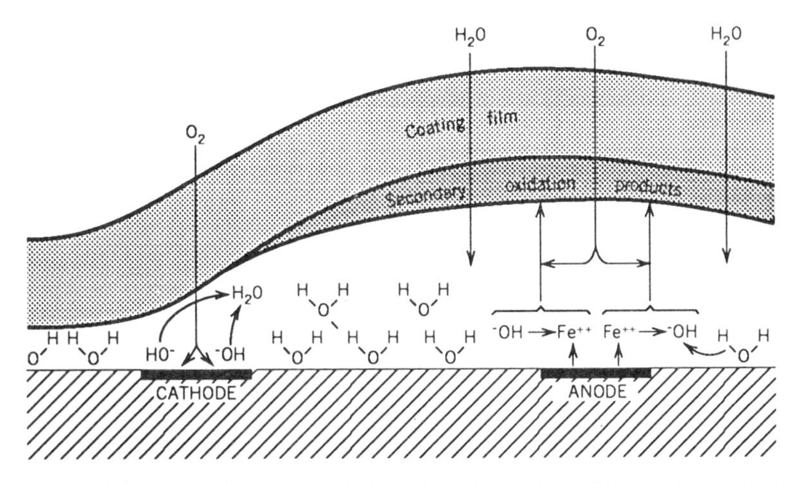

Figure 27.5. Schematic illustration of blistering. (From Ref. [12], with permission.)

Some still believe that conductivity of coatings is at least a factor in corrosion protection [5,6].

Current understanding of the protection of steel against corrosion by intact films is based to a significant degree on the work of Funke [7–14]. He found that an important factor that had not been given sufficient emphasis in earlier work was adhesion of the coating to the steel in the presence of water. Funke proposed that water permeating through an intact film could displace some area of the film from the steel. In such cases, the film shows poor *wet adhesion*. Water and oxygen dissolved in the water would then be in direct contact with the steel surface; hence corrosion would start. As corrosion proceeds, ferrous and hydroxide ions are generated, leading to formation of an osmotic cell under the coating film. Osmotic pressure can provide a force to remove more coating from the substrate. Osmotic pressure can be expected to range between 2500 and 3000 kPa, whereas the resistance of organic coatings to deformational forces is lower, ranging from 6 to 40 kPa [9]. Thus, blisters form and expand, exposing more unprotected steel surface. A schematic representation of blistering is shown in Figure 27.5.

It has also been proposed that blisters can grow by a nonosmotic mechanism [15]. The suggestion is made that water absorbed by a coating induces in-plane compressive stress within the coating and elastically extends the interfacial bonds between the coating and the steel substrate. At a point of weak adhesion between the coating and the substrate, the stress can lead to disbondment.

In either osmotic or nonosmotic mechanisms, the key to maintaining corrosion protection by the coating is sufficient adhesion to resist displacement forces. Both mechanisms predict that if the coating completely covered the entire surface of the steel, on a microscopic as well as macroscopic scale, and if perfect wet adhesion could be achieved at all areas of the interface, the coating would indefinitely protect the steel completely against corrosion. Unfortunately, it is difficult to achieve both of these requirements in the practical application of coatings, so a high level of wet adhesion becomes an important, but not the only, factor affecting corrosion protection by coatings. Funke has found that in addition to wet adhesion, low water and oxygen permeability also help increase corrosion protection [13]. If the wet adhesion is poor, corrosion protection will be poor in any case. However, if the wet adhesion is fairly good, a low rate of water and oxygen permeation may delay the loss of adhesion long enough that there will be adequate corrosion protection for many practical conditions.

It is widely recognized that primers made with saponification resistant vehicles commonly give better corrosion protection than primers made with vehicles that saponify readily [16,17]. When water and oxygen permeate through the film and water displaces some of the adsorbed groups of the coating resin from the surface of the steel, corrosion starts. Hydroxide ions are generated at the cathodic areas. Hydroxide ions catalyze the hydrolysis (saponification) of such groups as esters. If the backbone of the vehicle resin is connected by ester groups, hydrolysis results in polymer network degradation, leading to still poorer wet adhesion and ultimately to catastrophic failure.

27.3.2. Adhesion for Corrosion Protection

Chapter 26 deals broadly with adhesion. In view of the critical importance of adhesion in achieving corrosion protection, wet adhesion, which is especially critical to corrosion protection, is reviewed here. Obviously, good dry adhesion must be taken as a given in achieving corrosion protection. If there is no coating left on the substrate, it cannot protect the steel. It has not been so obvious, however, that good wet adhesion is required. Good wet adhesion means that the adsorbed layer of the coating will not desorb when water permeates through the film and approaches the interface.

The first step to obtain good wet adhesion is to clean the steel surface, especially to remove any oils and salts. Application of phosphate conversion coatings gives further advantages. Various types of steel and coated steel may require different cleaning and treatment methods [18]. After cleaning and treating, the surface should not be touched and should be coated as soon as possible. Fingerprints will leave oil and salt on the surface. After exposure to high humidity, fine blisters can form, disclosing the identity of the miscreant by his finger prints. A rusty handprint was observed on a ship after only one ocean and lake passage [3]. In painting surfaces near the ocean, it is critical to avoid having any salt on the metal surface when the paint is applied.

It is also critical to achieve as nearly complete penetration into the micropores and irregularities in the surface of the steel as possible. If any steel is left uncoated, when water and oxygen reach that surface, corrosion will start, generating an osmotic cell that can lead to blistering. As discussed in Section 26.1, the most important factor in achieving penetration is to have the viscosity of the external phase be as low as possible and remain so long enough to permit complete penetration. It is preferable to use slow evaporating solvents, slow cross-linking systems, and, whenever possible, baking primers.

Wet adhesion requires that the coating not only be adsorbed strongly on the surface of the steel but that it not be desorbed by water that will permeate through the coating. Empirically, it is found that wet adhesion may be enhanced by having several absorbing groups scattered along the resin chain with parts of the resin backbone being flexible to permit relatively easy orientation and parts rigid to assure that there will be loops and tails sticking up from the surface for interaction with the rest of the coating. Another reason that baking primers commonly provide superior corrosion protection is that at the higher temperature there may be greater opportunity for orientation of resin molecules at the steel interface.

Amine groups are particularly effective polar substituents for promoting wet adhesion. Perhaps water is less likely to displace amines than other groups from the surface. Phosphate groups have also been suggested as groups that promote adhesion and, particularly, wet adhesion. For example, epoxy phosphates have been used to enhance the adhesion of epoxy coatings on steel [19]. There is need for further research to understand the relationships between resin structure and wet adhesion.

Saponification resistance is another important factor. Corrosion generates hydroxide ions at the cathode, raising pH levels as high as 14. Ester groups in the backbone of a binder can be saponified, degrading the polymer near the interface and reducing wet adhesion. Epoxy-phenolic primers are an example of high-bake primers that are completely resistant to hydrolysis. In some epoxy-amine primers there are no hydrolyzable groups. The amine-terminated polyamides, which are widely used in air dry primers to react with epoxy resins, have amide groups in the backbone that can hydrolyze. However, amides are more resistant to base-catalyzed hydrolysis than are esters.

Although having polyester backbones, alkyd resins are used when only moderate corrosion protection is required and low cost is important. Epoxy ester primers show substantially greater resistance to saponification than do alkyd primers and are widely used.

Water-soluble components that may stay in the barrier primer films should be avoided because they can lead to blister formation. For example, zinc oxide is, generally, an undesirable pigment to use in primers. Its surface interacts with water and carbon dioxide to form zinc hydroxide and zinc carbonate, which are somewhat soluble in water and can lead to osmotic blistering. It should be noted that passivating pigments, discussed in Section 27.4.2, cannot function unless they are somewhat soluble in water; their presence in coating films can, therefore, lead to blistering. Funke [9] has shown that residual hydrophilic solvents, which become insoluble in the drying film as other solvents evaporate, can be retained as a separate phase and can lead to blister formation.

27.3.3. Factors Affecting Oxygen and Water Permeability

Many factors affect the permeability of coating films to water and oxygen [20]. Water and oxygen can permeate, to at least some extent, through any amorphous polymer film even though the film has no imperfections such as cracks or pores. Small molecules travel through the film by jumping from free volume hole to free volume hole. Free volume increases as temperature increases above T_g. Therefore, normally one wants to design coatings with a T_g above the temperature at which corrosion protection is desired. Since cross-linking reactions become slow as the increasing T_g of the cross-linking polymer approaches the temperature at which the reaction occurs, and become very slow at $T < T_g$, air dry films cannot have T_g values much above ambient temperatures. The higher T_g values that can be reached with baked coatings may be another factor in their generally superior corrosion protection. In general, higher cross-link density leads to lower permeability. Of course, both T_g and cross-link density affect other coatings properties; so some compromise in T_g and cross-link density must be accepted.

Permeability is also affected by the solubility of oxygen and water in the film. The variation in oxygen solubility is probably small, but variation in water solubility can be large. Salt groups on the polymer lead to high solubility of water in films. This fact makes it difficult to formulate high-performance air dry, water-reducible coatings that are solubilized in water by amine salts of carboxylic acids. Although to a lesser degree than salts, resins made with polyethylene oxide backbones are likely to give high water permeabilities, as are silicone resins. On the other side, water has low solubility in halogenated polymers, hence vinyl chloride and vinylidene chloride copolymers and chlorinated rubber are commonly used in formulating top coats for corrosion-resistant systems.

Pigmentation can have significant effects on water and oxygen permeability. Oxygen or water molecules cannot pass through pigment particles; therefore, permeability decreases as PVC increases. However, if the PVC exceeds CPVC, there are voids in the film and passage of water and oxygen through the film is facilitated. Some pigments have high-polarity surfaces that adsorb water and, in cases where water can displace polymer adsorbed on such surfaces, water permeability can be expected to increase with increasing pigment content. Pigments should be used that are as free as possible of water-soluble impurities, and use of hydrophilic pigment dispersants should be avoided or at least minimized.

Pigments with platelet-shaped particles are particularly effective, reducing permeability rates as much as fivefold when they are aligned parallel to the coating surface [11,21]. A factor favoring alignment is shrinkage during solvent evaporation. Since oxygen and water vapor cannot pass through the pigment particles, the presence of aligned platelets can substantially reduce the rate of vapor permeation through a film. Mica, talc, micaceous iron oxide, and metal flakes are examples of such pigments.

Aluminum flake is widely used; stainless steel and nickel platelets, while more expensive, have greater resistance to extremes of pH. When the appearance requirements permit, the use of leafing aluminum pigment in the top coat is particularly effective. The particles of leafing aluminum are surface treated so that their surface tension is very low. As a result, the platelets come to the surface during film formation, creating an almost continuous barrier. In formulating coatings with leafing aluminum, it is necessary to avoid resins and solvents that could displace the surface treatment from the flakes.

A Monte Carlo simulation model of the effect of several variables on diffusion through pigmented coatings has been devised [22]. The model indicates, as would be expected, that finely dispersed, lamellar pigment particles at a concentration near but below CPVC give the best barrier performance. The model's basic assumptions must be experimentally tested, but the model shows promise as a basis for considering the effect of variations in pigmentation on diffusion.

There are advantages in applying multiple layers of coatings. First, the primer can be designed so that it has excellent penetration into the substrate surface and has excellent wet adhesion without particular concern about other properties. The top coat(s) can provide for minimum permeability and other required properties. The primer film does not need to be thick as long as the top coat is providing barrier properties; the lower limit is probably controlled by the need to assure coverage of the entire surface. Funke [14] has reported good results with 0.2 μm primer thickness. Another advantage of multiple coats is the corresponding decrease in probability that any areas of the substrate will escape having any coating applied.

Film thickness affects the time necessary for permeation through the films. Thicker films would be expected to delay somewhat the arrival of water and oxygen at the interface, but would not be expected to affect the equilibrium condition. The corrosion protection afforded by intact films would be expected to be essentially independent of film thickness. Since film thickness affects the mechanical performance of films, there may well be some optimum film thickness for the maintenance of an intact film. For example, erosion losses would take longer to expose bare metal as film thickness increases; on the other hand, the probability of cracking on bending increases as film thickness increases.

However, in the case of air dry, heavy duty maintenance coatings, it has been the general experience that there is a film thickness, dependent on the particular coating, that provides a more than proportional increase in corrosion protection relative to thinner films. Commonly this film thickness will be as much as 400 μm or more. Funke [23] suggests that below certain coating thicknesses there may be microscopic defects extending down through the film all the way to the substrate. The film looks intact but there may be microscopic defects that are large compared to the free volume holes through which ordinary permeation in fully intact films occurs. A potential source of such defects is cracks resulting from internal stresses introduced as the last solvent is lost from a coating with T_g of the solvent-free system around ambient temperature. Funke suggests that if the film thickness is great enough, such defects may not reach the substrate, hence reducing the passage of water and oxygen substantially. This hypothesis is consistent with the general observation that greater protection is achieved by applying more coats to reach the same film thickness. In line with this proposal, the use of barrier platelet pigments permits a substantial reduction in the required film thickness without the loss of protection. The platelets may minimize the probability of defects propagating through the film to the substrate. Such defects are less likely to occur in baked films, and this may be another factor in the generally superior corrosion protection afforded by baked films even though thinner films are used.

27.4. CORROSION PROTECTION WITH NONINTACT FILMS

Even with coatings designed to minimize the probability of mechanical failure, in many end uses there will be breaks in the films during the service life. Furthermore, there are situations where it is not possible to have full coverage of all of the steel surface. In such cases it is generally desirable to design the coatings to suppress electrochemical reactions rather than primarily for their barrier properties.

27.4.1. Minimizing Growth of Imperfections

If there are gouges through the film down to bare metal, water and oxygen will reach the metal and corrosion will start. If the wet adhesion of the primer to the metal is

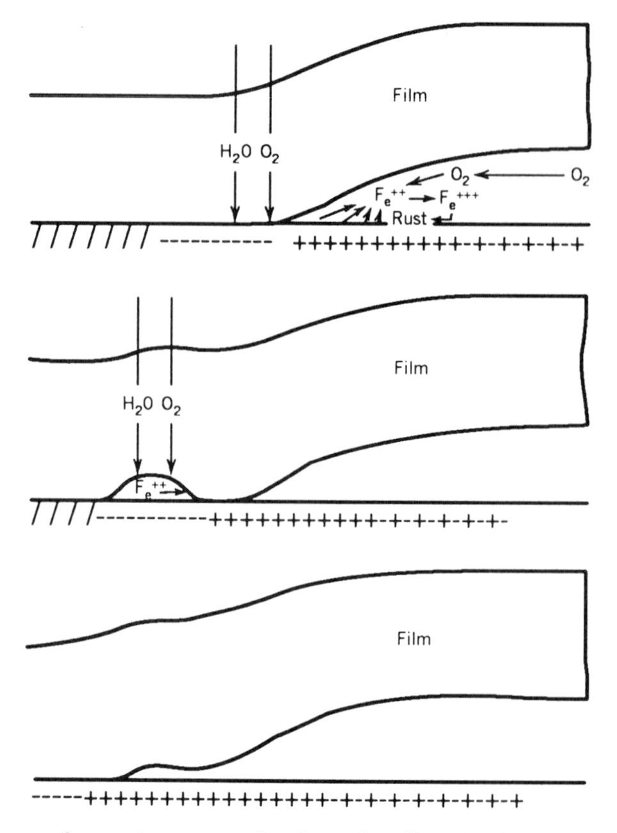

Figure 27.6. Stages of corrosion at a scribe through a film showing initial corrosion at the scribe and blistering adjacent to the scribe. (From Ref. [11], with permission.)

not adequate, water will creep under the coating and the coating will come loose from the metal over a wider and wider area. Furthermore, poor hydrolytic stability can be expected to exacerbate the situation. Even when wet adhesion is reasonably good, it has been shown that blisters are likely to develop under the film near the location of a gouge [3,11]. This mode of failure is called *cathodic delamination*, as illustrated in the idealized picture in Figure 27.6. Control of cathodic delamination requires adequate wet adhesion and saponification resistance.

When wet adhesion varies on a local scale, *filiform corrosion* can occur [24]. It is characterized by development of thin threads of corrosion wandering randomly under the film but never crossing another track. Formation of these threads often starts from the edge of a scratch. At the head of the thread, oxygen permeates through the film and cathodic delamination occurs. The head grows following the directions of poorest wet adhesion. Behind the head, oxygen is consumed by oxidation of the ferrous ions and ferric hydroxide precipitates, passivating the area (explaining why threads never cross). Since the ion concentration decreases, osmotic pressure drops, and the thread collapses, but it leaves a discernible rust track. Filiform corrosion can be difficult to see through pigmented films.

27.4.2. Primers with Passivating Pigments

Passivating pigments promote the formation of a barrier layer over the anodic areas, passivating the surface (see Section 27.2.1). In order to be effective, such pigments must have some minimum solubility. However, if the solubility is too high the pigment will leach out of the coating film too rapidly, limiting the time it is available to inhibit corrosion. Furthermore, for the pigment to be effective, the binder must permit diffusion of water to dissolve the pigment. Therefore, the use of passivating pigments can be expected to lead to blistering or other coating failures after exposure to humid conditions. Such pigments are most useful in applications where the need to protect the steel substrate after film rupture has occurred outweighs the desirability of minimizing the probability of blistering. They are also useful when it is not possible to remove all surface contamination (blistering will probably occur anyway) or when it is not possible to achieve complete coverage of the steel substrate surface by the coating.

Red lead pigment Pb_3O_4, containing 2–15% PbO, has been used as a passivating pigment since the mid-19th century. Red lead in oil primers are widely used for air dry application over rusty, oily steel (see Section 36.1.3). The mechanisms of action are not fully understood [3]. They presumably include oxidation of ferrous ions to ferric ions followed by coprecipitation of mixed iron–lead salts or oxides. The somewhat soluble PbO raises the pH and neutralizes any fatty acids formed over time by hydrolysis of the drying oil. Toxic hazards of red lead restrict its use to certain industrial and special purpose applications.

The utility of chromate pigments for passivation is well established. Various mechanisms have been proposed to explain their effectiveness [3]. All the proposed mechanisms require that the chromate ions be in aqueous solution. Like all passivators, chromate ions accelerate corrosion at low concentrations. The critical minimum concentration for passivation at 25°C is approximately 10^{-3} mols CrO_4^{2-} L^{-1}. The critical concentration increases with increasing temperature and increasing NaCl concentration.

Sodium dichromate is an effective passivating agent but would be a poor passivating pigment; its solubility in water (3.3 mols CrO_4^{2-} L^{-1}) is too high. It would be rapidly leached out of a film and would probably cause massive blistering. At the other extreme, lead chromate (chrome yellow) is so insoluble (5×10^{-7} mols CrO_4^{2-} L^{-1}) [25] that it has no electrochemical action and is used only for its color.

"Zinc chromates" have been widely used as passivating pigments. The terminology is poor since zinc chromate itself is too insoluble and could promote corrosion rather than passivate. Zinc yellow pigment [$K_2CrO_4 \cdot 3ZnCrO_4 \cdot Zn(OH)_2 \cdot 2H_2O$] (alternative ways of giving the same composition are $Zn_4K_2Cr_4O_{20}H_6$ and $4ZnO \cdot K_2O \cdot 4CrO_3 \cdot 3H_2O$ [25]) has an appropriate solubility (1.1×10^{-2} mols CrO_4^{2-} L^{-} at 25°C) and has been widely used in primers. Zinc tetroxychromate [$ZnCrO_4 \cdot 4Zn(OH)_2$ also given as $Zn_5CrO_{12}H_8$] has a solubility lower than desirable (2×10^{-4} mols CrO_4^{2-} L^{-1}), but is used in *wash primers* (see Section 36.2). Phosphoric acid is added to wash primers before application; it may be that this changes the solubility so that the chromate ion concentration is raised to an appropriate level. Strontium chromate ($SrCrO_4$) has an appropriate solubility in water (5×10^{-3} mols CrO_4^{2-} L^{-1}), and is sometimes used in primers, especially latex paint primers where the more soluble zinc yellow can cause problems of package stability.

It has been established that zinc chromates, and presumably other soluble chromates, are carcinogenic to humans. They must, therefore, be handled with appropriate caution. In some countries their use has been prohibited. Substantial efforts have been undertaken to develop less hazardous passivating pigments. It is difficult to conclude from the available literature and supplier technical bulletins how these newer pigments compare with each other and with zinc yellow. In some cases, a formulation that has been optimized for one pigment is compared to a formulation containing another pigment that may not be the optimum formulation for that pigment. (A common example is the substitution of one pigment for another on an equal weight basis rather than formulating to the same ratio of PVC to CPVC; results could be very misleading since primer performance is quite sensitive to PVC/CPVC.) Much of the published data is based on comparing corrosion resistance in salt fog chamber tests (or other laboratory tests) rather than actual field experience. As discussed in Section 27.5, there is no laboratory test available that provides reliable predictions of field performance.

Basic zinc and zinc–calcium molybdates are said to act as passivating agents in the presence of oxygen, apparently leading to the precipitation of a ferric molybdic oxide barrier layer on the anodic areas. Barium metaborate is the salt of a strong base and a weak acid. It may act by increasing the pH, thus lowering the critical concentration of oxygen required for passivation. To reduce its solubility in water, the pigment grade is coated with silicon dioxide. Even then, some workers feel that the solubility is too high for use in long-term exposure conditions. Zinc phosphate, $Zn_3(PO_4)_2 \cdot 2H_2O$, has been used in corrosion protective primers and may act by forming barrier precipitates on the anodic areas. There is considerable difference of opinion as to its effectiveness. Calcium and barium phosphosilicates and borosilicates are being used increasingly; they may act by increasing pH.

All of these pigments are inorganic pigments; a much wider range of potential oxidizing agents would seem to be available if organic pigments were used. The search for effective organic passivating pigments is a long-standing research activity. An example of a commercially available organic pigment is the zinc salt of nitrophthalic acid. It is said to be as effective as zinc yellow at lower pigment levels. (Since over a long time any effective passivating pigment will be lost by leaching, it seems doubtful if equal lifetime could be achieved at substantially lower pigment content.) 2-Benzothiazoylthiosuccinic acid has also been recommended as a passivating agent.

It is probable that all of these pigments (and others that are offered commercially [26]) offer significant corrosion protection in primers; however, many years of field experience are needed to assess how effective they may be as compared to zinc yellow and red lead.

27.4.3. Cathodic Protection by Zinc-Rich Primers

Zinc-rich primers are another approach to protecting steel with nonintact coatings. They are designed to provide the protection of galvanized steel, but to be applied to a steel structure after fabrication [27]. The primers contain high levels of powdered zinc—over 90% by weight is usual. On a volume basis, the zinc content exceeds CPVC to assure good electrical contact between the zinc particles and with the steel. Furthermore, when PVC is greater than CPVC, the film will be porous, permitting water to enter, thereby completing the electrical circuit. The CPVCs of zinc powders vary depending primarily on particle size and particle size distribution; values on the order of 65% have been reported [28]. The zinc serves as a sacrificial anode and zinc hydroxide is generated in the pores.

It should be evident that vehicles for zinc-rich primers must be saponification resistant. Alkyds are not appropriate resins for this application. Among organic binders, epoxies are the most commonly used. However, probably the most widely used vehicles are tetraethyl orthosilicate and oligomers derived from it by controlled partial hydrolysis with a small amount of water (see Section 13.3.2). Ethyl alcohol is used as the principle solvent since it helps maintain package stability. After application, ethyl alcohol evaporates and water from the air completes the hydrolysis of the oligomer to yield a film of polysilicic acid partially converted to zinc salts. Such a primer is referred to as an *inorganic zinc-rich primer* or sometimes as a *zinc silicate primer*.

Properly formulated and applied, zinc-rich primers are very effective in protecting steel against corrosion. Their useful lifetime is not completely limited by the amount of zinc present, as one might at first assume. It has been found that initially the amount of free zinc decreases as would be expected from the electrochemical reaction. Later, the loss of zinc metal becomes slow, but the primer continues to protect the steel. Possibly the partially hydrated zinc oxide formed in the initial stages of corrosion of the zinc fills the pores, and, together with the remaining zinc, acts as a barrier coating [29]. It is also possible that the basic zinc hydroxide raises the pH to the level where oxygen can passivate the steel.

Zinc is expensive, especially on a volume basis. Early attempts to replace even 10% of the zinc with low-cost inert pigment caused a serious decrease in performance, presumably owing to a decrease in metal to metal contact even though the PVC was above CPVC. A relatively conductive inert pigment, iron phosphide (Fe_2P), has shown some promise [30]. It has been reported [31] that in ethyl silicate based coatings up to 25% of the zinc can be replaced with Fe_2P, but that with epoxy/polyamide coatings replacement of part of the Zn with Fe_2P leads to a reduction in protection. The Steel Structures Painting Council is comparing field exposure performance of Fe_2P extended primers to conventional zinc-rich primers.

Zinc-rich primers are usually topcoated to minimize corrosion of the zinc to protect against physical damage and to improve appearance. Formulation and application of top coats must be done with care. If the vehicle of the top coat penetrates into the pores in the primer film, the conductivity of the primer may be substantially reduced, rendering it ineffective. This critical problem is discussed in Section 36.1.2.

27.5. EVALUATION AND TESTING

There is no laboratory test available that can be used to predict corrosion protection performance of a new coating system. This unfortunate situation is an enormous obstacle to research and development of new coatings, but it must be recognized and accommodated.

Use testing is the only reliable test of a coating system, that is, to apply it and then observe its condition over years of actual use. The major suppliers and end users of coatings for such applications as bridges, ships, chemical plants, and automobiles have systematically collected data correlating the performance of different systems over many years. These data provide a basis for selection of current coatings systems for particular applications. They also provide insight into how new coatings could be formulated to improve performance.

Simulated tests are the next most reliable tool for predicting performance. One common approach is to expose laboratory prepared panels on test fences in inland south Florida or on beaches in south Florida or North Carolina. The difficulties in developing tests to simulate corrosion in marine environments are discussed in Ref. [32]. Test conditions must simulate actual use conditions as closely as possible. For example, exposure at higher temperatures may accelerate corrosion reactions; however, oxygen and water permeability are strongly affected by $(T - T_g)$. If actual use will be at temperatures below T_g and the tests are run above T_g, no correlations should be expected.

The many variables in the preparation of test panels are frequently underestimated. The steel used is a critical variable [33]. Also significant are how the steel is prepared for coating and how the coating is applied. Film thickness, evenness of application, flash-off time, baking time and temperature, and many other variables affect performance. Results obtained on carefully prepared and standardized laboratory panels can be quite different than results on actual products. In view of these problems, it is desirable, when possible, to paint test sections on ships, bridges, chemical storage tanks, and so forth, and to observe their condition over the years. The long times required for evaluation are, of course, undesirable, but the results can be expected to correlate reasonably with actual use.

To obtain results more quickly, there have been many attempts to develop improved laboratory tests, and these efforts continue. Available tests have limited reliability in predicting performance. Nonetheless they are widely used.

The most widely used test method for corrosion resistance is the salt spray (fog) test [ASTM Method B117-73 (1979)]. Coated steel panels are scribed (cut) through the coating in a standardized fashion exposing bare steel and then hung in a chamber where they are exposed to a mist of 5% salt solution at 100% relative humidity at 35°C. Periodically, the nonscribed areas are examined for blistering and the scribe is examined to see how far from the scribe mark the coating has been undercut or has lost adhesion. It has been repeatedly shown that there is little, if any, correlation between results in salt spray tests and actual performance of coatings in use [34–37]. Reversals among different coatings are common in comparison to actual performance results.

Many factors are probably involved in the unreliability of the salt spray test. Outdoor weathering can have a significant effect on film properties. Environmental factors such as acid rain vary substantially from location to location. The application of the scribe mark can be an important variable—narrow cuts generally affect corrosion less than broader ones. Also, if the scribe mark is cut rapidly there may be chattering of cracks out from the main cut, whereas slow cutting may lead to a smooth cut. A passivating pigment with high solubility might be very effective in a laboratory test but may provide protection for only a limited time in field conditions owing to loss of passivating pigment by leaching.

Since with intact films it is common for the first type of failure to be blister formation, humidity resistance tests are widely used (ASTM Method D 2247). The face of the panel is exposed to 100% relative humidity at 38°C while the back of the panel is exposed to room temperature air. Thus, water continuously condenses on the coating surface. This humidity test is a more severe test for blistering than the salt fog test because pure water on the film will generate higher osmotic pressures with osmotic cells under the film than will the salt solution used in the salt fog test. It is common to run the test at 60°C "because it is a more severe test." The pitfalls of this approach should be obvious in view of the discussion above of the importance of $(T - T_g)$. Humidity tests will not provide a prediction of the life of corrosion protection, but may provide useful comparisons of wet adhesion. Funke [23] recommends testing for wet adhesion by scribing panels after various exposure times in a humidity chamber immediately followed by applying pressure-sensitive tape across the scribe mark and then pulling the tape off the panel. Wet adhesion can also be checked after storing panels in water [38].

It is often observed that alternating high and low humidity causes faster blistering than continuous exposure to high humidity. A large number of humidity cycling tests have been described, commonly involving repeated immersion in warm water and removal for several hours. In some industries such tests have become accepted methods of screening coatings, although their predictive value is questionable. Simply correlating them with salt fog tests proves little.

A testing regimen called "Prohesion" has been reported to correlate better with actual performance than the standard salt spray test [39]. The procedure combines care in selection of substrates that will reflect real products, use of thin films, emphasis on adhesion checks, and a modified salt mist exposure procedure. Instead of 5% NaCl, a solution of 0.4% ammonium sulfate and 0.05% NaCl is used. Scribed panels are sprayed with the mixed salt solution cycling over 24 hours, six 3-hour periods alternating with six 1-hour drying periods using ambient air. Much more extensive comparisons with field performance will be needed to evaluate the predictive value of the procedure. A variety of other cycling tests are also being investigated as alternatives to salt spray tests [40].

Neither salt fog nor humidity tests have good reproducibility. It is common for differences between duplicate panels to be larger than differences between panels with different coatings. Precision can be improved by testing 10 or more replicate panels of each system. (Commonly, decisions are based on the results from testing two or three panels.) The problem is further complicated by the difficulty of rating the degree of severity of failure. A new rating system and an approach to statistical analysis of data has been published [41]. The study was based on panels with an acrylic clear coating and a pigmented alkyd coating exposed to 95% relative humidity at 60, 70, and 80°C. The times to failure were extrapolated down to ambient temperatures. (Neither of the coatings tested would be expected to have good corrosion protection properties.) In light of the effect of T_g on permeability, the validity of such extrapolations is doubtful.

A further problem in evaluating panels for corrosion protection is the difficulty of detecting small blisters and rust areas underneath a pigmented coating film without removing the film. Infrared thermography has been recommended as a nondestructive testing procedure [42].

A great deal of effort has been expended on electrical conductivity tests of paint films and electrochemical tests of coated panels. While there are papers that support the utility of such tests [3,43,44], there are also papers that warn against the uncritical use of data from such tests [45].

A variety of cathodic disbonding tests specifically for testing pipeline coatings has been established by ASTM,—ASTM: G8–79, G19–83, G42–80, and G80–83. In these tests a hole is made through the coating and the pipe is made the cathode of a cell in water with dissolved salts and at a basic pH. Disbonding (loss of adhesion) as a function of time is followed. While there is considerable variability inherent in such tests, and their utility for predicting field performance is doubtful, useful guidance in following progress in modifying wet adhesion may be obtained. Such tests may be useful for more than pipeline coatings. For a discussion of research on cathodic delamination, including investigation of the migration of cations through or under coating films, see Ref. [46].

Another approach to testing for delamination is the use of *electrochemical impedance spectroscopy* (EIS) [47,48]. Impedance is the apparent opposition to flow of an alternating electrical current, and is the inverse of apparent capacitance. When a coating film begins to delaminate there is an increase in apparent capacitance. The rate of increase of capacitance is proportional to the amount of area delaminated by wet adhesion loss. High-performance systems show slow rates of increase of capacitance, so tests must be continued for long time periods. EIS can be a powerful tool to study the effect of variables on delamination.

Appleman [49] has reported the results of an extensive survey of accelerated test methods for anticorrosive coating performance. The need for everyone in the industry to become aware of the current testing situation and to work cooperatively to develop more meaningful methods of testing is emphasized.

The lack of laboratory test methods that reliably predict performance puts a premium on collection of data bases permitting analysis of interactions between actual performance and application and formulation variables. It is especially critical to incorporate data on premature failures in the data base. The availability of such a data base can be a powerful tool for a formulator and may be especially useful in testing the validity of theories about factors controlling corrosion. In time, it may be possible to predict performance better from a knowledge of the underlying theories than from laboratory tests. Many workers feel this is already true in comparison with salt fog chamber tests.

GENERAL REFERENCES

H. H. Uhlig, *Corrosion and Corrosion Control*, 2nd ed., Wiley, New York, 1971.

Z. W. Wicks, Jr., *Corrosion Control by Coatings*, Federation of Societies for Coatings Technology, Blue Bell, PA, 1987.

A. Smith, *Inorganic Primer Pigments*, Federation of Societies for Coatings Technology, Blue Bell, PA, 1988.

REFERENCES

1. R. A. Dickie and F. L. Floyd, Eds., "Polymeric Materials for Corrosion Control," *ACS Symposium Series*, No. 332, American Chemical Society, Washington, DC, 1986.

2. H. H. Uhlig, *Corrosion and Corrosion Control*, 2nd ed., Wiley, New York, 1971.
3. H. Leidheiser, Jr., *J. Coat. Technol.*, **53** (678), 29 (1981).
4. J. E. O. Mayne, *Official Digest*, **24** (325), 127 (1952).
5. J. E. O. Mayne in *Corrosion*, Vol. 2, L. L. Shreir, Ed., Butterworth, Boston, 1976, pp. 15:24–15:37.
6. H. Leidheiser, Jr., *Prog. Org. Coat.*, **7**, 79 (1979).
7. W. Funke, U. Zorll, and B. G. K. Murth, *J. Coat. Technol.*, **41** (530), 210 (1969).
8. W. Funke and H. Haagen, *Ind. Eng. Chem. Prod. Res. Dev.*, **17**, 50 (1978).
9. W. Funke, *J. Oil Col. Chem. Assoc.*, **62**, 63 (1979).
10. W. Funke, *Prog. Org. Coat.*, **9**, 29 (1981).
11. W. Funke, *J. Coat. Technol.*, **55** (705), 31 (1983).
12. W. Funke, *Ind. Eng. Chem. Prod. Res. Dev.*, **24**, 343 (1985).
13. W. Funke, *J. Oil Col. Chem. Assoc.*, **68**, 229 (1985).
14. W. Funke, *Farbe Lack*, **93**, 721 (1987).
15. J. W. Martin, E. Embree, and W. Tsao, *J. Coat. Technol.*, **62** (790), 25 (1990).
16. J. W. Holubka, J. S. Hammond, J. E. DeVries, and R. A. Dickie, *J. Coat. Technol.*, **52** (670), 63 (1980).
17. J. W. Holubka and R. A. Dickie, *J. Coat. Technol.*, **56** (714), 43 (1984).
18. B. Perfetti, *J. Coat. Technol.*, **63** (795), 43 (1991).
19. J. L. Massingill, P. S. Sheih, R. C. Whiteside, D. E. Benton, and D. K. Morisse-Arnold, *J. Coat. Technol.*, **62** (781), 31 (1990).
20. N. L. Thomas, *Prog. Org. Coat.*, **19**, 101 (1991).
21. B. Bieganska, M. Zubielewicz, and E. Smieszek, *Prog. Org. Coat.*, **16**, 219 (1988).
22. D. P. Bentz and T. Nguyen, *J. Coat. Technol.*, **62** (783), 57 (1990).
23. W. Funke, private communication, 1986.
24. R. T. Ruggeri and T. R. Beck, *Corrosion-NACE*, **39**, 453 (1983).
25. E. Lalor, "Zinc and Strontium Chromates," in *Pigment Handbook*, Vol. 1, T. C. Patton, Ed., Wiley, New York, 1972, pp. 847–859.
26. A. Smith, *Inorganic Primer Pigments*, Federation of Societies for Coatings Technology, Blue Bell, PA, 1988.
27. J. E. O. Mayne and U. R. Evans, *Chem. Ind.*, **63**, 109 (1944).
28. M. Leclercq, *Materials Technology*, March, 57 (1990).
29. S. Feliu, R. Barajas, J. M. Bastidas, and M. Morcillo, *J. Coat. Technol.*, **61** (775), 63, 71 (1989).
30. N. C. Fawcett, C. E. Stearns, and B. G. Bufkin, *J. Coat. Technol.*, **56** (714), 49 (1984).
31. S. Feliu, Jr., M. Morcillo, J. M. Bastidas, and S. Feliu, *J. Coat. Technol.*, **63** (793), 31 (1991).
32. T. S. Lee and K. L. Money, *Mater. Perform.*, **23**, 28 (1984).
33. R. G. Groseclose, C. M. Frey, and F. L. Floyd, *J. Coat. Technol.*, **56** (714), 31 (1984).
34. R. Athey, R. Duncan, E. Harmon, D. Izak, K. Nakabe, J. Ochoa, P. Shaw, T. Specht, P. Tostenson, and R. Warness, *J. Coat. Technol.*, **57** (726), 71 (1985).
35. J. Mazia, *Met. Finish.*, **75** (5), 77 (1977).
36. R. D. Wyvill, *Met. Finish.*, **80** (1), 21 (1982).
37. W. Funke, in *Corrosion Control by Coatings*, H. Leidheiser, Jr., Ed., Science Press, Princeton, NJ, 1979, pp. 35–45.
38. M. Hemmelrath and W. Funke, *IXXth FATIPEC Congress Book*, Vol. IV, 137 (1988).
39. F. D. Timmins, *J. Oil Colour Chem. Assoc.*, **62**, 131 (1979).
40. B. R. Appleman, *J. Protective Coatings & Linings*, Nov., 71 (1989).
41. J. W. Martin and M. E. McKnight, *J. Coat. Technol.*, **57** (724), 31, 39, 49 (1985).
42. M. E. McKnight and J. W. Martin, *J. Coat. Technol.*, **61** (775), 57 (1989).
43. B. S. Skerry and D. A. Eden, *Prog. Org. Coat.*, **15**, 269 (1987).

44. H. Leidheiser, Jr., *J. Coat. Technol.*, **63** (802), 21 (1991).
45. W. Funke, *Fitture e Vernici*, **7**, 42 (1984).
46. J. Parks and H. Leidheiser, Jr., *XVIIth FATIPEC Congress Book*, Vol. II, 317 (1984).
47. W. S. Tait, *J. Coat. Technol.*, **61** (768), 57 (1989).
48. J. R. Scully, *Electrochemical Impedance Spectroscopy for Evaluation of Organic Coating Deterioration and Under Film Corrosion—A State of the Art Technical Review*, Report No. DTNSRDC/SME-86/006, D. W. Taylor Naval Ship Research and Development Center, Bethesda, MD, 1986.
49. B. R. Appleman, *J. Coat. Technol.*, **62** (787), 57 (1990).

CHAPTER XXVIII

Solvent-Borne and High Solids Coatings

In this chapter and Chapters 29–32, we discuss several of the principal classes of coatings: solvent-borne, water-borne, electrodeposition, powder, and radiation cure coatings. The intent is to discuss the principles involved in these classes. In Chapters 33–36 end use applications for these coatings are discussed.

Historically almost all coatings were solvent-borne; the other four classes of coatings have been developed in large measure to reduce solvent usage. The original motivation to reduce solvent usage was a desire to reduce fire hazards and odor, and to permit cleanup with water. However, since the 1960s a major driving force has been the need to reduce VOC emissions into the atmosphere. While there have been significant reductions in solvent usage by shifting to the other classes of coatings, another important approach has been to decrease the solvent content of solvent-borne coatings by formulating so-called high solids coatings. In this chapter we discuss solvent-borne coatings with particular emphasis on high solids coatings.

In some end uses, especially where cost is particularly important, a single coating is adequate; however, in many more end uses, performance requirements can only be met by applying at least two coats over the substrate. Almost always when more than one coat is to be applied, it is preferable to have a coating, generally called a *primer*, specifically designed to be the first coat, and a different coating for the *top coat*. The primer is designed to adhere strongly to the substrate and to provide a surface to which the top coat will adhere well. It is not generally necessary for the primer to meet such other requirements as exterior durability, which are critical for top coat performance. Commonly, primers are less expensive than top coats. Usually primers need not be resistant to UV; however, if the top coat does not give complete hiding, UV can penetrate to the primer/top coat interface and may lead to degradation and loss of intercoat adhesion. Incorporation of UV absorber, particularly one that strongly absorbs radiation in the range of 280–300 nm (see Section 25.2.1), can minimize the problem. In some cases, it is necessary to formulate the primer with UV resistant epoxies instead of BPA epoxies or aliphatic isocyanates instead of aromatic isocyanates.

28.1. PRIMERS

The first consideration in formulating a primer is to achieve adequate adhesion to the substrate. Adhesion is discussed in Chapter 26; here the conclusions of those discussions are briefly summarized. The substrate should be clean and preferably have a rough surface. The surface tension of the primer must be lower than that of the substrate. The viscosity of the continuous phase of the primer should be as low as possible to promote penetration of the vehicle into pores and crevices in the surface of the substrate. Penetration is also promoted by the use of as slow evaporating solvents as possible, slow cross-linking systems, and, whenever feasible, baking primers. The primer binder should have polar groups scattered along the backbone of the resin that can interact with the substrate surface. The interaction between the binder and the substrate should be such that it will not be displaced by water when water molecules permeate through the coating toward the interface. In the case of primers for use over metal substrates as well as alkaline surfaces, such as masonry, saponification resistance of the primer is an important criterion in binder selection.

28.1.1. Binders for Primers

The binders in a large fraction of all primers used on metal substrates are BPA epoxy resins and their derivatives (see Sections 11.1–11.6). While a full theoretical explanation is not available, experience has shown that, in general, BPA epoxy resin-based coatings tend to exhibit superior adhesion. Use of a small amount of epoxy phosphate esters (see Section 11.6) can further enhance adhesion, especially to reduce loss of adhesion by displacement by water to provide maximum corrosion resistance. For baked coatings, epoxy-phenolic coatings are particularly appropriate since films show not only good adhesion but also good saponification resistance. For air dry coatings, epoxy-amine coatings are commonly selected. Epoxy ester-based primers provide adhesion and with saponification resistance somewhat inferior to epoxy-phenolic and epoxy-amine primers at lower costs.

For applications where moderate resistance properties are adequate, alkyd resins are used because of their lower cost. Alkyds also have the advantage of low surface tension so that cleanliness of the steel is not as critical to obtain wetting. However, saponification resistance is limited.

The lowest cost vehicles used in metal primers are styrenated alkyds (see Section 10.3). They are generally used in air dry primers. The higher T_g resulting from the large fraction of aromatic rings leads to primers that give dry-to-touch films or even dry-to-handle films much more rapidly than the corresponding nonstyrenated alkyd. The styrenation reduces the fraction of ester groups by dilution, and hence may increase saponification resistance. However, because of the reduction in the available methylene groups between double bonds resulting from the free radical reactions during styrenation, styrenated alkyds give cross-linked films more slowly than do their nonstyrenated counterparts. As a result, the films develop solvent resistance slowly, and have a time window within which they should not be top coated. If little or no cross-linking has occurred or, if the cross-linking has proceeded far enough so that the whole film is sufficiently cross-linked to be solvent resistant, the primer can be top coated without problems. However, at intermediate times the films have poor and nonuniform solvent resistance; if the primer is top coated within this time interval, *lifting* is likely to occur. Lifting leads to puckered areas in the top coat surface due to the nonuniform degrees of swelling by the top coat solvent, which leads to uneven film shrinkage when the film finally dries.

As noted above, in designing primers for steel, saponification resistance is an important criterion; in the case of galvanized steel it is critical. The surface of galvanized steel is coated with a layer of zinc hydroxide, zinc oxide, and zinc carbonate; all are somewhat water-soluble, strong bases. Especially if the metal is not phosphate treated shortly before coating, the binder in the primer must be very resistant to saponification. Alkyds have been used in primers on galvanized steel, sometimes with satisfactory initial results. However, commonly there are sporadic serious adhesion failures. The differences in performance have been attributed to variations in the condition of the surface of the galvanized steel. Fresh galvanized steel surfaces or surfaces that have been carefully protected from dampness are relatively easy to adhere to. However, if the galvanized metal has been stored where it comes in contact with water, then the surface becomes coated with zinc hydroxide and its reaction products. The resulting alkaline conditions can lead to poor adhesion with a vehicle like an alkyd that is susceptible to saponification of its backbone ester groups. For air dry coatings, acrylic latex vehicles are more appropriate.

Primers are also required for other than metal substrates. Adhesion to some plastics can be difficult to achieve (see Sections 26.5 and 34.3).

Primers for wood substrates have somewhat different functions. Adhesion to new wood surfaces is almost never a problem, since the substrate is porous, permitting penetration into the wood surface. However, the porosity is uneven. If one applies a solvent-borne high-gloss coating directly on a wood surface, more vehicle penetrates into the more porous areas, resulting in an increase in PVC of the coating above these areas of the surface. As a result, the gloss of the coating varies substantially from relatively high gloss over the least porous areas to relatively low gloss over the most porous areas. If one applies a solvent-borne low-gloss coating on wood, over the porous areas the penetration of vehicle can lead to an increase of PVC so that the local PVC is greater than CPVC. Latex paints minimize these problems since the latex particles are large compared to the pores in the wood surface and do not penetrate far into the wood; hence the PVC remains about the same over both high- and low-porosity sections of the surface.

Concrete surfaces are alkaline and often require special surface treatments, most commonly washing with hydrochloric acid. The acid wash not only neutralizes the surface alkalinity, it also etches the surface. Saponification resistant primers such as epoxy-amine or latex primers provide longest life. Concrete blocks have very porous surfaces and penetration of solvent-borne paints into the surface requires a relatively large volume of coating for coverage. Substantially better coverage can be achieved by using latex paints as discussed in Section 35.1.

28.1.2. Pigmentation of Primers

Selection of pigments and the amount of pigment are among the most critical decisions in primer formulation.

Pigmentation level affects adhesion of top coats to the primer film. Formulations with a high ratio of PVC/CPVC are low-gloss coatings; the roughness and increased surface area of low-gloss films give improved intercoat adhesion. In some cases, it is desirable to formulate primers with PVC > CPVC. The resultant primer film is somewhat porous, permitting penetration of top coat vehicle into the pores, assuring excellent intercoat adhesion. The PVC should be only slightly higher than CPVC or the loss of vehicle from the top coat may be enough to raise the PVC of the top coat film sufficiently to decrease the gloss of the top coat. Primers with PVC > CPVC can be sanded more easily and are less likely to clog the sand paper than primers with PVC < CPVC. Since the inert pigments are generally the least expensive components of the dry film, high PVC minimizes cost. See Section 21.2 for a discussion of factors controlling CPVC; in general terms, CPVC is higher with pigments having a broad principle size distribution.

Since primers almost always are low gloss coatings, inexperienced formulators may not think it is necessary to have good pigment dispersions for primers. This is not so! Since CPVC is affected by the degree of dispersion, particularly by the extent of flocculation, and since primers are usually formulated to be either slightly above or slightly below CPVC, pigment dispersion can be critical. If the pigment dispersion is not properly stabilized, CPVC decreases; instead of having a primer with a PVC slightly less then CPVC, its PVC could be greater than CPVC. Or, in the case of a primer designed to have a PVC slightly greater than CPVC, the PVC may become much greater than CPVC, leading to loss of gloss of the top coat.

Pigment selection, obviously, affects hiding and color of the primer. If only one color top coat is to be applied over the primer, it is usually desirable for the primer to have a similar color since this minimizes the effect of the primer color on the final top coat color. However, in many cases, various color top coats are applied over the same primer. Then it is usually desirable to use a light gray primer. Gray primers have better hiding than white primers, or equal hiding at lower cost, since not as much TiO_2 is needed to hide the substrate when a pigment like lamp black that strongly absorbs light is also present. Light gray primers have relatively little affect on top coat color. With a white primer, dark color top coats may give hiding problems and with dark colored primers, light color top coats may have hiding problems. Primers pigmented with red iron oxide provide good hiding at low cost but are more likely than gray primers to affect the color of the top coat.

With the exception of pigments for passivating steel against corrosion, pigments should be selected that are completely insoluble in water. If, for example, zinc oxide is used as a pigment in a primer formulation, some of the zinc oxide could dissolve in water, permeating through the film, and resulting in the establishment of an osmotic cell that can cause blistering.

As discussed in Section 27.3.3, water and oxygen permeabilities are important factors in corrosion protection of metals by barrier coatings. The pigments in the film can strongly affect its oxygen and water permeability. In general, the higher the PVC of the film, the lower the gas and vapor permeability. In order to occupy the largest possible volume of the dry film with impermeable pigment, it is desirable to select a combination of pigments that give a high CPVC and to formulate with PVC close to CPVC. Platelet-shaped pigment particles tend to provide better barriers to oxygen and water permeability. Mica and micaceous iron oxide are widely used in primers for metals because of their platelet form.

One must also use resin/pigment combinations in which the resin is strongly adsorbed on the pigment surface. If the pigment has a polar surface with a weakly adsorbed resin, water permeating through the film may displace the resin from the pigment surface, leading to an increase in water permeability.

As discussed in Section 27.4.2, passivating pigments can be useful in protecting steel against corrosion when the coatings on the steel substrate have been ruptured. However, passivating pigments must be somewhat soluble in water in order to passivate the steel. Furthermore, dissolving the pigments requires that the binder be able to swell with water to a degree. These characteristics mean that blistering is more likely than with a primer not containing a passivating pigment. In most coatings for OEM products, it is preferable to avoid the use of passivating pigments and rely on the barrier properties of the coating films to provide corrosion protection. This is especially likely to be the case when baking coatings are used. However, applications such as coatings for bridges, storage tanks, ships, and offshore drilling platforms are examples of end uses where the coating cannot be baked and in many cases film rupture must be anticipated. As discussed in Sections 36.1.2 and 36.1.3, primers for such applications containing passivating pigments are commonly used as are zinc-rich primers.

28.1.3. High Solids Primers

In common with all other coatings, there is considerable regulatory pressure to reduce the VOC emissions from primers. In conventional low solids primers, the effect of pigmentation on the viscosity of the coating is relatively small if the pigment is not flocculated. However, as solids are increased, the volume of dispersed phase, including both the volume of the pigment and the volume of the adsorbed layer on the surfaces of the pigment particles, increases and becomes an important factor controlling the viscosity of the coating and hence limits the solids at which the coating can be applied. See Section 20.2.3 for a discussion of the factors affecting the viscosity of high solids pigmented coatings.

In coatings with a nonvolatile volume (NVV) content below about 70, the effect of pigmentation levels less than required for a PVC of 20 in the dry film is small. But pigmentation levels such as encountered in low gloss coatings with PVC of 45 or higher substantially increase the viscosity of the wet coating, requiring reduction of solids for application. Figure 28.1, based on model calculations, shows plots of viscosity as a function of volume solids for three sets of calculations: one for the unpigmented coating, one for a 20-PVC (i.e., high gloss) coating, and one for a 45-PVC (i.e., low gloss) coating. The assumptions made in carrying out the calculations are provided in Ref. [1].

With the present status of our knowledge, somewhere in the neighborhood of 60 NVV is probably an upper limit for a primer with a PVC near to or above CPVC. A great challenge to increasing this level to a higher solids contents is development of a reasonable cost means of stabilizing the pigment dispersion in a primer with thinner adsorbed layer thicknesses. However, even assuming stabilization with an adsorbed layer thickness of 5 nm, the upper limit of volume solids is probably not over 70 or at most 75 NVV.

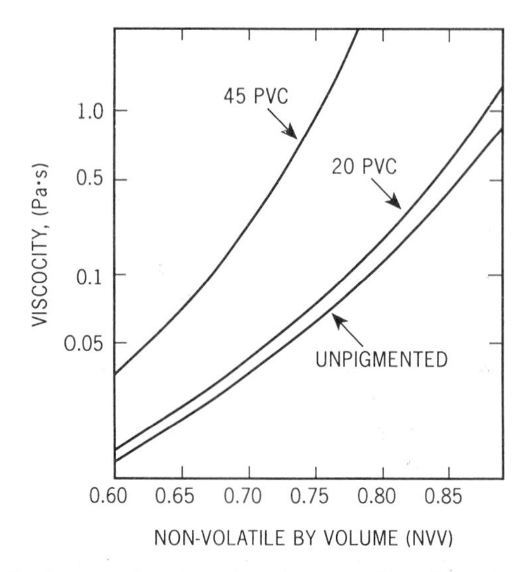

Figure 28.1. Effect of pigmentation on viscosity as a function of volume solids for an unpigmented coating and for two pigmented coatings based on the same binder with pigment loadings sufficient to give 20% PVC and 45% PVC in the dry films. See Ref. [1] for the assumptions made in the calculations. (Adapted from Ref. [1], with permission.)

As a result of such limitations, the major thrust of development of reduced VOC primers has been on water-borne primers. These are discussed in Chapter 29 and in Chapter 30, which covers electrodeposition coatings.

28.2. TOP COATS

Top coats include coatings for use both over primers and directly on the substrate. In the former case, the primer provides the adhesion to the substrate and a major part of the corrosion protection for coatings on metal. The top coat need only adhere well to the primer and provide the appearance and other desired properties. Coatings used as a single coat must combine both functions. In general, it is preferable to use a primer/top coat system. However, one-coat application can be fully functional for many applications and is usually substantially less expensive than multiple-coat systems. Single coats are widely used on products that need little corrosion protection and where the need for maintaining adhesion in the presence of water is not critical. In other cases where appearance and exterior durability requirements are minimal, a primer with excellent corrosion protection properties may be used without a top coat, for example, inside ballast tanks of ships and for the interior of structural components of aircraft.

28.2.1. Binders for Top Coats

To an important degree the properties of top coats are controlled by the class of resins used as the principal binder in the top coat. The chemistry of these various binders is discussed in Chapters 4–13; in this chapter, the advantages and disadvantages of the various classes of resins for top coats especially for OEM product coatings on metal are compared.

28.2.1.1. Alkyds

In the 1930s and 1940s, alkyds (see Chapter 10) were the major binders for coatings. While being increasingly replaced by other binders, alkyds are still used on a large scale. It seems appropriate to use them as a base to which to compare all other binder systems. A major advantage of alkyds in many applications is lower cost. Not only are the resin solids generally low in cost but also relatively low cost hydrocarbon solvents can be used as the only, or at least the major, solvent. A second major advantage is that application of alkyd coatings is, in general, the most foolproof. Solvent-borne alkyd coatings are least subject to film defects of all the classes of coatings. This advantage results from the low surface tension of most alkyd coatings. Therefore, there are seldom problems with crawling, cratering, and other defects resulting from surface tension driven flows or surface tension differential driven flows (see Section 23.3). Also, in general, it is relatively easy to make pigment dispersions in alkyds that do not flocculate. A further advantage is their ability to cross-link by autoxidation. This makes air dry or low temperature baking readily possible and avoids the need for cross-linking agents with potential toxic hazards.

The major limitations, which vary in degree depending on the alkyd and other coating components, are comparatively poor color retention on baking and limited exterior durability. It is also difficult to achieve very high solids in alkyd solvent-borne coatings (see Section 10.2). While water-borne alkyd coatings (see Section 10.3) are available, their hydrolytic stability is limited and air dried films generally show relatively poor water resistance. A further problem can be generation of smoke in baking ovens that can cause visual air pollution problems in some cases.

As discussed in Chapter 10, there are two broad classes of alkyd coatings: those that cross-link only through oxidation of unsaturated fatty acid ester moieties and those in which at least part of the cross-linking occurs by the reaction of free hydroxyl groups on the alkyd with a cross-linker such as a melamine–formaldehyde (MF) resin or a polyisocyanate.

With metal salt driers, oxidizing alkyd coating films air dry in a few hours, force dry at 60–80°C in an hour, or bake at 120–130°C in half an hour. Baking time is shortest with an oil length of about 60 with highly unsaturated drying oil derived alkyds. Tung oil modified alkyds give the fastest cross-linking, followed in turn by mixed tung/linseed oil alkyds, and then linseed oil modified alkyds. Cured films from these alkyds are quite yellow and turn yellow brown on overbaking. Generally, they are only used for dark color coatings. Since the cured films still contain substantial amounts of unsaturation and metal driers, the films embrittle with age. Somewhat better color and color retention can be obtained using alkyds made with fatty acids or oils such as soy or tall oil fatty acids that are less unsaturated. These alkyds are also less expensive, but cure somewhat more slowly than the more highly unsaturated alkyds.

Oxidizing alkyd binders are used for low performance, low cost coatings such as for steel shelving, machinery, coat hangers, and exterior coatings for drums. Individual applications may be relatively small, but there are a host of products where the performance of these coatings is adequate. They are also used for a significant fraction of architectural gloss enamels (see Section 35.3).

Medium oil alkyds have a substantial number of hydroxyl groups on each molecule; these functional groups can be cross-linked with a wide variety of cross-linkers. Since MF cross-linkers are generally of lowest cost, they are the most widely used. In order to achieve compatibility, butylated, isobutylated, or octylated MF resins are used, not methylated ones. Generally, the more reactive, Class II MF resins with high NH content are used (see Chapter 6). The solvent system should contain some butyl alcohol to promote package stability and to reduce viscosity as compared to straight hydrocarbon solvents. Usually no catalyst need be added since the residual unreacted carboxylic acid groups on the alkyd are sufficient catalyst for the class II MF resin. If needed, relatively weak acid catalysts such as an alkylphosphoric acid can be used.

Speed of cure and color is affected by the type of fatty acid used in the alkyd. Highly unsaturated alkyds are seldom used in top coats because of their dark color but soy, tall oil, and dehydrated castor alkyds are widely used. Some of the cross-linking results from the unsaturation and the balance from the reactions of the MF resin. The color, color retention, and resistance to embrittlement are significantly better than obtained using the oxidizing alkyd without the MF resin since the MF cross-linking minimizes or eliminates the need for metal driers to catalyze the oxidation reactions.

The best color, color retention, and exterior durability are obtained using non-oxidizing, saturated fatty acid based alkyds. For example, coconut oil based alkyds with MF resin are still used by one automobile manufacturer for solid color exterior top coats. To minimize degradation due to hydrolysis under acidic conditions, alkyds made with isophthalic acid (or with hexahydrophthalic anhydride) are used rather than those from phthalic anhydride. While the exterior durability is not as good as the best acrylic coatings, it is still adequate. The resulting coating is said to exhibit a greater appearance of depth than is obtained with acrylic/MF coatings. The cost of these coatings is similar to, or even higher than, polyester–melamine coatings, but film defects during application are less of a problem.

The hydroxyl groups on alkyds can also be cross-linked at room temperature or under force dry conditions with polyisocyanates. Both IPDI isocyanurate and TMXDI/TMP prepolymers are often used for this purpose since the color stability and exterior durability are superior to TDI derivatives (see Section 12.3.2). In many cases, the use of the polyisocyanate with an oxidizing alkyd permits sufficiently rapid development of tack-free coatings so that metal driers are not needed; this further improves the exterior durability.

28.2.1.2. Polyesters

Polyester resins (see Chapter 8) have been one of the major classes of resins replacing alkyd resins in MF cross-linked baking enamels. Polyesters are also widely used in urethane coatings. Their costs are generally somewhat higher than oxidizing alkyds, but in some cases less than nonoxidizing alkyds. Color, color retention, exterior durability, and resistance to embrittlement are better than obtained with most alkyds, but exterior durability and resistance to saponification are generally not as good as obtained with acrylics. Adhesion of polyester-based coatings without primers over clean, treated steel and aluminum substrates is comparable to alkyds and generally superior to acrylics. Polyester coatings generally have higher surface tensions than alkyd coatings, and hence are more subject to crawling and surface tension differential driven flow defects, such as cratering.

Most polyesters are hydroxy-terminated and are cross-linked with MF resins or with isocyanates. The MF resins are less expensive. Most commonly methylated or mixed methylated/butylated MF resins are used. Class I MF resins, that is, those with high degrees of alkoxymethylation, are the most widely used. For lower temperature cure systems or air dry systems, aliphatic polyisocyanates are used.

A major advantage of polyesters over alkyds and most acrylic resins is the relative ease of preparing polyester resins suitable for very high solids coatings. Even low molecular weight hydroxy-terminated polyesters can be synthesized that have at least two hydroxyl groups on virtually every molecule (see Sections 8.2 and 28.2.2).

Other polyesters are carboxylic acid-terminated. These can be cross-linked with MF resins, but are more commonly cross-linked with epoxies or with hydroxy-alkylamides. The largest uses for these polyesters are in powder coatings (see Chapter 31).

28.2.1.3. Acrylics

In very general terms, the major advantages of acrylic binders are their low color, excellent color retention, resistance to embrittlement, and exterior durability at relatively modest costs. They are quite photochemically stable and very hydrolytically stable. In general, their surface tensions are intermediate between alkyds and polyesters and, as a result, the susceptibility of acrylic based coatings to film defects is intermediate. Generally, their adhesion to metal surfaces is inferior to both alkyds and polyesters; therefore, they are usually used over a primer.

Thermoplastic solution and NAD acrylic resins (see Section 4.1) were widely used in OEM automotive coatings but have had to be replaced with thermosetting acrylics (TSAs) (see Chapter 7) in order to reduce VOC emissions. Most commonly TSAs are hydroxy-functional with perhaps a minor amount of carboxylic acid functionality. They are cross-linked with MF resins or with polyfunctional isocyanates. Either Class I or Class II MF resins can be used; the choice usually depends on curing temperature requirements. Aliphatic isocyanate cross-linkers are more expensive than MF resins and present greater toxic hazards, but cure at lower temperatures and, with HALS light stabilizers, usually provide somewhat greater exterior durability and frequently have better acid rain spot resistance.

As discussed in Section 7.2, it is difficult to make very high solids acrylic resins because of the difficulty of assuring that substantially all of the low molecular weight molecules have at least two hydroxyl groups. While progress is being made in increasing solids, it is unlikely that acrylic resins will ever be made with as high solids as polyesters.

28.2.1.4. Epoxies and Epoxy Esters

While the major uses for epoxy resins (see Chapter 11) in solvent-borne coatings are in primers, significant volumes of epoxy-based top coats are used. Since epoxy-based coatings generally exhibit excellent adhesion to metal, especially in the presence of water vapor, as well as resistance to saponification, they are commonly used as single coats. For BPA and novolac based epoxy resins, a major limitation is their poor exterior durability. However, many applications, for example, beer and soft drink can linings, do not require exterior durability. Other epoxy resins are available which give films with greater exterior durability (see Section 11.1.2), and the use of a HALS additive to increase durability of a BPA based coating has been reported (see Section 25.1).

Epoxy esters offer properties intermediate between alkyds and epoxies. An example of their use in top coats is in coating bottle caps and crowns. They exhibit the requisite combination of hardness, formability, adhesion, and resistance to water. They are more resistant to saponification than alkyds and share their low surface tension and, hence, reduced probability of film defects during application.

28.2.1.5. Urethanes

Polyisocyanate cross-linkers have been mentioned in connection with alkyds, polyesters, and acrylics. They are widely used as cross-linkers because of low temperature curing and to achieve abrasion resistance combined with resistance to swelling of cured films with solvent. They are mentioned further here because urethanes can also be major backbone links in resins having reactive groups other than isocyanates. Hydroxy-terminated urethane resins (see Section 12.4.6) can be cross-linked with polyisocyanates or with MF resins. Hydroxy-terminated urethane resins can also be prepared by reactions not involving the use of isocyanates [2]. Compared to acrylics, urethanes show superior abrasion resistance and solvent swelling resistance, but, in some cases, they increase moisture sensitivity. The cost of light stable isocyanates is a limiting factor, but when MF resins are used for cross-linking, these systems have the advantages of being one package coatings without the concern for toxic hazards due to free isocyanate. Further, there is no difficulty making low molecular weight resins where substantially all of the molecules have a minimum of two hydroxyl groups. Compared to polyesters, the cost is higher but the hydrolytic stability and, therefore, exterior durability can be superior. However, the intermolecular hydrogen bonding of urethane groups, even in solution in hydrogen-bond acceptor solvents, is such that viscosity of urethanes is higher at equal concentration solutions of equal molecular weight resins.

28.2.1.6. Silicone and Fluorinated Resins

The highest oxidation resistance both to thermal degradation and photooxidation is obtained with silicones (see Section 13.3) and fluorinated resins (see Sections 4.2.3 and 13.4.3). The cost is high, especially for the fluorinated resins. Silicone alkyds give greater exterior durability than alkyds for air dry coatings. In a compromise to obtain excellent outdoor durability at an intermediate cost, silicone-modified polyesters and acrylics (see Section 13.3.2) are relatively widely used. The effects of the susceptibility of the cross-links to reversible hydrolysis can be minimized by using some MF resin as a supplemental cross-linker. Fluorinated copolymer resins cross-linked with MF resin or aliphatic polyisocyanates have outstanding durability.

28.2.2. VOC Considerations

Until the late 1950s virtually all industrial and special purpose coatings were solvent-borne as were a significant fraction of architectural coatings. As of 1992, a significant fraction of industrial and special purpose coatings and some architectural coatings are still solvent-borne. The purpose of this section is to consider the various types of top coats with a view of comparing them for the potential for reduction of solvent usage while still achieving the required level of performance.

Thermoplastic solution resins were formerly used in industrial top coats on a large scale. The coatings were based on nitrocellulose, thermoplastic acrylic resins, or thermoplastic chlorinated resins (vinyl chloride copolymers and chlorinated rubber). Thermoplastic coatings have many advantages. They can be dried under ambient conditions or baked. Since no cross-linking is required, package stability is seldom a problem. The viscosity of the lacquers is Newtonian and hence less subject to variation with minor variations in composition. The solvent mixtures can be adjusted so that there is seldom a problem in controlling leveling, sagging, popping, or wrinkling. The film properties are not significantly affected by relatively large variations in baking temperatures. While lacquer films are commonly baked to permit rapid and essentially complete removal of solvent and to achieve leveling, no cross-linking reactions are occurring that could vary with temperature. The applied coatings can be readily repaired if they are damaged since it is easy to achieve adhesion of a lacquer to a dried lacquer film. However, lacquers are applied at volume solids less than 15%, commonly 10–12%, corresponding to weight solids of less than 20%.

Nitrocellulose lacquers are still used in wood furniture finishing as discussed in Section 34.1. Some acrylic lacquers are still used in automotive refinishing as discussed in Section 36.3. Thermoplastic solution vinyl and chlorinated rubber coatings continue to be used to some extent in swimming pool paints and heavy duty maintenance coatings discussed in Section 36.1.

Nonaqueous dispersion (NAD) acrylic resin (see Section 4.1.2) lacquers were used extensively for automotive top coats starting in the early 1970s until they were phased out in the mid 1980s. These permitted application at NVVs of 27, or somewhat higher—a significant reduction in VOC emissions as compared to solution lacquers but still high compared to thermosetting high solids solution or water-borne coatings.

A third type of solvent-borne thermoplastic coating system is organosols (commonly, but not entirely properly, called plastisols). As discussed briefly in Section 4.2.2, organosols are composed of dispersions of vinyl chloride copolymer particles in a solvent solution of plasticizer. Weight percent solids of organosols are high, but the high density of the chlorinated resin means that the grams of VOC emission per liter of coating is still fairly high. If the T_g of the polymer is too low or if the solvent or plasticizer is too readily soluble in the polymer particles, the particles slowly swell. The resulting increase in the volume fraction of internal phase increases the viscosity. To maintain application viscosity, the T_g of the copolymer should be high enough so that the particles are not significantly swollen by the plasticizer or solvent. Swelling may also be retarded by the partially crystalline structure of the vinyl chloride copolymers. However, the high T_g and slow dissolving in plasticizer mean that the baking temperature must be quite high. The thrust to lower baking temperatures has now limited the use of organosols almost entirely to coil coating applications where the baking temperatures must be high anyway (see Section 33.4).

The majority of solvent-borne coatings being applied and to be expected in the future are solution thermosetting coatings. Conventional thermosetting coatings have volume solids on the order of 25–35%. Technical effort on *high solids coatings* since around 1970 has accomplished varying degrees of increases in solids of various types of coatings for various applications.

People ask for a definition of high solids coatings; there is no single definition. For automotive top coats in 1992, high solids means about 45 NVV. For a highly pigmented primer, high solids might be 50 NVV. For clear or high gloss pigmented coatings with modest performance requirements, 75 NVV or even higher may be possible. The situation is further complicated by the difficulty of exactly defining or determining the VOC of a coating. For example, in some cases solvents with functional groups may partially react with the cross-linker and hence not be evolved and there can also be volatile byproducts of cross-linking which must be included in VOC; the extent to which such emissions occur may vary with baking conditions. See Section 15.7.1 for a discussion of determination of VOC.

One limitation on solids content is the increasing difficulty in achieving desired mechanical properties as molecular weight and average functionality per molecule, $\bar{f}_n$, are decreased and molecular weight distribution is narrowed. As discussed in Section 7.1, in a conventional thermosetting acrylic coating, one uses a resin with a $\overline{M}_w/\overline{M}_n$ of the order of 35,000/15,000 and, depending on the end use, perhaps $\bar{f}_n$ of 10–20 hydroxyl groups per molecule. The cross-linker might have a $\overline{M}_w/\overline{M}_n$ of the order of 2000/800 and an $\bar{f}_n$ of 3–7 functional groups per molecule. An acrylic resin that permits 45 NVV might have a $\overline{M}_w/\overline{M}_n$ on the order of 8000/3000 with $\bar{f}_n$ of 3–6 hydroxyl groups per molecule. As $\bar{f}_n$ is reduced it becomes increasingly critical and difficult to control formulations and cure conditions to keep all film properties within the desired ranges.

As one goes to still lower molecular weights and functionalities for still higher solids coatings, achieving high performance properties becomes still more difficult. For NVV of 70, the $\overline{M}_w/\overline{M}_n$ must be 2000/800 or less with an average of little over two hydroxyl groups per molecule. It is critical that essentially all molecules have at least two functional groups per molecule. Any molecules with only one functional group cannot cross-link and will leave dangling ends in the network; any molecules with no functional groups will be plasticizers and, if low enough in molecular weight, may be partially volatilized in a baking oven. It is difficult to synthesize the required acrylic resins for very high solids coatings (see Section 7.2). On the other hand, it is relatively easy to make polyester or polyurethane resins with $\overline{M}_w/\overline{M}_n$ of 2000/800, or even lower, and an average of a little over two hydroxyl groups per molecule where essentially all of the molecules have at least two hydroxyl groups (see Sections 8.2 and 12.4.6).

In developing a formulation for an MF cross-linked, very high solids coating, the formulator would evaluate a series of ratios of one or more MF resins to polyester to arrive at an appropriate ratio, type of MF resin, catalyst, and catalyst concentration which provides the hardness and flexibility required for the application in the baking time and temperature specified by the customer. This is the same procedure used in conventional coatings. However, in conventional coatings, the *window of cure* is relatively large. That is, it makes little difference if the baking temperature is off by ±10°C, or the baking time is off by ±10%, or if the customer is a little sloppy in adding the catalyst and gets 10% over or under the recommended amount.

It has been found in the case of high solids coatings that the window of cure is narrower [3]. If there are a large number of hydroxyl groups on each resin molecule and 10% are not reacted, the change in properties may be small. If, however, there are only a little over two hydroxyl groups per average molecule and 10% are left unreacted, a significant fraction of the molecules will only be tied into the network once and there can be important effects on mechanical properties.

The problem can be minimized by using cross-linkers with higher $\bar{f}_n$; at least the problem only arises then from the polyester (or other resin) not from both. Class I MF resins, because of their greater average functionality compared to Class II MF resins, generally offer broader cure windows, as discussed in Section 6.3.1.4. The extent of self-condensation reactions of MF is particularly dependent on time, temperature, and catalyst concentration.

In using high solids coatings, the applicator should be more careful in controlling the time and temperature in baking ovens and in following the coating supplier's recommendations. The formulator must be more careful in checking the film properties when the temperature is about 10°C above and below the standard temperature. In making recommendations of cure cycles to a customer, one should use pieces of the customer's metal for establishing the baking schedule. The critical temperature in baking is that of the coating itself, not that of the air in the oven. Coatings applied to a heavy piece of metal will heat up more slowly in an oven than coatings on light gauge sheet metal. On sheet metal, the coating over a place where the sheet metal has been welded to a supporting member will heat up more slowly than on the rest of the surface. In conventional coatings, variations resulted from such differences are generally small; high solids coatings are much more likely to be subject to variation in properties owing to differences in coating temperature and time.

There is increasing evidence that breadth of distribution of molecular weight and composition can be important factors affecting the mechanical properties of films [4]. This is presumably related to the breadth of the transition region, as exhibited for example in the breadth of a tan delta peak (see Section 24.2). To increase solids, polydispersity of resins can be decreased. However, this may exacerbate the effect of low $\overline{M}_n$ and low $\bar{f}_n$ on mechanical properties. An approach to overcome this problem may be to blend resins with differences in composition, but enough similarity to be compatible. For example, polyester/acrylic blends using low molecular weight hydroxy-terminated polyesters reduce the viscosity and thereby increase the solids [5]. Very low molecular weight components must be avoided since they can volatilize in a baking oven.

Resins and cross-linkers for high solids coatings generally have $\overline{M}_n$ values well below 5000. In contrast to high molecular weight polymers, the entropy of mixing of different low molecular weight resins is large enough to be a significant factor favoring compatibility (see Section 2.1.3). The effect of molecular weight is illustrated by the compatibility of 50/50 blends of several acrylic and methacrylic homopolymers when their $\overline{M}_n$ values are less than 5000, but incompatibility when $\overline{M}_n$ values are above 10,000 [6]. The broader compatibility of low $\overline{M}_n$ resins permits formulation of high solids coatings based on mixtures of different generic type resins. Of course, not all blends of low $\overline{M}_n$ resins are compatible and there is the possibility that phase separation could occur as molecular weight increases in the early stages of cross-linking; compatibility should be checked by examining films made from unpigmented coatings.

In the case of high solids, two package urethane coatings, substantial research is being done. One approach has been to synthesize isocyanurates and biuret polyisocyanates (see Section 12.3.2) with lower molecular weights and hence lower viscosities. It has been shown, however, that the lower viscosity of these polyisocyanates is only one factor affecting the VOC of coatings [7]. In some cases, the lower viscosity cross-linker also has a lower equivalent weight; this means that the ratio of low viscosity cross-linker to much higher viscosity polyol is decreased. It was demonstrated that the lowest VOC could be obtained using a somewhat higher equivalent weight polyisocyanate even though it has a somewhat higher viscosity. Current research is aimed at synthesizing polyisocyanates having both low viscosity and higher equivalent weight. Another approach to decreasing VOC of two package polyurethane coatings is to replace some of the hydroxy-terminated acrylate or polyester with a reactive diluent [7]. Examples of such coreactants are aliphatic diols, low molecular weight ester or urethane diols, ketimines, and hindered diamines.

Recent studies of the direct reaction of ketimines with isocyanates have identified another product of the reaction different from those reported earlier in Section 12.1. It has been shown that the major product of the reaction of butyl isocyanate and the ketimine derivative of cyclohexylamine and methyl isobutyl ketone is an azetidinone [8].

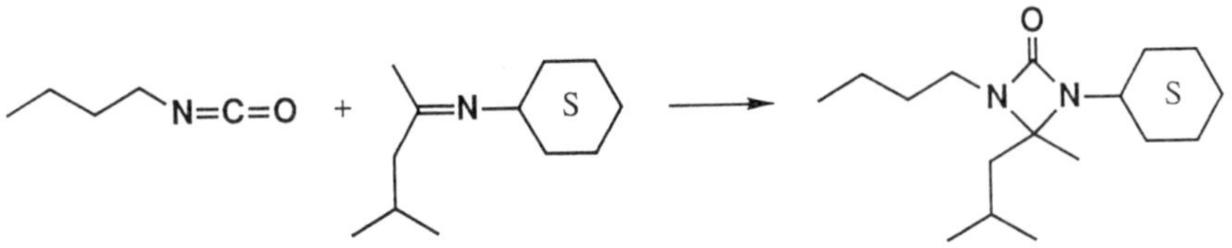

Further it has been found that the direct reaction of isocyanates with ketimines is catalyzed by carboxylic acids and other weak acids [8]. This catalysis makes it feasible to design coatings with adequate pot life combined with sufficiently rapid cure for applications such as refinish automobile top coats. It was shown that the volatile losses from the films were intermediate between those predicted from only direct ketimine-isocyanate reaction and those predicted from hydrolysis of the ketimine with water followed by reaction between the amine generated and isocyanate. Presumably both reactions occur during the drying of coating films.

As discussed in Section 12.1, most amines are too reactive with isocyanates to be useful in coatings. However, hindered amines have been developed that react at slow enough rates to be useful in two package coatings, as exemplified by aspartic acid esters [9, 10]. The esters are synthesized by reaction of difunctional primary amines with maleic or fumaric esters.

The reactivity of the aspartic esters depends upon the structure of the diamine from which they are derived. Viscosities and gel times, when mixed with equal

Table 28.1. Viscosities (mPa·s at 23°C) and Gel Times of Substituted Aspartic Acid Ethyl Esters[a]

R	Viscosity	Gel Time
$(CH_2)_6$	150	<5 min
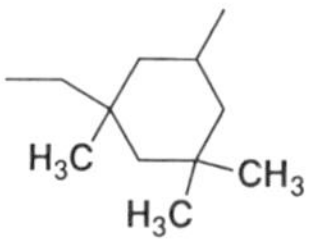 H_3C, H_3C, CH_3 (trimethylcyclohexyl-methylene structure)	800	2–3 hr
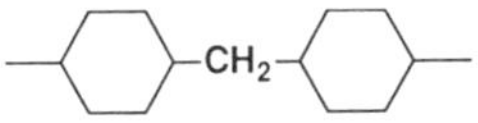 $-C_6H_{10}-CH_2-C_6H_{10}-$	1000	2–3 hr
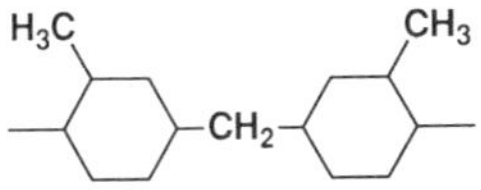 H_3C, CH_3; $-CH_2-$ (dimethyl-substituted structure)	1500	>24 hr

[a]The values given are when the esters are mixed with equal equivalents of a hexamethylene diisocyanate isocyanurate cross-linker at 65% solids in butyl acetate and hiflash naphtha.

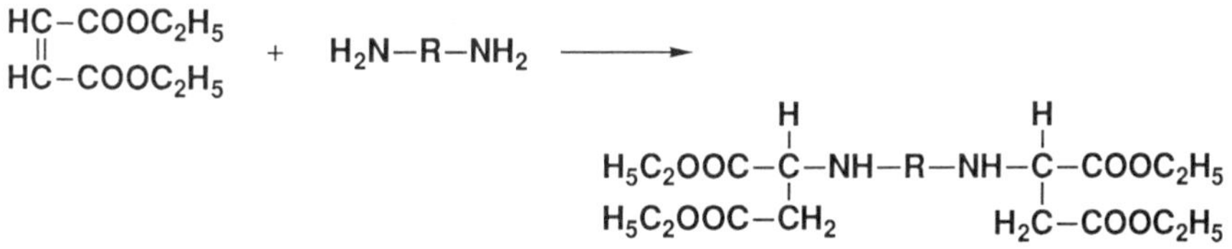

$$\begin{matrix} HC-COOC_2H_5 \\ \| \\ HC-COOC_2H_5 \end{matrix} + H_2N-R-NH_2 \longrightarrow \begin{matrix} & H & & H & \\ H_5C_2OOC- & C & -NH-R-NH- & C & -COOC_2H_5 \\ H_5C_2OOC- & CH_2 & & H_2C & -COOC_2H_5 \end{matrix}$$

equivalents of a hexamethylene diisocyanate isocyanurate cross-linker, of a series of substituted aspartic acid ethyl esters made with different diamines are given in Table 28.1 [9]. It has also been found that addition of dibutyltin dilaurate increases the pot life without a significant effect on the time required to obtain a hard film [10]. No mechanism to explain this unusual discovery has yet been published.

Pigment flocculation is more commonly encountered in high solids coatings; as molecular weight is decreased, the probability of pigment flocculation increases [11]. As discussed in Section 20.1.3, the primary factor controlling stabilization of pigment dispersions is the thickness of the adsorbed layer on the surface of the pigment particles. As molecular weight goes down, the adsorbed layer thickness should be expected to decrease. Furthermore, as the number of functional groups per molecule decreases, solvent can compete more effectively for adsorption sites on the pigment surface promoting flocculation. There is greater need for either surface treatment of pigment or of carefully designed additives that are strongly or irreversibly adsorbed on the pigment surface so as to stabilize against flocculation with the thinnest possible layer with little of the additive in the solution phase (see Section 20.3).

Another limiting factor for some high solids coatings is surface tension effects. Generally speaking, as molecular weight gets lower, the equivalent weight must

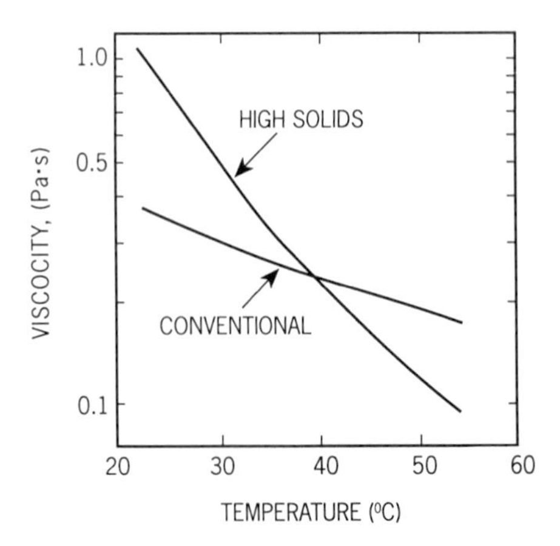

Figure 28.2. Viscosity as a function of temperature for a conventional and a high solids resin solution. The high solids solution is of a 1500 molecular weight polyester at 90% solids in ethylene glycol monoethyl ether acetate. The conventional solution is of a 20,000 molecular weight polyester at 25% solids in methyl ethyl ketone and ethylene glycol monoethyl ether acetate. (Adapted from Ref. [1], with permission.)

be lowered even further since the number of reactions required to achieve high molecular weight will be greater. In most coatings, the functional groups are highly polar groups, hydroxyl and carboxylic acid groups. Increased levels of such groups give higher surface tensions. Furthermore, achieving high solids at a given viscosity with such resins generally requires using hydrogen-bond acceptor solvents rather than hydrocarbon solvents. Again, this results in higher surface tensions than with most conventional coatings, and, therefore, increased probability of film defects during application.

In coating metal, it becomes more critical to have clean surfaces. In coating plastics, one must be careful to remove mold release, and a broader range of plastic materials have to be surface treated so as to avoid crawling and permit achieving adhesion (see Sections 26.5 and 34.3). Since the surface tension of freshly applied high solids coatings is generally higher than with conventional coatings, a larger fraction of contaminating particles floating in the air have lower surface tension than that of the wet high solids coating. Cratering is, thus, more probable (see Section 23.3).

Control of sagging is more of a problem with high solids than with conventional coatings [12]. As discussed in Section 23.2, this results from the slower rate of evaporation of solvents from high solids coatings as compared with conventional coatings. While the reasons for this difference have not been completely elucidated, the consequences pose significant problems. Sagging of spray applied high solids coatings cannot be easily controlled by adjusting the evaporation rate of the solvent in the coating or by changing the distance between the spray gun and the substrate. The sagging can be minimized by using hot spray (see Section 22.2.4) or supercritical fluid spray (see Section 22.2.5).

For many applications, thixotropic flow properties must be built into a high solids coating for spray application. Fine particle size SiO_2, bentonite clay pigments, zinc stearate, and polyamide gel thixotropes are additives that can be useful. The problem of sagging in metallic coatings used on automobiles is particularly serious; microgel particles have been developed which are effective (see Sections 23.2 and 33.1.3.1).

Even when sagging is not encountered during application, it can sometimes occur with high solids coatings when the coated object is put in the baking oven. The temperature dependence of viscosity of high solids coatings is greater than for conventional coatings, as illustrated in Figure 28.2. Before the cross-linking reaction has proceeded far enough to increase the viscosity, the temperature has decreased the viscosity sufficiently to cause sagging [1,12]. The phenomenon is called *oven sagging*.

Thus, while solids can be significantly increased, in many cases there are limitations in spite of two decades of intensive effort. Further progress is to be expected, but it will probably come slowly.

GENERAL REFERENCE

L. W. Hill and Z. W. Wicks, Jr., *Prog. Org. Coat.*, **10**, 55 (1982).

REFERENCES

1. L. W. Hill and Z. W. Wicks, Jr., *Prog. Org. Coat.*, **10**, 55 (1982).
2. W. J. Blank, *Prog. Org. Coat.*, **20**, 235 (1992).
3. D. R. Bauer and R. A. Dickie, *J. Coat. Technol.*, **54** (685), 101 (1979).
4. S. L. Kangas and F. N. Jones, *J. Coat. Technol.*, **59** (744), 99 (1987).
5. L. W. Hill and K. Kozlowski, *J. Coat. Technol.*, **59** (751), 63 (1987).
6. J. M. G. Cowie, R. Ferguson, M. D. Fernandez, M. J. Fernandez, and I. J. McEwen, *Macromolecules*, **25**, 3170 (1992).
7. S. A. Jorissen, R. W. Rumer, and D. A. Wicks, *Proc. Water-Borne Higher-Solids Powder Coat. Symp.*, New Orleans, 1992, p. 182.
8. E. P. Squiller and D. A. Wicks, *Proc. Water-Borne, Higher-Solids, Powder Coat. Symp.*, New Orleans, 1993, p. 15.
9. C. Zweiner, L. Schmalstieg, M. Sonntag, K. Nachtkamp, J. Pedain, *Farbe Lack*, **97**, 1052 (1991).
10. D. A. Wicks and P. E. Yeske, *Proc. Water-Borne, Higher-Solids, Powder Coat. Symp.*, New Orleans, 1993, p. 49.
11. S. Hochberg, *Proc. Water-Borne, Higher-Solids Coat. Symp.*, New Orleans, 1982, p. 143.
12. D. R. Bauer and L. M. Briggs, *J. Coat. Technol.*, **56** (716), 87 (1984).

CHAPTER XXIX

Water-Borne Coatings

Before about 1950, almost all coatings were solvent-borne. The introduction of latex paints for architectural uses was the first major step away from solvent-borne coatings. Since 1970, there has been a further trend away from solvent-borne coatings because of higher solvent prices and particularly the need to reduce VOC emissions to meet air quality standards. Powder coatings (Chapter 31) and radiation cure coatings (Chapter 32) have reached significant commercial volumes. Broader use of water-borne coatings has been the major trend in coatings. It is probable that the volume of water-borne coatings used in the United States in 1992 is similar to the volume of solvent-borne coatings. The use of water-borne coatings can be expected to increase further as restrictions on VOC get more stringent. As of 1992, almost all water-borne coatings contain some volatile organic solvents. The solvents play a variety of important roles in resin manufacture, coating production, application, and film formation. Substantial research efforts are being exerted to reduce and, in at least some cases, eliminate the need for solvents.

The two largest classes of water-borne coatings are *water-reducible* coatings (Section 29.1) and *latex* coatings (Section 29.2). In addition, modest amounts of *emulsion* coatings (Section 29.3) are used. There is also growing use of blends of different types and of binders that are not easily classified. The terminology is not uniform. Some authors call latex coatings, emulsion coatings; we recommend against this practice to avoid confusion of latexes (dispersions of solid polymer particles in water) with coatings that really are emulsions (dispersions of liquids in liquids). Many authors speak of *water-soluble* resins, but because these resins are not really soluble in water, we are using the term *water-reducible* resins (see Section 7.3 for a discussion of the morphology of these resins after dilution with water). These coatings are sometimes called *microemulsion* coatings. Sometimes the term *aqueous dispersion resins* is used, and may cover the same type of resins we include in the water-reducible category, but some kinds of latexes are also called aqueous dispersion resins.

The properties of water are very different from those of organic solvents; this leads to distinct differences in characteristics of water-borne coatings as compared to solvent-borne coatings. Some of these differences are advantageous. For example, water presents no toxic hazard and it is odor-free. Water is not flammable; this reduces risks and insurance costs. The lack of flammability permits the use of less make-up air in baking ovens and hence reduces energy consumption in some cases. There are no emission or disposal problems directly attributable to the use of water. With some formulations cleanup of personnel and equipment is easier with water-borne coatings; however, in other cases cleanup is more difficult. The cost of water is low—it does not, however, follow necessarily that the cost of water-borne coatings is low.

On the other hand, there are some important disadvantages to the use of water. Major problems with water-borne coatings of any kind result from the singular evaporation characteristics of water. Whereas a wide range of solvents with different evaporation characteristics enables the formulator to fine tune evaporation rates from solvent-borne coatings (see Section 15.1), there is only one kind of water. The heat capacity and heat of vaporization of water are high, resulting in high energy requirements for evaporation. With a given amount of energy available for evaporation, water evaporates more slowly than solvents with similar vapor pressures. At 25°C the relative evaporation rate of water is low; however, the vapor pressure increases with increasing temperature relatively rapidly. A problem, unique to water, is that the evaporation of water is affected by relative humidity (RH). As discussed in Section 29.1, variations in RH when coatings are applied can lead to major problems.

The surface tension of water is higher than that of any organic solvent. In latex paints surfactants must be used to reduce surface tension so as to wet pigments or so that a coating can wet many kinds of surfaces. The presence of surfactants tends to give films with poorer water resistance. In the case of water-reducible coatings, pigments can be dispersed in solvent solutions of the resins before addition of water and the surface tension can be reduced by the use of some organic solvent such as butyl alcohol or a butyl ether of ethylene or propylene glycol in the formulation.

A further problem with water-borne coatings is that the water tends to increase corrosion of storage tanks, paint lines, ovens, and so forth. This requires that corrosion resistant equipment be used in factories applying water-borne coatings, increasing the capital cost. For example, mild steel lines may have to be replaced with stainless steel.

The majority of water-reducible resins are used in OEM product coatings with limited applications in special purpose coatings. One class of water-reducible coatings—electrodeposition coatings—has such different formulation and application procedures from other coatings that we have chosen to discuss them separately in Chapter 30. Latex systems are used in a large majority of architectural coatings; while the general principles of latex coatings are covered in Section 29.2, detailed discussion is postponed to Chapter 35 on architectural coatings. Small but increasing fractions of both OEM product coatings and special purpose coatings are latex based.

29.1. WATER-REDUCIBLE COATINGS

As has been discussed in Section 7.3 on water-reducible acrylics, the resins in these water-reducible coatings systems are not soluble in water, although they are sometimes called water-soluble resins. Most are made as high solids solutions in water-miscible solvent(s) of resins with either carboxylic acid groups that are at least partially neutralized with low molecular weight amines or with amine groups that are partly neutralized with low molecular weight acids. Pigment is dispersed in the partly neutralized resin solution, cross-linker is added along with additives such as catalysts, and the coating is diluted with water to application viscosity.

The resin is not soluble in the aqueous solution of the solvent that results. However, the presence of the salt groups prevents precipitation of the resin in a macrophase. Rather, aggregates form with the salt groups oriented at the water–particle interfaces and with the low polarity parts of the resin molecules in the interior of the particles. The solvent partitions between the water phase and the particles. The particles are swollen by this solvent as well as by substantial amounts of water that associate with the salt groups and dissolve in the solvent. Generally the cross-linker is either partitioned or is entirely dissolved in the resin–solvent particles. The pigment particles are also inside the aggregate particles. Not only are acrylics used, but also a variety of other water-reducible binders: alkyds (see Section 10.3), polyesters (see Section 8.4), epoxy resins (see Sections 11.2.6, 11.4.1, and 11.6), and urethanes (see Section 12.4.7).

When solutions or coatings of these neutralized resins are diluted with water, the change of viscosity with concentration is abnormal. See Figures 7.1–7.3 for typical relationships between resin concentration and log viscosity when water-reducible acrylic resin solutions are diluted with water. As water is added, there is an initial rapid drop in viscosity. As more water is added, the viscosity plateaus and, frequently with further addition of water, the viscosity increases. Still further addition of water leads to a very rapid drop in viscosity. The solids volume of the system is quite low at application viscosity. While the viscosity of the organic solvent solution of resin is Newtonian, the viscosity of the water-diluted system in the region of the peak or plateau is highly shear thinning. The system as diluted to application viscosity is usually only slightly shear thinning and may be Newtonian. As explained in Section 7.3, this behavior is consistent with the formation of swollen aggregates.

As also explained in Section 7.3, the pH of these systems is also abnormal. In the case of carboxylic acid-functional resins neutralized with a low molecular weight amine, the pH is basic even though less than stoichiometric amounts of amine are generally used. The viscosity at application solids decreases rapidly as water is added and is very dependent on the ratio of amine to carboxylic acid. If a small excess of water is inadvertently added, the viscosity may be too low. However, this problem can usually be remedied by adding a small further increment of amine, which will increase the viscosity.

Selection of the amine for neutralization is an important formulation consideration. Amines are expensive and they add to VOC emissions after application of the coating. If there is insufficient amine, there will be macrophase separation of the coating. Therefore, it is desirable to select amines that provide for the necessary stability of the aggregate dispersion at the lowest possible concentration. The principal factor controlling the ratio of amine to carboxylic acid needed appears to be water solubility (polarity). For example, the amount of tripropyl amine required is substantially larger (on an equivalent weight basis as well as a weight basis) than the amount of triethyl amine required. Still more efficient are aminoalcohols; *N*,*N*-dimethylaminoethanol (DMAE) is probably the most widely used amine. Somewhat less efficient, but still more effective than trialkyl amines, are morpholine derivatives.

Since, as discussed in Section 28.1.3, reduction of VOC levels in solvent-borne primer formulations is particularly difficult, water-borne primers are particularly important. For example, maleated epoxy esters (see Section 11.4.1) are used for spray or dip applied primers for steel. All electrodeposition coatings, including the very large volume of primer for automobiles, are based on water-reducible resin systems (see Chapter 30). Interior linings for beverage cans are also water-reducible coatings, based on grafting acrylate ester, styrene, and acrylic acid onto BPA epoxy as described in Section 11.5. Water-borne coatings are especially suited for dip coating applications (see Section 22.3) because they eliminate the severe fire hazards of solvent-borne dip coatings. In dip coating, beads or tears tend to form at the lower edges or corners of objects as the coating drains after removing from the dip tank. To a degree, this can be minimized by electrostatic detearing.

Many kinds of water-reducible resins have been developed. The most widely used have been acrylics with both carboxylic and hydroxy functionality. While water-reducible alkyds and polyesters can be made and are used to a degree commercially, their use is limited because of the difficulty of achieving adequate saponification resistance for package stability. Water-reducible urethanes have excellent saponification resistance and give films with excellent properties, but they are generally more expensive than acrylic resins. In cationic electrodeposition primers discussed in Chapter 30, water-reducible amine-functional epoxy resins are the most widely used binders.

As in any other system, coatings made with water-reducible resins have advantages and limitations. An important advantage follows from the fact that the molecular weight of the resins used can be as high as that of resins used in conventional solution thermosetting coatings. This is true because the viscosity at application dilution is almost independent of molecular weight. For example, one can use water-reducible acrylics with $\overline{M}_w/\overline{M}_n$ on the order of 35,000/15,000 with an average of 10 or so hydroxyl groups and 5 or so carboxylic acid groups per molecule along with a Class I MF cross-linker and achieve properties essentially equal to a conventional solution acrylic enamel. The window of cure is comparable to that of the conventional thermoset acrylic also. The problems related to molecular weight and functionality of high solids coatings are not encountered.

A disadvantage is that the application solids are low. Typically such coatings are applied by spraying, roll coating, or curtain coating at around 20–30 NVV. The low solids mean that more wet film thickness has to be applied in order to achieve the same dry film thickness. On the other hand, in automotive metallic coatings the low solids is an advantage in that it permits better orientation of the aluminum pigment in the film. Since only about 20% of the volatiles are organic solvents, the VOC emitted per unit volume of coating excluding water is fairly low. In a gloss water-reducible acrylic coating, VOC is equivalent to the amount of solvent emitted by a solution acrylic with approximately 60 NVV.

Water-reducible coatings with 20% of the volatiles as organic solvents have usually satisfied air pollution regulations in the United States. However, air quality goals in the future will probably require still lower VOC emissions. Research is underway for further reductions in VOC. In the case of acrylics and polyesters a challenge is to reduce the amount of solvent required in the polymerization or initial reduction stages. This can be done but, with current technology, material handling problems are severe. Particular progress has been made in reducing solvent content of polyurethane dispersions [1]. Reductions in organic solvent content are likely to amplify the already difficult application problems described below.

The viscosity of water-reducible resin coatings depends strongly on the ratio of water to solvent. Depending on the humidity there can be substantial changes in the ratio of the two components remaining in the coating as volatiles evaporate. If, as discussed in Section 15.1.6, the humidity is above a critical level, water will evaporate more slowly than even slow evaporating solvents such as 2-butoxyethanol. In some cases, the viscosity of the coating during flash off after application can decrease rather than increase as evaporation continues, leading to delayed sagging [2].

Variations in humidity can be a major problem. If the RH is over 70%, the rate of evaporation of water is very slow and, of course, at 100% there is no evaporation of water. In the mid-RH range, the effect can be reduced by a relatively modest increase in temperature, since RH decreases with temperature. However, if the humidity is very high, the only recourses have been to cool the air to condense out some of the water and then rewarm it, an expensive expedient, or to wait until the RH is lower. At very high humidities, water evaporation is so slow that required flash off times can become excessive. It is generally desirable to formulate for best application at a relatively high RH, say 55%, since it is less expensive to increase RH than to decrease it.

Another major problem when applying relatively thick coatings can be popping (see Section 23.6). Popping during the baking of water-reducible coatings is more difficult to control than with solvent-borne coatings [3]. The probability of popping increases as film thickness increases. Since the solids are low and the rate of evaporation of water during spraying and flash off is low, the wet film thickness required to apply the same solids is higher than with solvent-borne coatings. Furthermore, water evaporates more slowly during spraying, flash off, and in the oven than most solvents. A factor is the high heat of vaporization of water as compared to solvents. This leads to a slower rate of heating of the coating, hence slower evaporation. Also, until the amine evaporates, there are polar salt groups that tend to retain water. When the coated article enters the oven, water evaporates fastest from the surface so that the viscosity of the surface layer increases. Later, when the water remaining in the lower layers of the film volatilizes, some of the bubbles of water vapor either cannot break through the surface layer, leaving a blister, or break through at a stage when the viscosity of the surface layer is so high that the crater produced cannot flow out.

Popping can be minimized by spraying the coating in more, thinner coats (often called passes) so that there is a greater chance for evaporation as the film thickness is built up; by having longer flash off times before entering the baking oven; and by zoning the oven, so that the first part of the oven is at lower temperature, permitting the water to diffuse out of the film before the viscosity at the surface increases unduly. However, all of these actions have relatively high capital or operating costs.

The degree of the problem, of course, depends on the film thickness and the method of application. Thin coatings applied by roll coating or curtain coating almost never give a popping problem. With reasonable care about flash off times, curtain coating of relatively thick films can be accomplished without major problems. Water-reducible coatings are used on a large scale for such applications as can and panel coatings. The greatest difficulties occur when spray applying relatively thick coats. One cannot apply completely uniform films by spray. In order to assure that sufficient coating is applied in all areas, there will be some areas with more than the average film thickness. Also there is a greater chance of air entrapment with spray applied coatings. If the air bubbles do not break, the air in the bubbles expands when the coating is baked, finally leading to popping. Water-reducible coatings have been successfully applied as automobile topcoats on a large scale, but the economics were such that solvent-borne high solids coatings were favored. As discussed in Section 33.1.3.2, the newer base coat/clear coat systems are being successfully applied with one or both of the coatings being water-borne.

The formulator can minimize the problems of popping by using as low a T_g resin as is consistent with required film properties. Use of some slow evaporating solvents such as 1-propoxy-2-propanol and the monobutyl ether of diethylene glycol in the formulation assists in reducing the probability of popping.

At first blush, one might think that the high surface tension of water would cause serious problems with crawling and cratering. This is not the case, presumably because orientation of nonpolar segments of solvent and resin molecules to the surface is rapid, so that the actual surface tension is low.

In air dry coatings, equivalency of properties has not been achieved. Water-reducible alkyds can be, and are, made, as are modified alkyds such as silicone-modified alkyds. Their hydrolytic instability leads to limited storage life. With careful inventory control, they can be used for some industrial applications, but they are generally unsatisfactory for trade sales paints, where a shelf life of one or more years is needed. More seriously, until the neutralizing amine required for water dilution evaporates from the film, the films are very water sensitive. Even after the amine has evaporated, the residual carboxylic acid moieties lead to water and, particularly, base sensitivity. The time required to lose amine is related to both the volatility and the base strength of the amine. Ammonia is the most volatile amine and is widely used. However, it must be remembered that the last of the amine is being lost at a stage when volatile loss is being controlled by diffusion rate through free volume holes, not by volatility. In such cases, the time for amine loss may be reduced by using a combination of ammonia and a less volatile but relatively weakly basic amine such as *N*-methylmorpholine.

It has been assumed that water-borne urethane coatings involving free isocyanate groups would not be feasible due to the reactivity of isocyanates with water. However, it has now been demonstrated that two package (2K) urethane coatings with adequate pot life for commercial use can be formulated [4]. A water-dispersible aliphatic "hydrophilically modified" polyisocyanate was used in one package with the second package including a hydroxy-terminated, water-reducible polyurethane with carboxylic acid modification with dimethylolpropionic acid (see Section 12.4.7). A 2:1 ratio of N=C=O/OH was used to offset possible reaction with water. Films cross-linked within a week at 25°C when the RH was 55% or lower; at high humidity, solvent resistance did not develop; elevating the temperature to 31°C permitted cross-linking even at 80% RH. Presumably water-reducible acrylic resins can also be used in 2K coatings with the water-dispersible polyisocyanate. Considerable formulation remains to be done, but the principle has been established.

29.2. LATEX-BASED COATINGS

Latexes have been widely used for many years in architectural coatings and are by far the major type of vehicle for this class of coating. For major household applications, such as flat wall paint, the advantages of latex paints over any solvent-borne paint are so large that solvent-borne paints are seldom marketed. Important advantages for interior latex paints include rapid drying, avoiding the odor of solvent and oxidation byproducts of drying oils and alkyds, easier cleanup, reduced fire hazard, and better long-term retention of mechanical properties. For exterior paints a major advantage is that exterior durability of high performance latex paints is far superior to that of drying oil or alkyd paints. On wood siding there is also a major advantage of reduced blistering since the latex films are more permeable to water vapor. On the other hand, adhesion of latex paints to chalky surfaces is inferior to solvent-borne paints (see Section 35.1).

An important advantage of acrylic, styrene/acrylic, and styrene/butadiene latexes over alkyds is their excellent resistance to saponification. Latex paints generally show much better adhesion to galvanized metal surfaces than do alkyd paints. They also generally show much better performance over cement and concrete surfaces than alkyds since these too are alkaline surfaces. Also, for reasons discussed in Section 35.2, latex paints give better coverage over porous cement surfaces.

Latex coatings form films by coalescence of the polymer particles (see Section 3.3.1). Coalescence can occur only if the film formation temperature is higher than the T_g of the polymer particles. While initial coalescence proceeds rapidly at temperatures just a few degrees above T_g, the rate of completion of coalescence is relatively slow unless the temperature is significantly higher than T_g. In the case of most architectural coatings, slow final coalescence is not a real problem so that the T_g need be only a little below the film formation temperature. In the case of baked industrial coatings, film formation should be complete by the time the coated article comes out of the baking oven, therefore, baking temperatures have to be significantly above the T_g.

There are limitations to latex paints, particularly on how low the temperature can be while still allowing proper coalescence for film formation. In order to have a final film with a high enough T_g to resist blocking, it is common to use *coalescing solvents* in the formulation. The coalescing solvent dissolves in the polymer particles, reducing the T_g, and thus permitting film formation at a lower temperature. After film formation, the coalescing solvent slowly diffuses out of the film and evaporates. Even with the use of coalescing solvents, however, there are limitations on the temperatures required for good film formation (see Section 35.1).

In 1992 latex paints that are said to contain no coalescing solvents or other volatile organics were introduced into the U. S. market in response to pressures for further reduction of VOC. Paints with no VOC have been marketed in Sweden for several years. The technology involves very small particle size latexes, possibly with a mixture of latexes having different T_g values; details are proprietary.

Another limitation in the formulation of latex coatings is the difficulty in formulating high gloss latex paints. The problem of formulating gloss coatings results from the random distribution of pigment and latex particles as the volatiles evaporate so that there is not the same chance of obtaining a pigment-free, or low pigment content, upper surface to the film as is true with solvent-borne coatings (see Section 17.3). This problem can be minimized, but not eliminated, by using a small particle size latex. Gloss is also limited by the presence of surfactants in the dry film, which can lead to haze and blooming. See Section 35.3 for further discussion of gloss latex paints.

A major disadvantage is that all stable latexes have at least some surfactant that remains in the films. Generally, it is not completely compatible with the polymer, leading to haze and, commonly, bloom. The residual surfactant also reduces the water resistance of the films.

A further problem with latex coatings is that they tend to be excessively shear thinning. When the viscosity at high shear rate is set appropriately, the viscosity at the low shear rates encountered in leveling tends to be high. This is one of several reasons that leveling of latex-based coatings tends to be poorer than solvent-borne coatings. This important topic is discussed further in Section 35.3. Substantial progress has been made in recent years by the use of *associative thickeners*, as discussed in Section 35.3.

The use of latexes has been more limited in industrial coatings. There have been many reasons for this. The problems of evaporation of water on a conveyor line and in ovens mentioned in the last section are certainly part of the problem. There can be a real problem of popping with latex coatings too. In this case, it is desirable to use a latex polymer with as high a T_g as possible to minimize the chances of coalescence of the latex particles at the surface of the film before the water has completely evaporated.

Probably the major limitation for industrial applications has been the flow properties of latex-based coatings. Leveling requirements for many industrial coatings are more rigorous than for architectural coatings. Most industrial coatings have been Newtonian fluids. In general, latex coatings have exhibited relatively high degrees of shear thinning and in many cases thixotropy. In some cases, the flow properties result from flocculation of latex particles. Flocculation of latex will increase the low shear viscosity to a major degree; furthermore, it exacerbates the gloss problem. The use of so-called associative thickeners minimizes this problem and relatively rapid progress in formulating latex coatings with more nearly Newtonian flow properties can be expected. Since the major part of the published work with associative thickeners has been done with latex paints for architectural end uses, discussion is deferred to Section 35.3.

A major advantage of latexes is that they have a high molecular weight so that excellent mechanical properties can be achieved without the need for cross-linking reactions. The viscosity of latexes is independent of molecular weight so that they can be applied at relatively high solids even though the molecular weight is high. In the case of highly pigmented coatings, the fraction of internal phase volume becomes so large that the solids have to be reduced, but the reduction is done primarily with water so that VOC emissions are minimal.

As the pressure to reduce VOC emission increases still further and the problems of high solids coatings become more and more apparent, it is to be anticipated that there will be substantial increases in the use of latexes as well as water-reducible resins in industrial applications of both OEM product coatings and special purpose coatings.

29.3 EMULSION COATINGS

While water-reducible and latex coatings account for a very high percentage of water-borne coatings, emulsion coatings are used in some applications and may have potential for wider use.

Two package coatings in which one package is a BPA epoxy resin solution and the second is an amine-terminated cross-linking agent (see Section 11.2.6) containing a nonionic surfactant have had significant commercial applications. The amine cross-linker package is diluted with water, and the epoxy resin solution package is added with vigorous stirring. The pot life is limited to a few hours since the epoxy resin can react not only with the amine groups but also, slowly, with water. Such emulsion epoxy paints are used where hard, easily cleaned wall coatings are needed, for example, in hospitals and food processing plants. The residual surfactant reduces the corrosion protective properties for application to metal surfaces. An ingenious approach to eliminating, or at least minimizing, the need for surfactants is the use of nitroethane as one of the solvents [5,6]. As discussed in Section 11.2.6, nitroethane forms a salt with an amine group of an amido-amine which then acts as a surfactant. When the film dries the nitroethane evaporates, leaving a less water-sensitive film. Coatings based on this principle are being used in high performance applications such as aircraft primers (see Section 36.4).

Another example of emulsion coatings is nitrocellulose lacquers emulsified into water for use as top coats for wood furniture [7] (see Section 34.1). The emulsions have significantly lower VOC than solvent-borne lacquers, but longer times are required to achieve print resistance.

Alkyd emulsion paints have been formulated for such applications as flat wall paint, but their performance is inferior to latex paints. However, there is at least one example of a large-scale use of alkyd emulsion coatings. Large quantities have been shipped to Saudi Arabia for application to sand dunes near highways. The coating stabilizes the dunes against drifting over highways. This application illustrates the situation that there are so many diverse possible uses for coatings that it is not safe to conclude that there is no market for coatings that are inferior for conventional applications. The largest use for alkyds emulsified into paints is in some types of latex paints. As discussed in Section 35.1, for certain applications there are advantages in using an alkyd emulsion to replace 15% of latex binder.

REFERENCES

1. J. W. Rosthauser and K. Nachtkamp, *Water-Borne Polyurethanes*, in K. C. Frisch and D. Klempner, Eds., *Advances in Urethane Science and Technology*, Vol. 10, 1987, pp. 121–162, Technomic Publishers, Westport, CT.
2. L. B. Brandenburger and L. W. Hill, *J. Coat. Technol.*, **51** (659), 57 (1979).
3. B. C. Watson and Z. W. Wicks, Jr., *J. Coat. Technol.*, **55** (732), 61 (1983).

4. P. B. Jacobs and P. C. Yu, *Proc. 19th Water-Borne, Higher-Solids, and Powder Coat. Symp.*, New Orleans, LA, 1992, p. 363.
5. R. Albers, *Proc. 10th Water-borne Higher-Solids Coat. Symp.*, New Orleans, LA, 1983, p. 130; U. S. Patent 4,352,898 (1982).
6. J. A. Lopez, U. S. Patent 4,816,502 (1989).
7. C. M. Winchester, *J. Coat. Technol.*, **63** (803), 47 (1991).

CHAPTER XXX

Electrodeposition Coatings

For many end uses, electrodeposition can be a very efficient method of applying high performance coatings. The largest volume uses are for primers, but there are also significant uses for single-coat applications. The general principle is relatively simple; development and long-term production use are very complex [1,2]. There are two types of electrodeposition coating systems: anionic and cationic. There is a wide variety of terminology used: E-coat, electrocoat, electropaint, ED, and ELPO are fairly common synonyms for electrodeposition coatings; and some authors call anionic and cationic coatings, anodic and cathodic coatings, respectively.

In anionic systems, negatively charged particles of coating in an aqueous dispersion are electrophoretically attracted to a substrate which is the anode of an electrochemical cell. The paint particles are precipitated by the hydrogen ions generated there by the electrolysis of water. In cationic systems, the object is made the cathode and positively charged particles of coating are attracted to the cathode and precipitated on its surface by the hydroxide ions generated there. In both types, the coatings are baked; thermosetting binders are generally used.

The systems must be designed so that all coating components are attracted to the electrode at the same rate; otherwise the composition will change with time. It is possible to electrodeposit polymer films from solutions of the salt of a polymer in water, but it is not possible to use a dissolved polymer as the binder for pigmented coatings. With a soluble resin, it would not be possible to design a coating in which all the resin, pigment, cross-linker, and other components would be deposited in the same ratio constantly over a long time. Similarly, one can electrodeposit polymer films from a latex. For some purposes, electrodeposited latex films can be useful, but they are not appropriate for the binder for pigmented coatings. For useful pigmented E-coats, it is necessary to use a vehicle in which pigments can be dispersed and cross-linkers dissolved that will then form a stable, electrically charged dispersion of aggregate particles when diluted with water. The pigment must be preferentially wet by the resin so that it will not migrate out of the resin aggregates. Similarly, the cross-linker must dissolve in the aggregates, with essentially none dissolved in the water phase.

The coating is diluted to 10–20% solids with water. This low solids is necessary for two reasons. First, especially for coating products like automobiles and appliances, the electrodeposition tanks are very large—up to 400,000 L. The tank is kept full at all times; therefore, part of the capital investment is the cost of a tank full of coating—the investment is lower with 10–20% solids than it would be with 50% solids. Second, when the coated object is brought out of the tank, it carries a layer of the bath liquid with it that must be rinsed off. Losses are less and rinsing is easier if the solids are low.

A critical requirement for electrodeposition coating formulations is that they be indefinitely stable after dilution to the concentration at which they will be run in production. As the coating is applied, coating solids are removed from the bath, they have to be continually replaced to maintain the same composition in the tank. Ideally, the tank would never be emptied. Some of the material from the original loading will be in the tank for very long times and, therefore, must be very stable to hydrolysis and mechanical agitation. The dispersion must have a high level of stability against coalescence during the continuous stirring and recirculation of the bath. The cross-linker must be stable in the system at a pH over 7 in the case of anionic systems and under 7 in the case of cationic systems. Stability to oxidation is also critical since air will be continually mixed into the bath by the agitation. If an oxidizing type vehicle is to be used, an antioxidant that will volatilize in the baking oven is an essential additive in the formulation.

30.1. ANIONIC ELECTRODEPOSITION COATINGS

The resins used in anionic systems are substituted with carboxylic acid groups so that they have an acid number in the range of 50–80 mg KOH/g resin. The pigments and other components of the coatings are dispersed in the resin and the carboxylic acid groups are partially neutralized with an amine such as 2-(*N*,*N*-dimethyamino)ethanol (DMAE). To load the tank, the coating is diluted to about 10% solids with water. The degree of substitution with salt groups is designed to be such that the resin is not soluble in water but rather that it forms aggregates on dilution. The aggregates are stabilized as a dispersion in water with salt groups on the outer surface of the particles. Even with less than the theoretical amount of amine to neutralize the carboxylic acid groups, these systems have a pH above 7 due to the entrapment of unneutralized COOH groups in the center of aggregate particles (see Section 7.3).

In early work on electrodeposition primers for automobiles, maleated linseed oil (see Section 9.3.5) was used as a vehicle. The anhydride moiety was bonded to the linseed oil molecules by carbon–carbon bonds that cannot be hydrolyzed and hence the system showed reasonable stability in the electrodeposition bath. However, the adhesion to steel was relatively poor and maleated linseed oil was soon replaced with maleated epoxy esters (see Section 11.4.1) which had still better hydrolytic stability and provided superior adhesion to steel. The cross-linking obtained through the drying oil fatty acid esters in the epoxy ester was supplemented by using some melamine–formaldehyde (MF) resin as a cross-linker. Mixed methyl ethyl ether Class I MF resins (see Section 6.2) are the most appropriate for electrodeposition purposes. The mixed ether MF resin has sufficient solubility in water to permit easy incorporation, but is more soluble in the resin aggregates than in water so that it deposits in constant ratio as the bath is used over long times. Carboxylic acid-substituted resins made by reacting moderate molecular weight polybutadiene with maleic anhydride have also been used as vehicles for anionic primers. Since the backbone linkages are all carbon–carbon bonds, there is no problem of hydrolysis in the bath.

Maleated epoxy esters are not appropriate for many top coat applications because their color stability and chalking resistance are poor. The most widely used resins for anionic electrodeposition top coats are acrylic copolymers made using some acrylic (or methacrylic) acid and usually some 2-hydroxyethyl methacrylate as comonomers. The carboxylic acid groups are required to permit formation of stable aqueous dispersions with electronegatively charged particles; the hydroxyl groups as well as the carboxylic acid groups serve as sites for cross-linking with MF resin.

The primary reaction occurring at the anode in anionic E-coats is the electrolysis of water to yield hydrogen ions. These hydrogen ions neutralize the carboxylate ions on the resin at the surface of the anode. This neutralization removes the charge that has helped stabilize the aggregates

$$2H_2O \rightarrow 4H^+ + O_2 + 4e^-$$
$$RCOO^- + H^+ \rightarrow RCOOH$$

against coalescence. Also, when the salt groups are neutralized, the surface is less polar; there is less swelling of the surface with water so that potential stabilization by entropic repulsion is also reduced, if not eliminated. Thus, the aggregate particles coalesce on the metal surface. Not all of the salt groups have to be neutralized in order for precipitation to occur. When the film is formed, some salt groups and, hence, ammonium ions are trapped within the film. Side reactions can also occur at the anode. Iron can dissolve to form ferrous ions that are oxidized to ferric ions. The ferric ions can form insoluble salts with the carboxylic acids on the resin; this side reaction leads to reddish brown discoloration. This is not a particular problem with primers but can be a limitation in making light colored or white electrodeposition top coats for steel objects. In some cases, zinc phosphate conversion coatings reduce discoloration to acceptable levels. Anionic top coats are commonly used on aluminum since aluminum salt formation does not affect the color of the coating.

An important side reaction in electrocoating of steel that has been phosphate conversion coated is the partial dissolution of the iron–zinc phosphate layer by the hydrogen ions generated at the anode surface. This partial removal

$$Zn_3(PO_4)_2 + 2H^+ \rightarrow 3Zn^{2+} + 2(HPO_4)^{2-}$$

of conversion coating has two potentially serious effects. Damage to the phosphate coating can lead to poorer adhesion to the steel surface and hence less corrosion protection. Also, the soluble ion concentration in the bath is increased, which leads to higher conductivity of the water phase. As discussed later in this chapter, maintaining constant, relatively low conductivity of the water phase is critical.

Another side product of the reaction occurring at the anode as a result of the electrolysis of water is the generation of oxygen gas. As also discussed later, in some cases the gas can be generated at the metal surface after it has been coated, leading to film rupture as the bubbles of oxygen escape through the film.

While, in theory, a Kolbe decarboxylation reaction could occur at the anode and some early work on anionic systems suggested that it might be an important reaction, later work showed that this reaction is probably unimportant.

$$2\text{R-COO}^- \rightarrow \text{R-R} + 2CO_2 + 2e^-$$

At the cathode, hydrogen gas and hydroxide ions are generated. The hydroxide ions neutralize the ammonium

$$4H_2O + 4e^- \rightarrow 4OH^- + 2H_2$$
$$R_3NH^+ + OH^- \rightarrow R_3H + H_2O$$

counterions. The amines that are formed are water-soluble and only a fraction of the amine is removed when the coated article being painted is taken out of the bath. This would, of course, lead to an unacceptable accumulation of amine in the bath. Methods of control of amine concentration are discussed in Section 30.4. An advantage of electrodeposition as compared to other water-borne systems is that only a small amount of counterion is deposited on the substrate with the rest of the coating. Some anionic electrodeposition tanks have been run using potassium hydroxide as the neutralizing base. This would clearly not be feasible with water-borne coatings applied by other methods.

30.2. CATIONIC ELECTRODEPOSITION COATINGS

Cationic coatings have positively charged aggregates that are attracted electrophoretically to the cathode. The resins have amine groups that are neutralized by a low molecular weight, water-soluble acid such as acetic or lactic acid. It is highly desirable to develop a coating that will be stable at a pH a little below 7, otherwise stainless steel or other expensive corrosion resistant piping and handling facilities must be used. Some suppliers recommend corrosion resistant equipment in all cases. Commercial cationic electrodeposition tanks are operated at a narrow range often within pH 5.8–6.2 but sometimes lower. The resins used in the largest scale application of cationic E-coats—automotive primers—are proprietary, but they are known to be based on BPA epoxy resins reacted with polyamines to yield a resin with both secondary and tertiary amine groups as well as hydroxy groups. Salts are formed with the amine groups with a low molecular weight carboxylic acid such as formic or lactic acid. Cross-linking is effected by alcohol-blocked toluene diisocyanate. This cross-linking agent is stable in the slightly acidic water system, whereas MF resins are not. During the baking of the coating, the blocked isocyanate reacts with a secondary amine group to form a substituted urea cross-link or with a hydroxyl group to form a urethane cross-link.

Pigmented coatings at about 20% solids made with such a resin system are widely used for automotive primers because the corrosion protection is significantly better than that provided by an anionic electrodeposition primer. As is generally the case, amine-substituted resin binders provide greater corrosion protection for steel, perhaps owing to strong interaction between the amine groups and the substrate surface, increasing the wet adhesion. As discussed in Section 27.3.2, wet adhesion is the most critical factor in corrosion protection. There is not the same problem of acid dissolving the phosphate conversion coating that there is in the case of anionic deposition. Zinc phosphate conversion coatings (hopeite) are somewhat alkali soluble so that Zn–Fe phosphate conversion coatings that generate phosphophyllite crystal layers must be used (see Section 26.4.1). On zinc coated steels, Zn–Mn–Ni phosphate conversion treatments are used [3].

Coatings based on BPA epoxies and TDI exhibit poor color retention and exterior durability. For top coats, use of blocked aliphatic diisocyanates with acrylic resins gives better color retention and exterior durability. Use of 2-(*N*,*N*-dimethylamino)ethyl methacrylate and hydroxyethyl methacrylate as comonomers provides the needed amine groups for salt formation and hydroxyl groups for cross-linking. The curing temperature required for alcohol blocked aliphatic isocyanates to react with hydroxyl groups is high. One can use oxime-blocked isocyanates that will cross-link at lower temperatures, but then the bath stability may be limited. Alternatively, one can make an acrylic resin using glycidyl methacrylate as a comonomer. Then the pendant epoxy groups can be reacted with amines to yield secondary amines or mixed secondary and tertiary amines. Alcohol-blocked aliphatic isocyanates can be used to cross-link the secondary amine groups.

Cationic top coats are desirable because they avoid the iron staining problem encountered with anionic top coats, but their usage has been limited because of the curing/bath stability situation and because it has been difficult to develop resin systems that will be stable at a pH near to neutral. Since amine groups remain in the final cured coating, the exterior durability of such coatings is inferior to the durability that can be obtained using anionically deposited acrylic coatings that have no amine groups remaining in the final film.

30.3. EFFECT OF VARIABLES ON ELECTRODEPOSITION

After the current is turned on, there is no instantaneous deposition of coating. Some time is required for the concentration of hydrogen ions (in the case of anionic coatings, hydroxide ions in the case of cationic coatings) to build up sufficiently so that neutralization of the charges on the aggregate particles will lead to precipitation. After this initial time interval, the rate of deposition is affected by the rate of electrophoresis of the aggregates. This rate is importantly affected by the impressed voltage; the higher the voltage, the faster the film deposition. In commercial operation, the system is designed to permit complete coating in 2–3 min with 250–400 V commonly used. The high voltage is not needed to electrolyze the water, but rather to increase the driving force for electrophoretic attraction of the coating particles to the electrode.

The first areas covered are the edges of the metal parts since the current density is highest there. As the coating film thickness increases, the electrical resistance increases, reducing the rate of deposition at those sections coated first. There will be a limiting film thickness beyond which deposition of further coating stops, or at least becomes very slow. After the edges are coated, the outer flat surfaces of the object are coated, followed by the recessed and enclosed areas. The further back in a recess, the later the area is coated. Particularly for corrosion protective primers, it is desirable to have the entire surface of the steel coated, so it becomes important to coat the furthest recessed areas in the 2- to 3-min dwell time in the tank.

The rate of deposition is also affected by the equivalent weight of the coating. The higher the equivalent weight the greater the amount of coating precipitated by each hydrogen (or hydroxide) ion and, therefore, the faster the buildup of film thickness. On the other hand, it is critical to have a low enough equivalent weight so that there are sufficient polar salt groups to maintain the stability of the dispersion of aggregates in the coating bath. The rate is also affected by the amount of soluble low molecular weight ions present in the bath. In effect, these ions are also attracted to the electrodes in competition with the aggregate particles. Since they are small, they can move more rapidly in the electrophoretic field. Thus, the concentration of soluble salts must be low and maintained at close to constant concentration.

A coating that will deposit in recessed areas quickly is said to have a high *throw power* (called *throwing power* by some authors). Standard tests have been developed for throw power that determine how far up into a pipe or hollow rectangular anode the coating is applied at a standard voltage in a standard time. The extent of throwing increases with higher impressed voltage and longer dwell times. However, if the voltage is increased to too high a level, there will be film ruptures of the coating applied to the outer surfaces of the object. At a sufficiently high voltage, the current will break through the film, leading to local generation of gas under the film (oxygen in anionic electrodeposition and hydrogen in cationic electrodeposition) and then bubbles of gas can blow out through the film, leaving film defects. It has been shown that at higher voltages electric discharges occur through the film during electrodeposition causing visible sparks [4]. These sparks may also be responsible for the film ruptures. Sparking is reported to occur at lower voltages, (240 V) when the substrate is galvanized steel, than when the substrate is steel, about 300 V or higher are used. As conductivity of the deposited film increases, film rupture tends to occur at lower voltages. Direct current electricity obtained by rectifying AC electricity has a relatively wide variation in voltage; the current is said to ripple. The effect of rippling can be to break through at lower (average) voltages [5].

Throw power is also strongly affected by the conductivity of the bath: the higher the conductivity, the greater the throw power. Again, however, there is a limitation, as the conductivity due to the presence of soluble salts increases, the rate of electrophoresis of coating aggregates decreases. Increased numbers of salt groups on the resin increase conductivity and hence throw power, but the correspondingly lower equivalent weight decreases the rate of deposition. Furthermore, entrapment of conductive material in the film is increased with a corresponding increase in the likelihood of film rupture. Therefore, a compromise on conductivity must be reached. Conductivities used are in the range of 1000–1800 microsiemens (μS), the older unit designation mho is still used; 1 μS = 1 μmho.

Film rupture, and hence throw power, is affected by variations in coatings composition. If the viscosity of the aggregates precipitated on the surface is high, they will not coalesce and a porous film will result; conductivity of the film will be high and throw power will be low. On the other hand, if the viscosity of the precipitated aggregates is very low, the films will be soft; then, if any electrolysis takes place below the surface, bubbles will break through easily and film rupture will be severe. As is common in all coating formulations, a compromise must be made between the extremes. The glass transition temperature of the resin is a controlling factor and, correspondingly, the temperature at which the electrodeposition is carried out is a factor. Temperature of the bath must be controlled within a fairly narrow range (typically 32–35°C). Higher T_g resins can be used if some solvent is included in the formula. Of course, solvents can affect the electrical conductivity of the deposited film so care must be taken in their selection. Excess solvent will lead to film rupture at lower voltage and hence lower throw power. Also, the partition coefficient must be such that essentially all of the solvent is dissolved in the aggregates and essentially none in the water or the solvent concentration in the bath will build up over time.

Pigment concentration also affects coalescence. If the PVC is near to or above CPVC and the amount of solvent, if any, is small, the film deposited will not coalesce. There is also a significant effect of the degree of pigmentation on the leveling of the film even if it is low enough to permit coalescence. Since the film as applied has a low solvent concentration, its viscosity is dependent on the pigment concentration. Unless the pigment concentration is quite low, in most systems probably less than half of CPVC, leveling will be poor because of the high viscosity. Because of the low level of pigmentation as compared to conventional primers, the gloss of electrodeposited primers is higher, especially if the pigment content is reduced to low enough levels to permit good leveling. In selecting pigments, any with a significant degree of solubility in water (at the pH involved) must be avoided. Among other things, this means that passivating pigments cannot be used in electrodeposition primers; corrosion protection results from a barrier coating with outstanding wet adhesion.

30.4. APPLICATION OF ELECTRODEPOSITION COATINGS

A schematic diagram of an E-coat system is shown in Figure 30.1. The object to be coated is hung from a conveyor and carried into the dip tank where the coating is electrodeposited. Both in order to return coating that is carried out of the tank by the object and to avoid local accumulation of excess paint on the object, the object must be rinsed as it comes out of the tank. At the rinsing stage, the coating has not been cross-linked, but it is of sufficiently high viscosity after being precipitated on the metal surface that rinsing will not displace the electrodeposited coating, only the bath liquid is washed from the surface of the coating.

Note the heat exchanger used to maintain temperature within a narrow range. Very importantly, the bath liquid is continuously recirculated through an ultrafiltration unit. The ultrafiltration membrane permits removal of excess water, and materials soluble in water, while not removing the aggregates containing the resin, pigment, and cross-linker. Ultrafiltration permits the concentration of soluble salts to be maintained essentially constant so that there will be constant electrical conductivity. It also permits removal of an amount of water equivalent to that being brought into the bath from the rinsing operation so that the concentration of coating solids is held constant.

The drawing also shows an electrolyte tank. This represents a system for controlling the concentration of solubilizing agent in the bath (amine for anionic systems, acid for cationic systems). As noted earlier, when the coating is precipitated

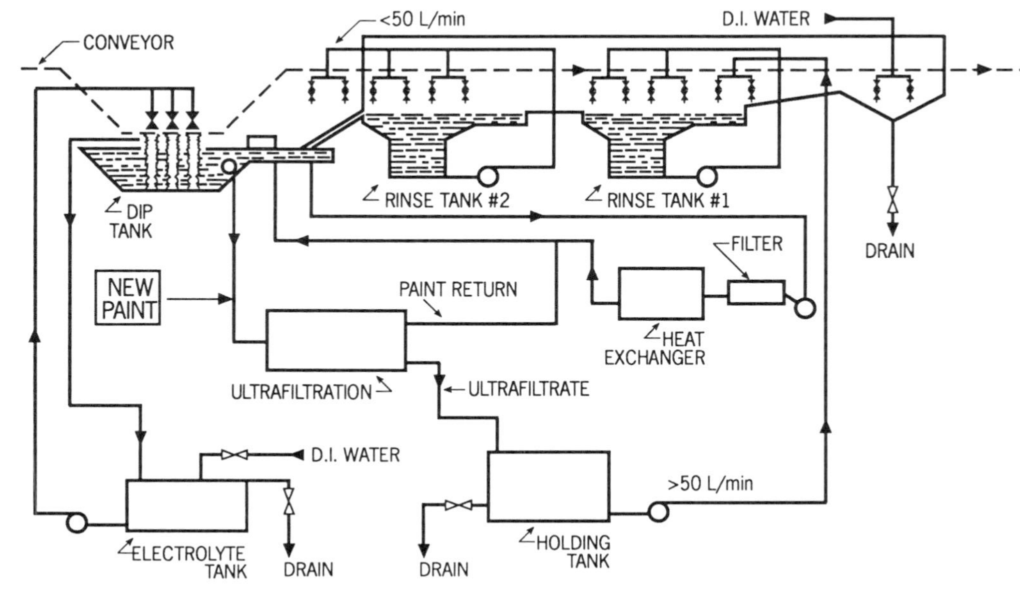

Figure 30.1. Schematic drawing of a cationic automotive primer electrodeposition installation. (Adapted from Ref. [2], with permission.)

on the substrate surface, a corresponding amount of solubilizing agent is released. The concentration of solubilizing agent in the bath must be maintained at a constant level. Some small amount of solubilizing agent is carried out of the bath with the coated product, and some is removed by ultrafiltration. But the sum of these losses of solubilizing agent is less than the amount being released by the electrocoating. There are two ways of maintaining the balance. In some cases it is feasible to use a sufficiently low level of solubilizing agent content in the make-up paint to be added to the bath, with the remainder of the required level of solubilizing agent being the excess left in the bath.

The other, more effective, approach is to have the counter electrode (cathode in the case of anionic systems and anode in the case of cationic systems) in a *membrane box*. For example, the anode in a cationic system can be located inside a microporous polypropylene box. The pore size of the membrane must be such that the aggregate particles can not pass through the membrane, but water and soluble materials such as the lactate or acetate ions can easily go through, especially since these ions are attracted by the anode whereas the positively charged aggregates are not. In some cases, it is necessary to use ion-selective membranes. The clear liquid in the box is recirculated to the electrolyte tank where the concentration can be continually monitored and corrected. Maintenance of the proper level of solubilizing agent is critical. It is common to hear that the pH of the bath must be kept constant. While this is true, the pH of these coatings is insensitive to the ratio of the weak acid and weak base (see Section 7.3). Conductivity is usually a more important control criterion. The application lines can be highly automated with feedback control of rate of addition of make up paint, solubilizing agent, and replacement of water.

30.5. ADVANTAGES AND LIMITATIONS OF ELECTRODEPOSITION

Electrodeposition has become a very widely used method for applying coatings to a wide variety of products. A large majority of all automobiles have electrodeposited primer coats. Many appliances also are primed by electrodeposition. Aluminum extrusions, drapery fixtures, metal toy trucks, and steel furniture are but a few of the many examples of single coat electrodeposition applications.

Electrodeposition can be a highly automated system with very low labor requirements, especially as compared to spray application. A startling example of manpower savings has been reported in the use of cationic electrodeposition to apply a single epoxy coating to air conditioners [6]. The former coating system of a flow coated primer and a spray applied acrylic top coat required 50 people, including those who did the required touch up and repair. The E-coat operation required only one operator. Further savings result from the elimination of paint losses from overspray. Paint utilizations in excess of 95% are reported. The economic advantage of the combination of these two factors is large in assembly line operations. Of course, the capital cost of the automated line is high, limiting the applicability of the highly automated line to large production operations. Simpler installations are used for applications like coating metal toys.

Solvent content of E-coats is low so that VOC emissions are low and the fire hazard is reduced. Another major environmental advantage as compared to spray applied coatings is that there is no overspray sludge disposal problem. (Unless, perish the thought, a poor job of formulating, coating production, bath maintenance, or control has been done and it is necessary to dispose of 250,000 L of bad paint.) Since the solids of the deposited film is very high, only 3–5 min flash off time is required before entering the oven, an advantage relative to spray applied coatings.

A major advantage (assuming adequate throw power) is complete coverage of surfaces. There can be differences in film thickness; for example, the recessed areas will generally have thinner deposits than the exposed face areas, but all of the surface will have some coating. Recessed and enclosed areas that cannot be coated by spray application can be coated by electrodeposition.

Sagging can occur when the coated part is heated in the oven. However, owing to the relatively high viscosity of the coating immediately on application, severe sagging is less likely than during spray or conventional dip application methods. Also, one is less likely to encounter the relatively large differences in film thickness between the top and bottom of the dipped object that are commonly experienced with conventional dip coating.

The locally uniform film thickness can lead to a problem, especially with relatively highly pigmented primers, the applied coating follows the surface contours of the metal closely, so that a rough metal surface gives a rough primer surface. Some authors refer to this phenomenon of replication of the substrate surface profile as *telegraphing*. Unfortunately, the term telegraphing is also used to describe other film defects, as discussed in Section 23.3. Some authors discuss the effect using the terminology *metal filling*. A primer is said to show good metal filling if there is minimal replication of metal scratches in the surface of the primer coat. Substantial variations in the smoothness of the primer surface can result from changes in E-coat primer composition [7]. The E-coat films are relatively thin, varying from 15–30 μm, depending on coating composition and application variables, hence it generally is not feasible to sand the surface smooth because bare metal would be exposed before the surface would be level. In some cases defects are repaired by sanding and patching with a conventional primer.

Leveling can be improved by reducing the pigment volume content of the primer. With a lower PVC, the viscosity of the coating after application and before cross-linking is lower. When the product is baked, some leveling will occur before cross-linking increases the viscosity such that no further flow can occur. The lower viscosity can also permit undesirable flow—in some cases, edge coverage is reduced by flow away from the edges of the coated product. Presumably, the edge heats up first, reducing the surface tension of the coating there and leading to flow over the adjoining higher surface tension area. Also, when the pigment content is decreased, the gloss of the coating films is increased.

There can be a problem in achieving adequate adhesion of top coats to the surface of electrodeposited primers. In many cases, in spray applied primers the PVC is slightly greater than CPVC, so that the dry film of primer is slightly porous, permitting a little of the vehicle of the top coat to penetrate in, thereby providing a mechanical anchor between the top coat and the primer. As noted above, one cannot make and apply an electrodeposition primer with a PVC > CPVC. In fact, the PVC/CPVC ratio is significantly less than in conventional primers; therefore, the gloss of electrodeposited primers is higher and the surface smoother, resulting in greater difficulty in achieving intercoat adhesion. Of course, when the PVC is further reduced to promote leveling, the surface is still smoother, resulting in higher gloss and an exacerbation of the intercoat adhesion problem.

An approach to improving intercoat adhesion is to apply a *sealer* over the electrodeposition primer. A sealer is usually a top coat that has been reduced to very low solids with relatively strong, slow evaporating solvents. Then the top coat is applied over the wet sealer surface. This procedure, in effect, permits more time for solvent to penetrate into the surface of the primer and swell it so that the sealer can penetrate into the surface of the primer. Especially for hoods, trunks, and other areas where smoothness is critical, one can spray apply a primer–surfacer on top of the E-coat primer. If the solvents are carefully selected, penetration into the electrodeposited film will be possible, promoting intercoat adhesion. There is the further advantage that the primer–surfacer can be sanded to provide the desirable smooth surface and also can have a PVC > CPVC so that adhesion of top coat to its surface will not be a problem. In some cases, an even more difficult aspect of this adhesion problem is the formulation of sealants and adhesives that adhere to the electrodeposited primer surface. Chip resistant coatings (see Section 33.1.2) are commonly applied to the lower parts of bodies before top coating.

The substrate must be conductive, but the cured electrodeposited coating is usually an insulator. Usually, therefore, one cannot apply more than one coat by electrodeposition. Most applications have been for metal primers or for one-coat metal coatings. Patents describe two-coat combinations in which the first coat forms a somewhat conductive film, and, as of 1992, at least one two-coat electrodeposition system is being used in production.

Another limitation that electrodeposition shares with any other dipping system is the difficulty of formulation changeovers. If it is decided to change color of an automotive primer, what do you do with 250,000 L of coating? At least once when this problem was faced, the coating supplier developed the new color primer such that it would be compatible with the old color primer so that the new coating could be introduced into the existing tank. Of course, for a considerable period there was a slow change of primer color from the old color to the new color. This was not serious for a primer but, obviously, would not be acceptable for a top coat. In effect, use of electrodeposition for top coats is limited to applications where long runs of the same color are made so that a line can be dedicated to a particular color. In some small installations, such as electrodeposition of coatings on metal toys, one coating bath can be pumped out of the electrodeposition tank into a holding tank, where it must be kept continuously agitated, and another color is pumped into the tank. But these are relatively small tanks; the procedure would not be economically feasible with large objects.

REFERENCES

1. F. Beck, *Prog. Org. Coat.*, **4**, 1 (1976).
2. M. Wismer, P. E. Pierce, J. F. Bosso, R. M. Christenson, R. D. Jerabek, and R. R. Zwack, *J. Coat. Technol.*, **54** (688), 35 (1982).

3. C. K. Schoff, *J. Coat. Technol.*, **62** (789), 115 (1990).
4. R. E. Smith and D. W. Boyd, *J. Coat. Technol.*, **60** (756), 77 (1988).
5. J. J. Vincent, *J. Coat. Technol.*, **62** (785), 51 (1990).
6. T. J. Miranda, *J. Coat. Technol.*, **60** (760), 47 (1988).
7. J. A. Gilbert, *J. Coat. Technol.*, **62** (782), 29 (1990).

CHAPTER XXXI

Powder Coatings

Powder coating is a segment of the industrial coatings industry that is undergoing rapid growth. An important factor leading to this growth is that VOC emissions are very low. Production of powder coatings in the United States in 1990 was estimated to be about 65 million kg [1]. Powder coatings are used even more extensively in other parts of the world, especially in Europe and Japan; worldwide production in 1990 was estimated at 290 million kg. The general principle is to formulate a coating from solid components, mix them, disperse pigments (and other insoluble components) in a matrix of the major binder components, and pulverize the formulation. In so far as possible, each particle should contain all ingredients in the formulation; the opposite extreme, a dry blended mixture of resin particles and pigment particles, is usually unsatisfactory. The powder is applied to the substrate, usually metal, and fused to a continuous film by baking, depending upon the coating, at temperatures in the range of 130–200°C. There are two broad categories of powder coatings—thermosetting and thermoplastic. By far the major portion of the market is for thermosetting powders.

Only an overview can be presented in this chapter. Further details can be found in the general references provided at the end of the chapter; the most extensive is that of Misev, which also provides many references.

31.1. BINDERS FOR THERMOSETTING POWDER COATINGS

The binders for thermosetting powder coatings consist of a mixture of a primary resin and a cross-linker, often called a *hardener*. The major types of binders are grouped somewhat arbitrarily into several classes, as shown in Table 31.1. The terminology used has grown historically and has become confusing. Epoxy coatings include only coatings based on BPA and novolac epoxy resins with amine, anhydride, and phenolic hardeners. Hybrid coatings also contain BPA epoxy resins but are cross-linked with carboxy-functional polyester resins. Polyester coatings contain polyesters with various cross-linkers other than BPA and novolac epoxies. The term is used only for coatings that exhibit good to excellent exterior durability. Acrylic coatings contain acrylic resins with various cross-linkers. In addition various blends, sometimes called *alloys*, of these classes are used. Among the factors in

Table 31.1. Classes of Thermosetting Powder Coatings

Common Name	Primary Resin	Cross-Linker
Epoxy	BPA (or Novolac) epoxy	Polyamines or phenolics
Hybrid	COOH-Functional polyester	BPA epoxy
Polyester	COOH-Functional polyester	Triglycidylisocyanurate or hydroxyalkylbisamides
	OH-Functional polyester	Blocked isocyanate or amino resin
Acrylic	Epoxy-Functional acrylic	Dibasic acid
	OH-Functional acrylic	Blocked isocyanate or amino resin

choosing a class of powder coating for a particular application are protective properties, exterior durability, and, of course, cost. Differences in exterior durability are illustrated in Figure 31.1.

In all these powder coatings, it is necessary to control the balance of binder T_g, $\bar{M}_n$, $\bar{f}_n$, and reactivity [3]. It must be possible to process the material without significant cross-linking, the resultant powder must not sinter (start coalescing) during storage, but must fuse during baking, level to form a desirable film, and cross-link to acquire desired film properties. In general, the primary resins are amorphous polymers with T_g values high enough to avoid sintering of the powder and with $\bar{M}_n$ values of a few thousand. Recommended minimum binder T_g values in Europe are 40°C and in the United States are 45–50°C, presumably reflecting the higher temperatures to be expected during shipment and storage in parts of

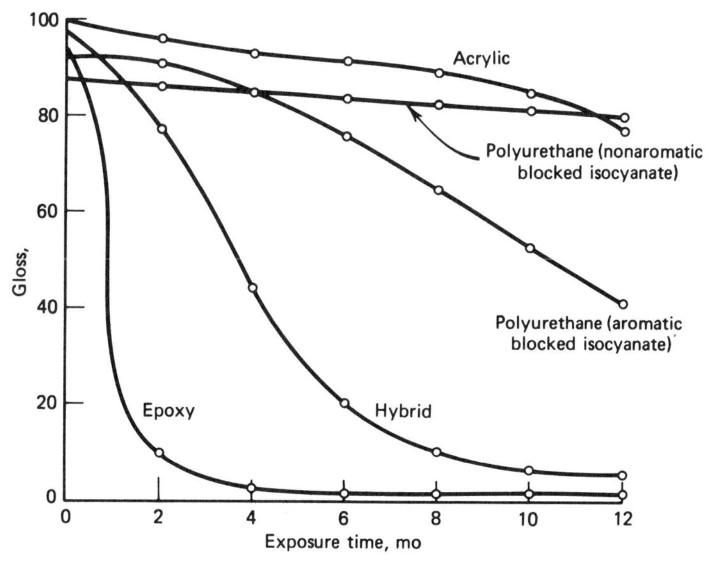

Figure 31.1. Florida outdoor exposure data on different types of powder coatings. (From Ref. [2], with permission.)

the United States. A typical powder coating might have a T_g of 55°C; could be melt processed at about 80°C; could be stored at temperatures up to about 40°C; during heating in the baking oven, the viscosity would briefly drop to about 10 Pa·s, allowing coalescence, flow, and leveling; and could be cross-linked by baking for 15 min at a temperature in the range of 130–200°C. It should be noted that the T_g values referred to are those of the combination of the primary resin plus the cross-linker. The required T_g for the primary resin alone varies depending on the cross-linker used with it.

31.1.1. Epoxy Binders

Epoxy powder coatings are the oldest and still one of the largest classes of thermosetting powder coatings. Decorative coatings have been based on BPA epoxy resins with n values of 3–5, usually made by the advancement process (see Section 11.1.1). The trend, however, is to lower molecular weights with some epoxy resins as low as $n = 2.5$ being adopted to provide better flow in thin film applications. For protective coatings, n values range up to 8. The most commonly used cross-linkers are dicyandiamide (dicy) (see Section 11.2.1) or a modified dicyandiamide. The curing reactions are complex and not completely understood. Modified dicys are more soluble in epoxy resins and tend to form uniform films more readily. 2-Methylimidazole is a widely used catalyst.

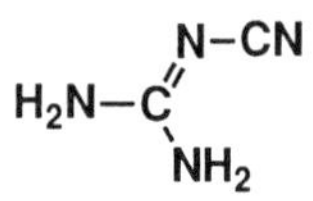

$$H_2N{-}C({=}N{-}CN){-}NH_2$$

Dicyandiamide

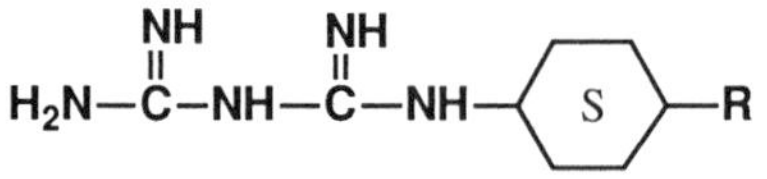

$$H_2N{-}\overset{\displaystyle NH}{\overset{\|}{C}}{-}NH{-}\overset{\displaystyle NH}{\overset{\|}{C}}{-}NH{-}\langle S \rangle{-}R$$

Modified Dicyandiamide

Epoxy powder coatings have good mechanical properties, adhesion, and corrosion protection; however, their exterior durability is poor (Fig. 31.1). Typical applications for decorative types include institutional furniture, shelving, and tools. Applications of protective epoxy coatings include pipe, rebars, electrical equipment, primers, and underbody automotive parts. Use of epoxy powder coated rebars can significantly extend the useful life of concrete bridges since an important cause of failure can be the rusting away of rebars within the concrete structure. Where enhanced chemical and corrosion resistance is needed, phenolic resins are used as cross-linkers (see Section 11.3.1), again with 2-methylimidazole as catalyst. Novolac epoxy resins (see Section 11.1.2) or blends of novolac and BPA epoxies give higher cross-link densities than BPA epoxies alone. All of these coatings discolor and chalk on exterior exposure. On the other hand, the coatings are tough and can be formulated and applied to steel to provide good corrosion protection.

Polycarboxylic acid anhydrides (see Section 11.3.3), such as trimellitic anhydride, are sometimes used with BPA epoxy resins in applications where greater resistance to yellowing and to acids and solvents is needed. These coatings are generally being replaced with hybrid coatings with somewhat better exterior durability and less question of toxic hazard.

31.1.2. Hybrid Binders

The BPA epoxy resins can be cross-linked with carboxylic acid-terminated polyester resins having $\bar{M}_n$s of a few thousand. Such formulations are called *hybrid powder coatings*, implying that they are intermediate between epoxy and polyester powder coatings. Hybrid powder coatings have better color retention and UV resistance than epoxy powder coatings but still do not have good exterior durability. Examples of end uses include water heaters, fire extinguishers, radiators, and transformer covers.

A variety of polyesters have been described. Most are derived from neopentyl glycol and terephthalic acid with smaller amounts of other monomers to reduce the T_g to the desired level and give branching to increase the $\bar{f}_n$ above two. (See Chapter 8 for discussion of polyester resin synthesis, functionality, and properties.) An example is a polyester from neopentyl glycol (NPG) (364 parts by weight, 3.5 mol), terephthalic acid (TPA) (423 parts, 2.55 mol), adipic acid (AA) (41 parts, 0.24 mol), and trimellitic anhydride (TMA) (141 parts, 0.74 mol) [4]. The acid number of the resin is given as 80 mg KOH/g of resin. The relatively high trimellitic anhydride content increases $\bar{f}_n$, compensating for the low (about 1.9) oxirane $\bar{f}_n$ of the BPA epoxy.

The primary cross-linking reaction is ring opening of the oxirane groups by carboxylic acids (see Section 11.3.2). Esterification and transesterification reactions involving hydroxyl groups of the epoxy resin and homopolymerization reactions of oxirane groups may also play a secondary role; to the extent that esterification of hydroxy groups occurs, the effective $\bar{f}_n$ of the epoxy resin is increased. Generally, a catalyst, such as an ammonium or phosphonium salt, for example, tetrabutylammonium bromide or choline chloride, is used to accelerate the reaction so that baking temperatures in the range of 160–200°C can be used. Often the polyester resins are supplied with the catalyst blended in—an example of what is called a *masterbatch*. Masterbatches are often used in powder coating production to permit uniform mixing of a minor component into the whole batch.

Flow properties of powder coatings containing carboxylic acid-terminated polyesters tend to be poorer than those made with hydroxy-terminated polyesters, perhaps owing to the stronger intermolecular hydrogen bonds between carboxylic acid groups. A proprietary, modified BPA epoxy has been reported that exhibits greater flow with comparable sintering resistance compared to conventional BPA epoxies [5].

31.1.3. Polyester Binders

Further improvement in exterior durability can be attained by replacing BPA epoxies with other cross-linkers. Depending upon the cross-linker, hydroxy-functional or carboxylic acid-functional polyester resins are used.

Triglycidyl isocyanurate (TGIC) (see Section 11.1.2) has been widely used as cross-linker for carboxylic acid-terminated polyesters with basic catalysts. The TGIC based powder coatings have good exterior durability (see Fig. 31.1) and mechanical properties. Examples of end uses are outdoor furniture, farm equipment, fence poles, and air conditioning units. There has been increasing concern about toxic hazards using TGIC.

While TGIC is expensive, the amounts required are relatively small because of its low equivalent weight. Typical binders contain 4–10 wt% TGIC and 90–96 wt% of carboxylic acid-terminated polyester. The polyesters used generally are less branched than those used in hybrid coatings because of the higher functionality of TGIC as compared to BPA epoxies. One polyester, for example, is made from NPG (530 parts by weight, 5 mol), TPA (711 parts, 4.3 mol), isophthalic acid (IPA) (88 parts, 0.47 mol), pelargonic acid (58 parts, 0.37 mol), and TMA (43 parts, 0.22 mol) with an acid number of 35 [4]. Such resins are usually prepared in a two-stage process to minimize the problems caused by the high melting point and low solubility of TPA (see Section 8.3) or using dimethyl terephthalate by transesterification (see Section 8.5). High equivalent weight (low acid number) is desirable since this reduces the required amount of TGIC, but cross-link density decreases with higher equivalent weight so there is an optimum for each application. The T_g of the polyester must be such that in combination with the amount of TGIC required for that polyester, the T_g of the combination will be appropriate.

Partly as a result of concern about toxic hazards with TGIC, use of tetra(2-hydroxyalkyl)bisamides (see Section 13.4) as cross-linking agents for carboxylic acid-functional polyesters in exterior durable coatings is being investigated [6]. These coatings also have good mechanical properties and flow. Accelerated exterior durability test results are excellent, but outdoor exposure test results have not yet been published. Water is evolved from the cross-linking reaction, which may limit the film thickness that can be applied without popping.

Another group of polyester coatings that is widely used, particularly in the United States, employs blocked aliphatic isocyanates (see Section 12.4.4) as cross-linkers with hydroxy-terminated polyesters. Typical coatings have exterior durabilities equal to, or somewhat better than, TGIC cross-linked polyesters and the excellent mechanical properties together with abrasion resistance typical of polyurethane coatings. Examples of their many end uses are automobile wheels, lighting fixtures, garden tractors, fence fittings, and playground equipment.

Probably the most widely used blocking agent is ε-caprolactam, but as the pressure to reduce curing temperatures increases, oxime blocking agents are also being used since they will react at lower temperatures. Isophorone diisocyanate (IPDI), bis(4-isocyanatocyclohexyl)methane ($H_{12}MDI$), and tetramethylxylidene diisocyanate (TMXDI), their isocyanurates, and/or low molecular weight prepolymers (see Section 12.3.2) are examples of isocyanates that give solid blocked isocyanates and are, therefore, appropriate for powder coatings. Blocked isocyanates from sterically crowded isocyanates such as TMXDI have the potential advantage of unblocking at a somewhat lower temperature [8]. Catalysts such as dibutyltin dilaurate (DBTDL) are used with all blocked isocyanates; they are often masterbatched in the polyester resin.

Polyesters are prepared with excess hydroxyl monomer utilizing a mixture of diol and triol. A representative polyester is made from NPG (436 parts by weight, 4.2 mol), TPA (552 parts, 3.33 mol), IPA (117 parts, 0.7 mol), sebacic acid (25 parts, 0.12 mol), and trimethylolpropane (TMP) (28 parts, 0.2 mol) with an hydroxyl number of 35 mg KOH/g resin [7]. The low ratio of TMP means that this polyester is only slightly branched and hence has a $\bar{f}_n$ only a little over 2. It would probably be used with a blocked isocyanate having a $\bar{f}_n$ of 3 or more.

Blocked isocyanate/polyester powder coatings generally show better flow than most powder coatings, perhaps because the polyesters are hydroxy not carboxylic acid-terminated. In the case of caprolactam blocked isocyanates, the low volatility of the caprolactam may mean that it acts as a solvent to reduce T_g after being released from the blocked isocyanate and before volatilizing. It has been reported that only about half of the caprolactam is volatilized, leading to speculation that the balance polymerizes to form nylon 6. The emission of blocking agent reduces the advantages resulting from the lack of VOC from powder coatings, but the total amount is only a few percent of the coating.

Tetramethoxymethylglycoluril (see Section 6.4.3) is also being investigated as a cross-linker for hydroxy-functional resins [9]. Since methyl alcohol is generated as a byproduct of the cross-linking reaction, film thickness may be limited by popping, that is, blister formation. Toluene sulfonamide-modified melamine–formaldehyde resins are also being studied as cross-linkers for hydroxy-functional polyesters [10]. The cross-linking reaction releases volatile byproducts with these resins also; it is reported that buffering by a cure rate regulator, 2-methylimidazole, minimizes popping [11].

31.1.4. Acrylic Binders

Two types of solid acrylic resins with appropriate T_g values are used commercially. Epoxy-functional acrylics, made with glycidyl methacrylate (see Section 11.1.2) as a comonomer, are cross-linked with dicarboxylic acids such as dodecanedioic acid [$HOOC(CH_2)_{10}COOH$]. Hydroxy-functional acrylic resins are cross-linked with blocked isocyanates (or other cross-linkers that will react with hydroxyl groups).

Acrylic powder coatings are not as widely used as polyester-based powder coatings. Acrylic coatings tend to have poorer impact resistance than polyester coatings [12]. On the other hand, they generally have superior detergent resistance and hence are used for applications such as washing machines. The ratio of polyester-based coatings to acrylic-based coatings is much higher in powder coatings than in liquid coatings. There are a variety of possible reasons for this in addition to impact resistance. Most solvent-borne acrylic coatings are applied over primer, which improves impact resistance and adhesion, whereas the powder coatings are generally applied directly on metal. The end uses for coatings that require especially good exterior durability, such as automobile exterior coats and some coil coated products, have not shifted significantly to powder. It is, comparatively speaking, easy to manufacture polyester powder resins since they can be synthesized without solvent. Acrylic resins generally must be polymerized in solvent and the solvent must be distilled out at the end of the reaction. It is also possible to make solid acrylic resins by polymerizing under pressure in methylene chloride and then removing and recovering the methylene chloride in a spray drier. The difficulty of synthesizing low molecular weight, low $\bar{f}_n$ acrylic resins with low levels of nonfunctional and monofunctional molecules as discussed in Section 7.2 may also be a limitation.

31.2. BINDERS FOR THERMOPLASTIC POWDER COATINGS

The first powder coatings were thermoplastic coatings that form films by coalescence without cross-linking. They now account for roughly 15% of the American market for powder coatings. Thermoplastic coatings have several disadvantages compared to thermosetting coatings. They are generally difficult to pulverize to small particle sizes; thus they can only be applied in relatively thick films. Owing to the high molecular weights of the binders required, even at high baking temperatures, they are viscous and give poor flow and leveling.

Vinyl chloride copolymers and, to a more limited extent, polyamides (nylons), fluoropolymers, and thermoplastic polyesters are used as binders. High vinyl chloride content copolymers (see Section 4.2.1) are formulated with stabilizers and with a limited amount of plasticizer, often a phthalate ester, so that the T_g will be above ambient temperature for storage of the powder. The partially crystalline nature of PVC may help stabilize the powder against sintering. The vinyl powders are generally applied as quite thick films, 0.2 mm and higher, by fluidized bed application (see Section 31.5.2). Coating of dishwasher racks, hand rails, and metal furniture are examples of end uses.

Nylon-11 and nylon-12 based powder coatings exhibit exceptional abrasion and wear resistance and are used as antifriction coatings, and coatings for hospital beds, clothes washer drums, and other applications where they must withstand frequent cleaning or sterilization and have good toughness and scratch resistance. Fluoropolymers, such as poly(vinylidene fluoride) and ethylene/chlorotrifluoroethylene copolymers are used for coatings requiring exceptional exterior durability, such as aluminum roofing and window frames and/or resistance to corrosive environments such as equipment for chemical plants. Thermoplastic polyester coatings are sometimes made using scrap or recycled poly(ethylene terephthalate).

Polyolefin-based powder coatings are used but the volume has been limited by generally inferior adhesion. Ethylene/acrylic acid (EAA) copolymer resins have recently been made available which give coatings with substantially better adhesion [13] (see Section 31.5.2).

31.3. FORMULATION OF POWDER COATINGS

A major challenge facing formulators is satisfying a combination of sometimes conflicting needs: (1) stability against sintering during storage, (2) coalescence and leveling at the lowest possible baking temperature, and (3) cross-linking at the lowest possible temperature in the least possible time. Further, the degree of flow and leveling must be balanced to achieve acceptable appearance and protective properties over the range of expected film thicknesses. Films that flow readily before cross-linking may have good appearance, but they may flow away from edges and corners, resulting in poor protection.

If the T_g of the coating is high enough, sintering can be avoided. However, coalescing and leveling at the lowest possible temperature are promoted by having the lowest possible T_g. Short baking times at low temperatures are possible if the resins are highly reactive and if the baking temperature is well above the T_g of the final cross-linked film. However, such compositions may cross-link prematurely

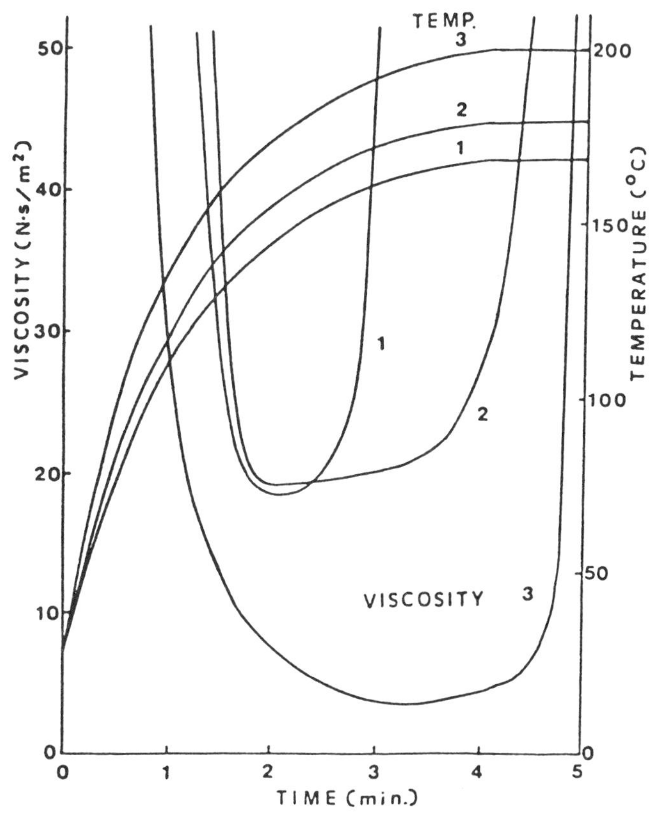

Figure 31.2. Nonisothermal viscosity behavior of powder coatings during film formation, as functions of time and panel temperature. (1) Acrylic/dibasic acid type. (2) Polyester/blocked isocyanate type. (3) Epoxy/dicy type. (From Ref. [14], with permission.)

during extrusion, and the rapid viscosity increase as the particles fuse in the oven limits the ability of the coating to coalesce and level. Clearly compromises are needed. With current technology, a crude rule of thumb is that the lowest feasible baking temperature is 50°C above the melt extrusion temperature and 70–80°C above the T_g of the uncured powder; that is, minimum baking temperatures of about 125–135°C are required for a powder with a T_g of 55°C. If users were willing to refrigerate the powder during storage or use it very soon after production, lower baking temperatures would be feasible.

Several studies have addressed the changes in viscosity during film formation [9,14–16]. Nakamichi [14] used a rolling ball viscometer to measure viscosities of powder coatings on a panel during heating. Results for three types of coatings are shown in Figure 31.2. In each case, viscosity is high immediately after fusion of the powder but drops off sharply with increasing temperature. Viscosity levels off as cross-linking reactions begin to increase the molecular weight and then increases rapidly as the coating approaches gelation. Flow is governed by the lowest viscosity attained and by the length of time that the coating stays near that viscosity—the *flow window*. In Figure 31.2, coating 2 will flow more than coating 1 even though the lowest viscosity is about the same because the reaction is slower and the flow window is longer. As baking temperatures are reduced, the problem becomes more difficult because more reactive resins have narrower flow windows. The problem is compounded when the same coating is to be applied at different film thicknesses and especially when thin films are desired. Recall that the rate of leveling increases with film thickness (see Section 23.1).

Some authors discuss the temperature dependence of viscosity in terms of Arrhenius-type relationships and discuss the activation energies for viscous flow. As indicated in Section 19.4.1, the dependency of viscosity on temperature does not actually follow Arrhenius relationships, but rather is dependent on free volume availability. The most important factor controlling free volume availability is $(T - T_g)$.

Differential scanning calorimetry (DSC) is widely used to characterize the T_g and cure characteristics of powder coatings. Dynamic mechanical analysis (DMA) (see Sections 24.1 and 24.4) is useful for characterizing cured films [10,12]. A DMA study [10] showed that the T_g values of cured films of a series of decorative powder coatings including a hybrid coating, a TGIC/polyester coating, and a blocked isocyanate/polyester coating were all in the range of 89–92°C. The average molecular weights between cross-links, $\overline{M}_c$, (see Section 24.2) for the cured films were in the narrow range of 2500 to 3000. It is noteworthy that years of trial and error in the formulation of such disparate compositions in different laboratories led to such similar T_g and $\overline{M}_c$ values. On the other hand, a protective epoxy powder coating with a modified dicy cross-linker gave cured films with a T_g of 117°C and an $\overline{M}_c$ of 2200. Pigmentation with TiO_2 had only a weak reinforcing effect and had essentially no effect on T_g, suggesting that pigment/binder interaction is weaker in powder coatings than in liquid coatings [10]. Dynamic mechanical analysis can be a powerful tool in helping suggest starting points for development work on new binders for powder coatings for similar applications.

It has been recognized that two quite different factors control the T_g of the resins used in powder coatings: the chemical composition and the molecular weight. It has been reported that it is advantageous to use higher molecular weight, more flexible resins since these can still have adequate package stability, but flow more easily at higher temperatures than lower molecular weight resins of similar T_g having more rigid chains [3]. Further research on the principles of resin design for optimum viscosity, flow window, and cured film properties is needed.

Another area needing more study is the important question of the driving forces for coalescence and leveling. In latex paints, coalescence is said to be promoted by capillary forces as the particles get close enough together; it is not evident how this mechanism could be a major factor in the coalescence of powder particles. It seems more plausible to suppose that the driving force could be the surface energy released with the reduction in surface area in the transition from small particles to a film. This would lead to the prediction that high surface tension might promote coalescence [15]. On the other hand, the effect of surface tension differences may be trivial as compared to the difference between the large surface area of the particles and the small surface area of the film.

While it is true that too high a surface tension could interfere with wetting and cause crawling and poor adhesion [3], in practice, powder coatings are only used over clean treated metal that has a higher surface energy than any powder coating.

In the case of leveling, it has been proposed that the driving force is surface tension, as reflected in the Orchard equation (see Section 23.1) [16]. Although the data obtained fit reasonably well for relatively thin and/or fluid films, thicker and/or somewhat more viscous films leveled better than predicted by the Orchard equation. The authors did not consider the possible effects of surface tension differential driven flow.

de Lange [15] points out that surface tension differential flow can cause cratering of powder coatings and that small amounts of additives such as poly(octyl acrylate) derivatives will overcome this problem. He refers to Overdiep's work [17] on surface tension differential driven flows in brush applied coatings, but does not seem to consider the possible implications in the leveling of powder coatings. de Lange indicates that powder coatings containing the acrylate additives have improved resistance to cratering but the coatings still show orange peel. He reports that, in some cases, other additives such as epoxidized soybean fatty acids and hydroabietyl alcohol reduce orange peel but do not reduce cratering. His statement that these additives increase the surface tension must be taken with considerable caution. As a general rule, one does not increase surface tension by adding something—the lowest surface tension material always comes to the surface.

Many powder coating formulations contain 0.1–1% of benzoin. This additive is said to act as an antipinholing agent and as a degassing aid, and it is generally believed that benzoin improves the appearance of cured powder coating films [1]. One study [9] showed that benzoin plasticizes the melt and increases the flow window of polyester/glycoluril formulations, indicating that it would improve leveling, but relatively high levels (1.4–2.4% of benzoin) were used. No complete account of the mechanism of the effect of adding benzoin has been published, and there is no indication of why benzoin is apparently unique in its action.

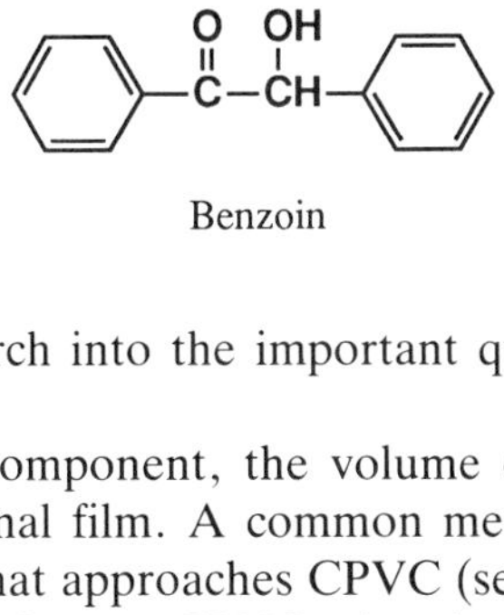

$$C_6H_5-\overset{O}{\overset{\|}{C}}-\overset{OH}{\overset{|}{CH}}-C_6H_5$$

Benzoin

The need for further research into the important questions of film formation and leveling is obvious.

Since there is no volatile component, the volume of pigment in the powder approaches the PVC of the final film. A common method for making low-gloss liquid coatings, using a PVC that approaches CPVC (see Section 21.1), cannot be used in powder coatings. At PVCs near CPVC, viscosity of the fused powder would be far too high for leveling. In fact, it has been shown that as pigmentation increases above a PVC of about 20, the problems of leveling increase due to the increase in melt viscosity [15]. If pigment flocculation would occur, it would make the problem even more severe.

Low gloss and semigloss powder coatings have been prepared using approaches other than high PVC. One approach is to use silica or polyolefin flatting pigments. Another approach is to blend two different primary resins or two different crosslinkers with substantially different reactivities or with poor compatibility. These methods work to a degree, but as a general rule it is more difficult to achieve low gloss powder coatings than with liquid coatings, and it is more difficult to control intermediate gloss levels reproducibly.

31.4. MANUFACTURING OF POWDER COATINGS

Manufacture of powder coatings poses production and quality control issues that are, in some respects, quite different from those of liquid coating manufacture.

31.4.1. Production

Most powder coatings are manufactured by the same general process—premixing, melt extrusion (Fig. 31.3), and pulverization (Fig. 31.4). All the major ingredients must be solids. Some liquid additives are used, but they must first be melted into one of the solid components to make a masterbatch that is granulated for use in the mix. The granulated ingredients, resins, cross-linkers, pigments, and additives, are weighed and mixed in a batch process. A variety of premixers can be used, but it is critical that they provide a uniform, intimate mixture of the solid ingredients. The premix is fed through an extruder in a continuous process. The barrel of the extruder is maintained at a temperature moderately above the T_g of the binder. In passing through the extruder, the primary resin and other low melting or low T_g materials are fused and the other components dispersed in the melt. The extruder operates at a high rate of shear so that it can be very effective in separating pigment aggregates. The melt can be extruded through a die—sometimes either a slot to produce a flat sheet or a series of round orifices to produce *spaghetti*. More commonly, so as to reduce heat exposure, the melt is extruded through a large bore die and the *sausage* is fed between chilled rollers to flatten it into a sheet and cool it. Often, but not always, the extruded material is deposited on a conveyor for further cooling. At the end of the conveyor, the material has hardened enough to be coarsely granulated and collected in bins for further processing. At this stage, it is a brittle, uncross-linked solid.

Extruders have developed into sophisticated and rather expensive pieces of equipment. Two types are commonly used—single screw and twin screw; in both types a powerful motor turns screws to drive the material through a barrel. The screws and barrel are configured to mix the material thoroughly and apply a high rate of shear. A popular single screw extruder uses a reciprocating action in addition to the radial turning of the screw to effect mixing and dispersion. Twin screw extruders use a combination of screw segments and kneading segments. Both types of extruders are capable of excellent dispersion of most pigments. They operate with relatively high viscosity formulations at high shear rates, and hence efficiently separate pigment aggregates (see Section 20.1.2). However, there is a trade-off between separating pigment aggregates and production rate. Production capacity can be increased by pushing more material through the extruder per unit time. Residence time in the extruder is sometimes reduced to 10 s or less, but at some point pigment dispersion, especially with some organic pigments, begins to suffer. Poor color development and color variability may result (see Section 31.4.2). There seems to be little problem of pigment flocculation. While there is no proof, it seems reasonable to speculate that the viscosity of the melt increases so rapidly, as the

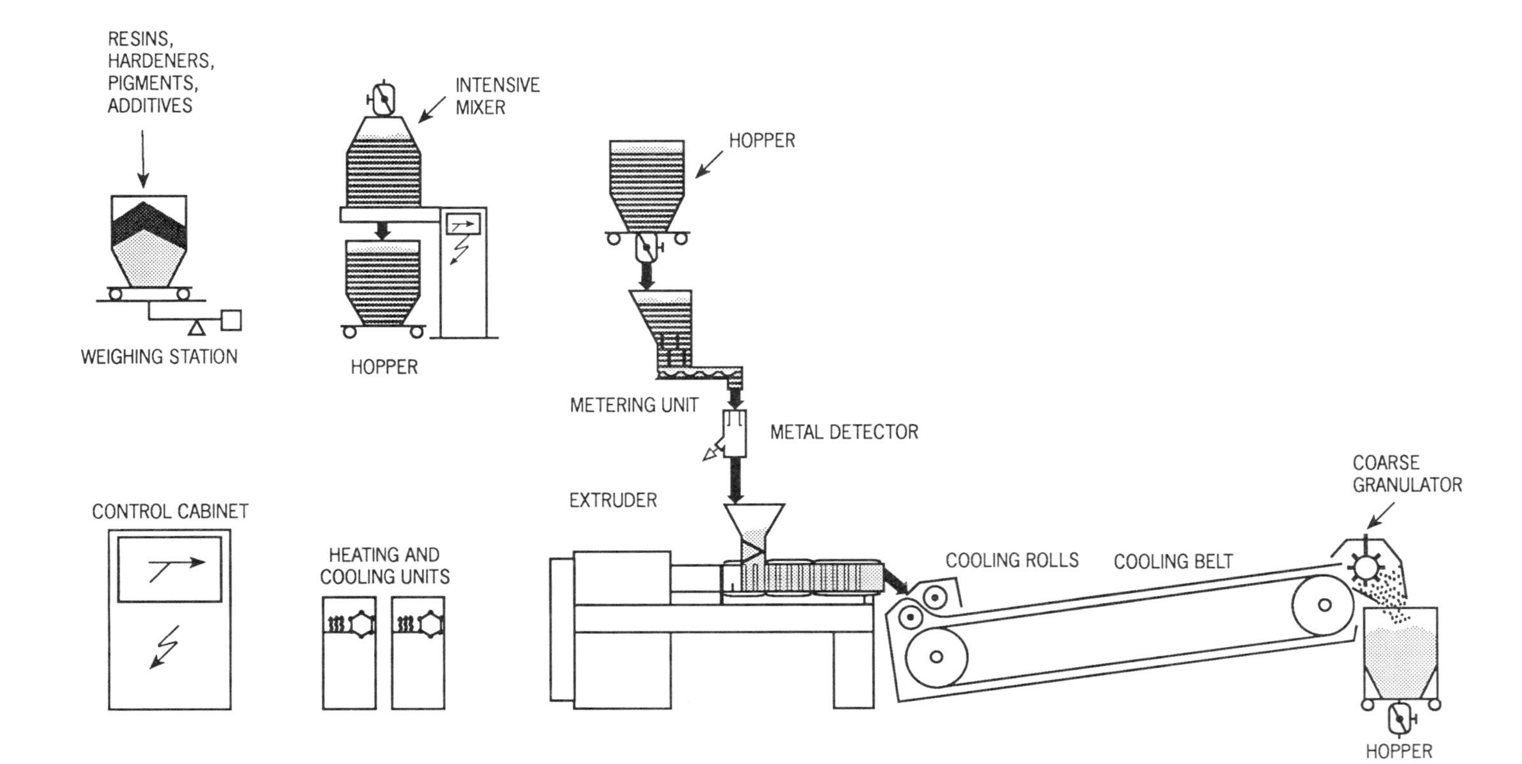

Figure 31.3. Schematic diagram of a line for premixing, melt extrusion, and granulation. (Adapted from Ref. [8], p. 242, with permission.)

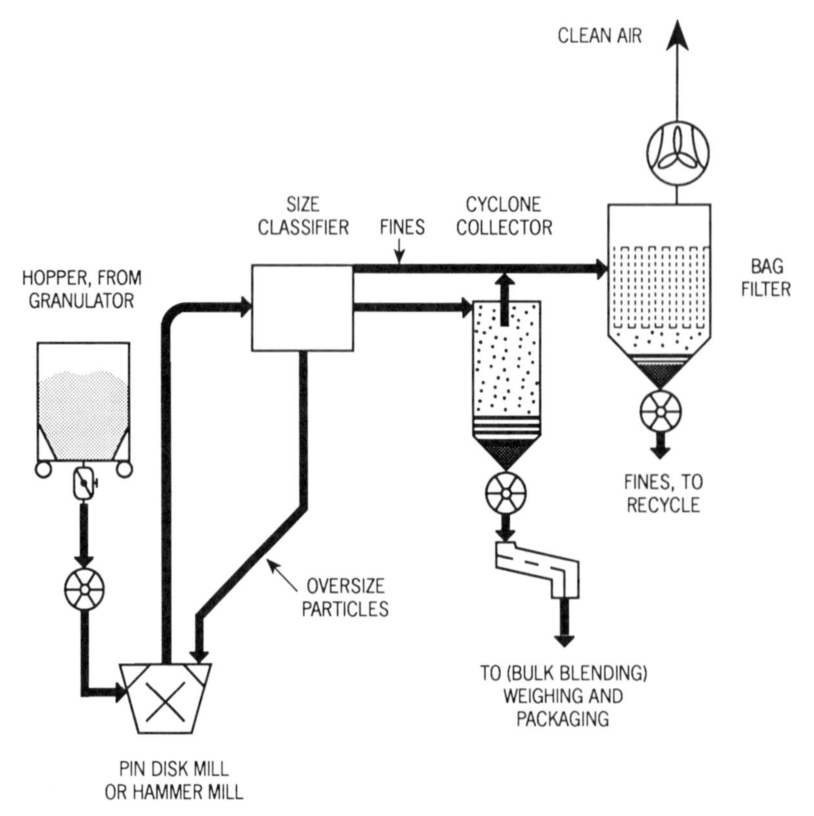

Figure 31.4. Schematic diagram of a line for pulverization and classification. (Adapted from Ref. [4], p. 252, with permission.)

coating comes out of the extruder that there is little opportunity for flocculation to occur.

The chopped granules are then pulverized. A variety of pulverizers are used. Some, such as *pin disk mills* and *hammer mills*, work on the principle of striking airborne granules with metal dowels or hammers mounted on a rapidly spinning disk. A newer type, *opposed jet mills*, work by causing high velocity collisions of granules with one another. Opposed jet mills perform well when small (<12 μm) diameter powders are needed for thin film application. Thermosetting extrudates are very brittle and are relatively easy to pulverize, but thermoplastics are generally quite tough and can be difficult to pulverize. In some cases, it is necessary either to cool the mill with liquid nitrogen or to grind dry ice along with the granules so that the temperature will be kept well below the T_g of the binder to offset the heating effect of the milling process. Even so, thermoplastic powders are generally available only in large particle sizes.

Some mills partially classify the powder, automatically returning oversize particles for further pulverization. Further size classification, by sieving and/or by an air classifier, is usually needed. The coarse particle fraction is sent back into the mill for further reduction in size. The fines are collected in a bag filter and recycled in the next batch of the same coating to be processed through the extruder. Finally, the classified powder is bulk blended for uniformity, packaged, and shipped to the customer.

31.4.2. Quality Control

Many of the quality control needs for powder coatings are similar to those for liquid coatings; those unique to powder coatings are covered in this section.

Close quality control of all resin components is required. In solvent-based coatings, the effect of small differences in molecular weight or molecular weight distribution can be readily adjusted for by small variations in the solids of the coating. In powder coatings, there is no solvent in the formula to make such adjustments. The only way of maintaining consistent quality in powder coatings is to assure that the raw materials have essentially no variation in such characteristics as molecular weight and molecular weight distribution as well as monomer composition.

In processing thermosetting compositions, considerable care must be exercised so that no more than a minimal amount of cross-linking occurs at the elevated temperatures in the extruder. In the extreme case, reprocessed material could gel in the extruder. More commonly, reprocessing can lead to increases in molecular weight due to some cross-linking reactions resulting in incomplete coalescence of the particles or poor leveling after application. The rate of travel through the extruder should be as rapid as possible, consistent with achieving the needed mixing and dispersion of the pigment aggregates. Reprocessing should be minimized. Only a limited amount of fines from the micropulverizer should be put in any one batch of coating. Reprocessing for color matching or other formulation changes should be avoided if at all possible. If reprocessing becomes essential, it is best to use a limited amount of the batch to be reprocessed in each of several new batches rather than reprocessing the batch alone.

This reprocessing limitation particularly affects color matching, the largest quality control issue in powder coating. One cannot blend batches of powder to carry out color matching, it must be done using the appropriate ratio of pigments in the extruder mix. In laboratory processing, it is feasible to make an extruder mix with your best estimate of the ratio of colorants needed to match a color and then check the color of the coating against a standard. Color can be adjusted to match the standard by mixing into the initial batch the estimated additional quantities of pigments and running the coating through the extruder and pulverizer again. In the lab, a third hit might be possible. The number of hits required is kept to a minimum by using computer color matching programs.

In production, however, it is almost essential to have the mix right the first time. Potential problems can be minimized by only extruding a small fraction of the production batch and then checking the color. If the color match is satisfactory, processing is continued. If the color needs adjusting, the initial fraction is granulated and returned to the hopper along with the required additional amounts of pigments. In pigment manufacture, there is some variation of color from batch to batch that cannot be avoided. The powder coating manufacturer who needs to make coatings with relatively close color matches must work out arrangements with the pigment manufacturer to be supplied with selected batches with narrow tolerance limits. In the case of large production runs, several extruder batches can be blended before pulverizing to average out batch to batch differences. With sufficient care, color reproducibility can be satisfactory for all but the most demanding end uses.

Another important quality control issue is particle size and particle size distribution. These important variables become increasingly critical as particle size is reduced for thin film (<50 μm) applications. Particle size distributions can be measured by passing the powder through a stack of graduated sieves and weighing the fraction retained on each sieve or by a variety of instrumental methods, such as laser diffraction particle sizing [18]. The important effects of particle size and its distribution on the handling and application characteristics of the powder are discussed in Section 31.5.1.

31.5. APPLICATION METHODS

Almost all thin film powder coatings are applied by *electrostatic spray*. Other application methods, important for protective and other thick-film powder coatings, include *fluidized bed*, *electrostatic fluidized bed*, and *flame spray*. See the general references listed at the end of the chapter for more extensive discussions.

31.5.1. Electrostatic Spray Application

Electrostatic spray is the major process for applying powder coatings. An electrostatic spray gun consists essentially of a tube to carry airborne powder to an orifice with an electrode located at the orifice. The electrode is connected to a high-voltage (40–100 kv), low-amperage power supply. Electrons are emitted by the electrode and react with molecules in the air, generating a cloud of ions, called a *corona*, in the air around the orifice. The corona probably consists predominantly of OH^- and perhaps O_2^- ions. As the powder particles come out of the orifice they pass through the corona and pick up electrons acquiring a negative electrostatic charge. The object to be coated is electrically grounded. The difference in potential attracts the powder particles to the surface of the part. They are attracted most strongly to areas that are not already covered, forming a reasonably uniform layer of powder even on irregularly shaped objects. The particles cling to the surface strongly enough and long enough for the object to be conveyed to a baking oven, where the powder particles fuse to form a continuous film, flow, and cross-link.

The powder particles that do not adhere to the object to be coated (*overspray*) are recovered and recycled by blending with virgin powder. Almost 100% is used eventually—a major advantage over spray-applied liquid coatings. A schematic diagram of electrostatic spray application of powder coatings including recovery of overspray is given in Figure 31.5. Production spray guns are usually mounted on automated reciprocators; such equipment functions smoothly with minimal worker attention as long as a single color and type of powder is being applied.

A limitation of powder coatings is the difficulty of changing colors. When spraying liquid coatings, one can simply flush the gun with solvent and shift the feed lines to the guns from those feeding one color to those feeding another color. In this way, successive objects on the conveyor line can easily be painted different colors. However, if changeover of powder coatings was attempted in this way, there would be sufficient dust in the air in the spray booth that one color would

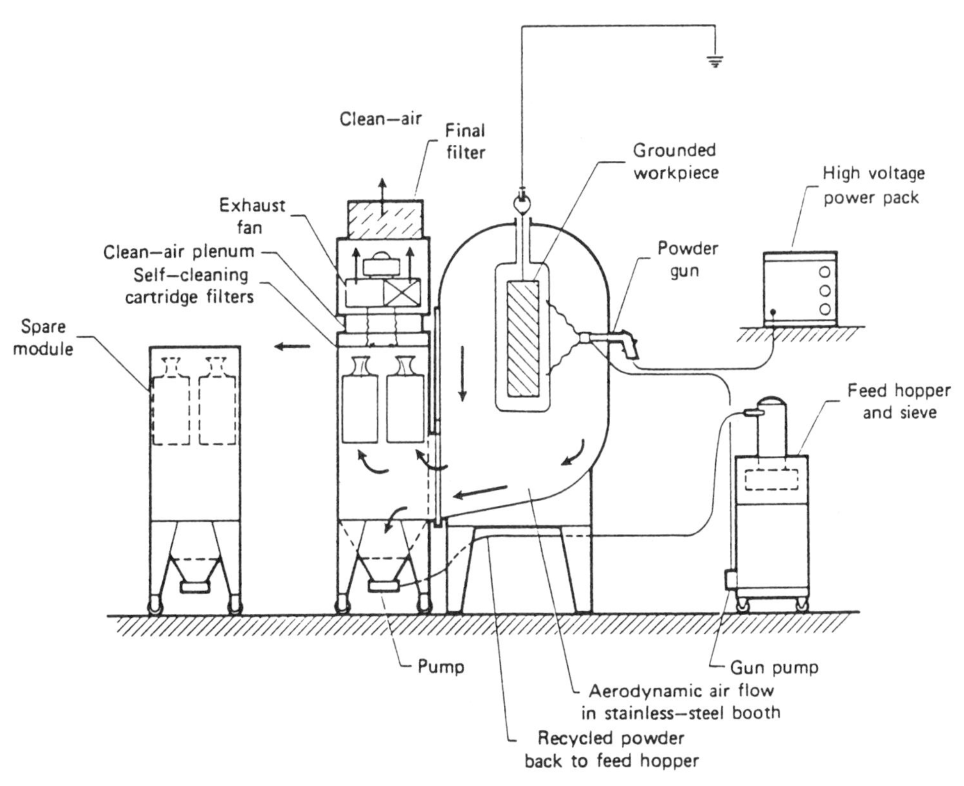

Figure 31.5. Production equipment for electrostatic spray application of powder coating showing the collection of overspray powder. (From Ref. [2], p. 18, with permission.)

contaminate another color both on the product and in the overspray collecting units. The operation must be closed down, the booth cleaned, and the overspray collecting units changed to collect the next color. While spray booths and overspray units have been designed to minimize cleanup time, it is still only economically feasible to make a reasonably long run of one color and then shift over to a reasonably long run of another color. Consequently, many of the important applications of powder coatings, such as fire extinguisher cases, are single color end uses, or are applications, such as metal furniture, where the runs of single colors are long enough to dedicate different spray booths to each of the limited number of colors involved.

In electrostatic spray, the charged powder particles wrap around the grounded object, to a degree, to coat exposed surfaces not in direct line with the spray gun. In production, it is often desirable to use spray guns on both sides of the object, making it possible to uniformly coat intricately shaped objects such as automobile wheels and tube-and-wire metal furniture. However, the process is strongly affected by the Faraday cage effect (see Section 22.2.3). As a result, it is difficult to get full coverage of areas such as interior corners of steel cabinets; the interior of pipes can only be coated by putting the spray gun inside the pipe.

Film thickness increases with increasing voltage and decreasing distance between the spray gun and the product being coated. Larger particle size powders tend to give increased film thickness. One can apply thicker coatings by heating the object to be coated before applying the powder. However, film thickness is limited because the powder coating acts as an insulator and does not attract further particles. The insulating properties of the powder coating mean that defective coated parts generally cannot be recoated and must be stripped to be recoated. Coating over other coatings and over plastics frequently requires application of a conductive primer coating.

There is considerable room for improvement of the process and considerable need for better understanding of how it works. For example, there is no satisfactory physical explanation of why the powder clings to the object as well as it does [19]. One opportunity for improvement is to increase the efficiency of charging by the corona of the electrostatic gun. It is estimated that only about 0.5% of the anions in the corona transfer electrons to powder particles [20]. The rest of the ions are attracted to the nearest grounded object, where they do no good at best and at worst may reduce deposition efficiency and increase the Faraday cage effect [21]. Research on ways to improve this situation is underway in several laboratories; some experts think that, if charging efficiency could be raised to 10%, deposition efficiency might become high enough to reduce overspray sufficiently that it would no longer be necessary to collect it. Such a development would make powder coating more attractive for applications where color is changed frequently.

A relatively recent innovation in electrostatic powder coating application is *triboelectric charging* of the powder particles. Instead of a high voltage source generating a corona at the gun orifice, the particles are charged by the friction of streaming through a poly(tetrafluoroethylene) spray tube in the gun. The mechanism is analogous to the buildup of static charge on a comb when combing your hair. Since there is not the large differential in charge between the gun and the grounded work being sprayed, no significant magnetic field lines are established and Faraday cage buildup is minimal, facilitating coating hollows in irregularly shaped objects. It is said that smoother coatings are obtained. On the other hand, throughput is slower and stray air currents can more easily deflect the particles between the gun and the object being coated. Tribo charging processes are widely used in Europe and are gaining popularity in North America. Guns are available which combine tribo and corona charging.

Particle size and particle size distribution have a critical effect on powder coating. The range of particle sizes must be limited; as a rule of thumb, the predominant particle diameter should be somewhat less than the intended film thickness and the largest particles should be no more than twice the film thickness. Figure 31.6 shows a typical distribution, presumably for a powder coating to be applied at a thickness of around 50 μm.

Very fine particles do not flow properly in hoppers and feed lines. In general only 6–8 wt% of the particles are smaller than 10 μm. Small particles have a higher ratio of surface to volume and hence acquire a higher charge/mass ratio as they pass through the corona of an electrostatic gun. After charging the particles are affected by three forces, the electrostatic field, air movements, and gravity. Theoretical calculations predict that gravity should be the predominant effect on very large particles and that air flow should predominate on the very small ones [23]. These theoretical predictions are confirmed by particle size analysis of overspray; with a typical powder, the overspray was rich in particle sizes below 20 and above

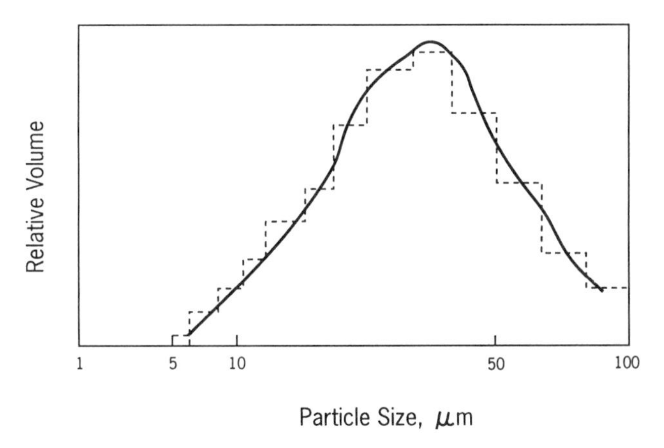

Figure 31.6. Particle size distribution of a typical white epoxy powder coating. (From Ref. [22], with permission.)

60 μm, indicating that the intermediate sizes were transferred most effectively. Small particles probably penetrate best into Faraday cages. Large particles flow better through the application system and retain their charge longer, and therefore cling to the object better between application and baking. Very small (<1 μm) dust particles may present a toxic hazard if they are inhaled. On balance, it appears that particle sizes in the 20–60 μm range are best for powder coatings that are to be applied at film thicknesses of 30–60 μm. Thinner films require smaller particle sizes, but handling and spraying characteristics are inferior, and difficulties increase as thinner and thinner films are sought.

31.5.2. Other Application Methods

The *fluidized bed* method is the oldest method of applying powder coatings. The equipment consists of a dip tank, the bottom of which is a porous plate. Air is forced through the porous plate and acts to suspend the powder in the dip tank in air. The flow behavior of air suspensions of powders resembles that of fluids, hence the term fluidized bed. The object to be coated is hung from a conveyor and heated in an oven to a temperature well above the T_g of the powder to be coated on it. The conveyor then carries the part into the fluidized bed tank. Powder particles fuse onto the surface of the object. As the thickness of fused particles builds up, the coating becomes a thermal insulating layer so that the temperature at the surface of the coating becomes lower, finally reaching the stage where further particles will not stick to the surface. The last particles that attach to the coated surface are not completely fused so the conveyor must now carry the object into another oven where the fusion is completed.

The film thickness depends on the temperature to which the part is preheated and the T_g of the powder. Thin films cannot be applied in this fashion. Most commonly, the method is used for applying thermoplastic coating materials.

Electrostatic fluidized beds are similar, but electrodes are added to generate ions in the air before it passes through the powder, and the object to be coated is grounded, as illustrated schematically in Figure 31.7 [24]. Powder is attracted to

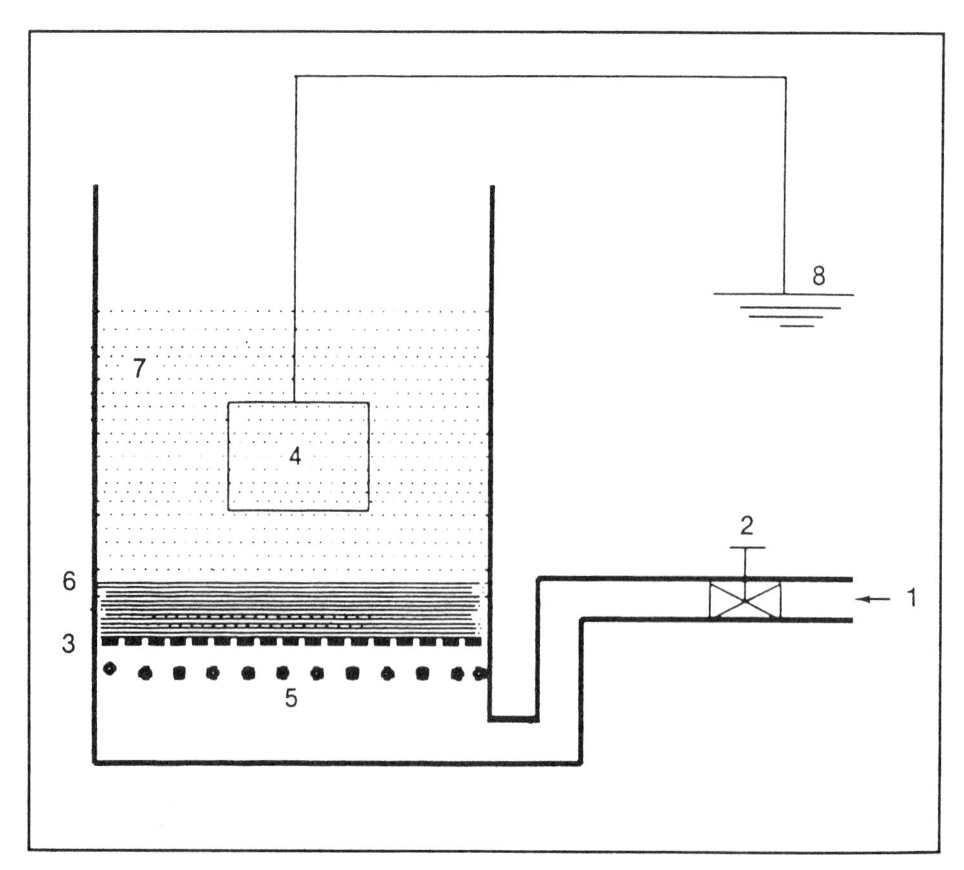

Figure 31.7. An electrostatic fluidized bed coating apparatus. 1, Air inlet; 2, air regulator; 3, porous membrane; 4, object to be coated; 5, electrodes; 6, fluidized powder; 7, cloud of charged powder; 8, ground. (From Ref. [24], with permission.)

the object by electrostatic force, as in electrostatic spray. The object can be heated when thick films are desired, but heating is not necessary. Thinner films can be applied than with the heated part, fluidized bed process. This method is used to apply thermoplastic and some thermosetting powders, for example, electrical insulation coatings. There is no overspray and powder losses are minimal; color changeover is also said to be easier. However, thin films are hard to apply; there is a strong Faraday cage effect; and it is difficult to coat large objects uniformly.

Flame spray is a developing technology for applying powder coatings [13,25]. In a flame spray gun, the powder is propelled through a flame, remaining just long enough to melt the powder, and then the molten powder particles are directed at the object to be coated. The flame heats and melts the polymer and also heats the substrate above the melting temperature of the polymer so that the coating can flow into irregularities in the surface to provide an anchor for additional adhesion. The combination of flame temperature (on the order of 800°C), residence time in the flame (small fractions of a second), T_g of the coating, particle size distribution, and substrate temperature must be carefully balanced. Particle size distribution must be fairly narrow; smaller particles will pyrolyze at 800°C before larger particles achieve melting temperatures.

In contrast to other methods of application, flame spray permits application in the field not just in a factory. Examples of commercial or experimental applications include drum linings, metal lamp poles, bridge rails, concrete slabs, grain rail cars, among others.

Since the application is not electrostatic, nonconductive substrates such as concrete, wood, and plastics can be coated. The coating need not be baked, so substrates that will withstand just the temperature from the impinging spray can be coated by this method. Since the coatings are thermoplastic and are not applied electrostatically, it is possible to repair damaged areas of coating, which is generally not possible with other powder coating methods.

Disadvantages of flame spray include limitation on the service temperature of objects coated with thermoplastic coatings (in the case of ethylene/acrylic acid (EAA) based coatings, 75°C) and the need for careful control of application variables [13]. Overheating the polymer can lead to thermal degradation and hence to poor coating performance with no visual indication of degradation until the polymer begins to pyrolyze.

31.6. ADVANTAGES, LIMITATIONS, AND USES OF POWDER COATINGS

In this section, the advantages and limitations of powder coatings are compared to solvent- or water-borne coatings. The primary advantages may be summarized as follows: VOC emissions approach zero. There is little waste—a major advantage compared to spray-applied fluid coatings. Flammability and toxicity hazards are substantially reduced. Thick films of 100–500 μm can be applied in a single application. Energy consumption is substantially reduced; at first blush, the low energy requirements of powder coatings are surprising since the baking temperatures are generally higher than for most baking solvent-borne coatings. However, with little or no volatiles being emitted into the oven, the air in the oven can be essentially completely recirculated with almost no make-up air. In contrast, with solvent-borne coatings the solvent concentration in the air in the oven must always be kept well below the lower explosive limit, resulting in the need to heat a substantial volume of make-up air. Similarly, the air flow through the spray booth can be substantially lower with powder coatings because it is no longer necessary to keep the solvent concentration in the air well below a safe concentration for people to be in the spray booth. In winter, the cost of heating air flowing at a high rate through a spray booth can be very high.

An important further factor in reducing overall cost of powder coatings is that the overspray powder can be collected in filter bags from the spray booth and reused. This not only increases paint utilization, it also essentially eliminates the cost and difficulty of disposing of the sludge obtained from the water wash spray booths that are used with liquid paints.

Important limitations of powder coatings are:

Explosive hazards. While the absence of solvents eliminates the flammability problem, suspensions of powder in air can explode. Consequently, the manufacturing and application facilities have to be designed to avoid powder explosions.

With good engineering and good housekeeping, the process can be operated quite safely. Triboelectric charge systems are said to be less likely to lead to explosive hazards.

Inability to coat large or heat-sensitive substrates. Proven application methods are limited to spray and fluidized bed methods; not all objects can be economically coated by these application methods. Only baking applications can be considered and, since the baking temperature must be fairly high, only substrates that can withstand the baking cycle can be used. This limits applications almost entirely to metal substrates. As noted in Section 31.5.2, flame spray can be used in some cases for such applications.

Appearance limitations. Broadly speaking, powder coatings can have good appearance, but some of the appearance effects attainable with liquid coatings are difficult or impossible to match with powders. Since there is no solvent loss, there is not the degree of shrinking during film formation required for metallic coatings showing the color flop typical of automotive coatings (see Section 33.1.3). One can make powder coatings containing aluminum flake pigments; they sparkle, but display little color flop.

As indicated in Section 31.3, a compromise between leveling and edge coverage is inevitable. The usual result is that the formulator must settle for a film that has a distinct orange peel to get acceptable edge coverage. The thinner the film, the more difficult the problem becomes. As also discussed in Section 31.3, there are problems of reproducibly controlling gloss of low- and medium-gloss powder coatings. Also as discussed in Section 31.4.2, close color matching is somewhat more difficult than with conventional types of coatings.

Materials limitations. Since all major components must be solids, there is a smaller range of raw materials available to the formulator. Furthermore, it is not possible to make thermosetting powder coatings where the T_g of the final film is low. This limits the range of mechanical properties that can be formulated into a powder coating, although excellent properties are attainable for most metal coating uses.

Limits of production flexibility. Economics of production and application suffer badly whenever frequent color changes are needed. Cleanup between color changes is time consuming. Powder coatings are best suited to reasonably long production runs of the same type and color of powder.

Some of the limitations of powder coatings can be overcome by making aqueous dispersions of powders. This eliminates the powder explosion problem; broadens the range of applicable application methods; and, particularly, reduces the storage stability problem. The T_g of the powder no longer needs to be high to avoid sintering and, therefore, more flexible coatings can be formulated and lower baking temperatures can be utilized. Of course, the new challenges are added of making stable aqueous dispersions without undue sacrifice of properties due to the presence of surfactant in the composition and controlling spray rheology.

An important application for powder coatings is pipe coatings. These are epoxy coatings with relatively large average particle size that are applied to give thick films, 300–450 μm. Other important applications are metal furniture, appliances, lawn mowers, fire extinguishers, rebars, and so on. Some trucks are painted with solid color powder top coats and some powder primer is used. While at one time, some believed that automobile and truck exterior body coatings would become important uses of powder coatings, the market has not developed broadly. The lack of flop in metallic coatings, difficulty of frequent color changes in the coating line, and the increasing use of heat-sensitive plastic body components are among the important factors limiting the use. Under the hood parts and wheels are examples of automotive components where these limitations do not apply and powder coatings are used on a large scale. New generations of powder coatings for automotive top coats are in development stages. Some predict that powder clear coats will be used with high solids or water-borne base coats, while others predict that powder coatings will eventually satisfy both uses.

GENERAL REFERENCES

J. H. Jilek, *Powder Coatings*, Federation of Societies for Coatings Technology, Blue Bell, PA, 1991.

T. E. Misev, *Powder Coatings Chemistry and Technology*, Wiley, New York, 1991.

REFERENCES

1. J. H. Jilek, *Powder Coatings*, Federation of Societies for Coatings Technology, Blue Bell, PA, 1991.
2. D. S. Richert, "Powder Coatings," in *Kirk-Othmer Encyclopedia of Chemical Science and Technology*, 3rd ed., Vol. 19, Wiley, New York, 1982, p. 11.
3. L. Kapilow and R. Samuel, *J. Coat. Technol.*, **59** (750), 39 (1987).
4. T. A. Misev, *Powder Coatings Chemistry and Technology*, Wiley, New York, 1991, p. 48.
5. G. C. Fischer and L. M. McKinney, *J. Coat. Technol.*, **60** (762), 39 (1988).
6. K. Kronberger, D. A. Hammerton, K. A. Wood, and M. Stodeman, *J. Oil Colour Chem. Assoc.*, **74**, 405 (1991).
7. Ref. [4], p. 146.
8. S. P. Pappas and E. H. Urruti, *Proc. Water-Borne Higher-Solids Coat. Symp.*, New Orleans, 1986, p. 146–161.
9. P. J. Achorn, W. Jacobs, W. E. Mealmaker, S. Sansur, and R. G. Lees, *Polym. Mater. Sci. Eng.*, **67**, 218 (1992).
10. H. P. Higginbottom, G. R. Bowers, J. S. Grande, and L. W. Hill, *Prog. Org. Coat.*, **20**, 301 (1992).
11. H. P. Higginbottom, G. R. Bowers, and L. W. Hill, *Polym. Mater. Sci. Eng.*, **67**, 175 (1992).
12. R. van der Linde and B. J. R. Scholtens, *Proc. 6th Annual Intern. Conf. on Cross-linked Polymers*, Noordwijk, Netherlands, June 1992, p. 131.
13. T. Glass and J. Depoy, *Society of Manufacturing Engineers, Finishing '91 Conference, FC91-384*, Cincinnati, OH, September 1991.
14. T. Nakamichi and M. Mashita, *Powder Coatings*, **6** (2), 2 (1984).
15. P. G. de Lange, *J. Coat. Technol.*, **56** (717), 23 (1984).
16. M. J. Hannon, D. Rhum, and K. D. Wissbrun, *J. Coat. Technol.*, **48** (621), 42 (1976).
17. W. S. Overdiep, *Prog. Org. Coat.*, **14**, 159 (1986).
18. G. H. Barth and S. T. Sun, *Anal. Chem.*, **61**, 143 (1989).
19. E. F. Meyer III, *Polym. Mater. Sci. Eng.*, **67**, 220 (1992).

20. J. F. Hughes, *Electrostatic Spraying of Powder Coatings*, Research Studies Press, Ltd., Letchworth, Hertfordshire, 1984.
21. J. F. Hughes, *Ytbehandlingdagar*, Stockholm, Sweden, May 1992, p. A-1.
22. Anon., *Technical Bulletin SC:586-82*, Shell Chemical Co., 1989.
23. H. Bauch, *Polym. Mater. Sci. Eng.*, **67**, 344 (1992).
24. Ref. [4], p. 346.
25. Ref. [4], pp. 347–349.

CHAPTER XXXII

Radiation Cure Coatings

Radiation cure coatings cross-link by reactions initiated by radiation rather than by heat. Such coatings have the potential major advantage of being indefinitely stable when stored in the absence of radiation. After application, cross-linking occurs rapidly at ambient temperature on exposure to radiation. Rapid cure at ambient temperature is particularly significant for heat sensitive substrates, including paper, some plastics, and wood.

There are two broad classes of radiation cure coatings: (1) UV cure coatings in which the initial step is excitation of a photoinitiator (or photosensitizer) by absorption of photons of UV-visible electromagnetic radiation, and (2) EB (electron beam) cure coatings in which the initial step is ionization and excitation of the coatings resins by high energy electrons (particle radiation). Cross-linking of polymerizable resins is initiated by the reactive intermediates that are generated from the photoexcited photoinitiator in UV curing and from the excited and ionized resins in EB curing. Infrared and microwave radiation are also used to cure coatings; but these systems are not included herein since the radiation is simply converted to heat which initiates thermal curing.

32.1. UV CURING

There are two broad classes of polymerization reactions used in UV curing: free radical initiated and acid initiated chain-growth polymerization reactions. Note that both classes are chain reactions. While there have been attempts to use photoreactions where the radiation leads just to the generation of a reactive functional group, this approach is not practical for coatings purposes. In such reactions, each photon absorbed can effect, at most, one cross-linking reaction. In chain reactions, absorption of a single photon can lead to the formation of many cross-links.

A key to successful use of UV curing is the availability of a UV source that reliably produces a relatively high intensity of UV radiation at low cost without generating excessive infrared radiation. The major sources in commercial use are medium pressure mercury vapor lamps. Such *electrode lamps* are tubes up to 2 m long; power outputs of 80W cm^{-1} are in wide use, and lamps with outputs up to 240 W cm^{-1} are available. The radiation consists of continuous wavelength distri-

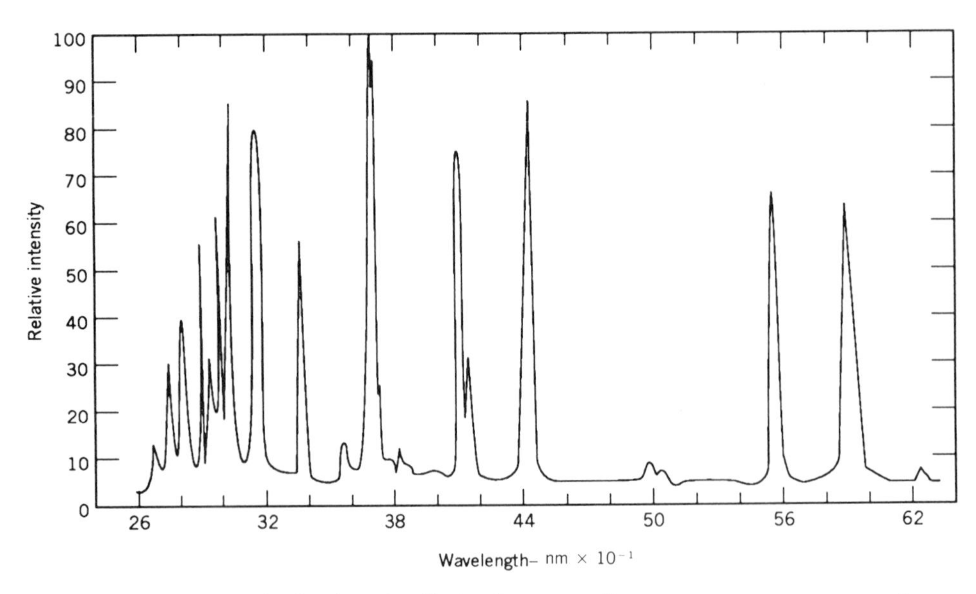

Figure 32.1. Energy Distribution of radiation from a medium pressure mercury vapor lamp. (Source: U.S. Patent 3 650 699 as cited in Ref. [1].)

bution radiation with major peaks at 313, 366, and 405 nm among others, including the 254-nm emission obtained from low-pressure mercury vapor sources. Substantial visible radiation and a minor, but not insignificant, amount of infrared radiation, that causes some heating, are also emitted. A typical wavelength distribution of an electrode lamp is shown in Figure 32.1.

In some cases there are advantages to other wavelength distributions, especially to those with increased fractions of the radiation in the very near UV-visible region. Changes in distribution can be made by doping the lamp with traces of other elements besides mercury or by having fluorescent coatings on the lamp tube that will absorb short UV and emit longer UV radiation.

Electrodeless lamps, powered by microwaves, also enjoy substantial commercial use. Electrodeless lamps are more suitable for doping since the lifetime of lamp electrodes is generally reduced by dopants. Electrodeless lamps have the further advantage of essentially instantaneous start-up and restart after the lamp is turned off. On the other hand, electrodeless lamps are more expensive than the electrode type. For further discussion of UV sources, see Ref. [1].

Radiation is emitted in all directions around the tubular-shaped lamp, and its intensity drops off with the square of the distance from the source. To increase the efficiency of use of the radiation, the lamps are mounted in an elliptical reflector with a focal length such that the maximum intensity of radiation is focused at the distance between the lamp and the surface being coated. Since thermal energy is also produced, the lamp housing must be water- and air-cooled.

UV radiation is hazardous and can lead to severe burns. It is essential to avoid exposure of eyes to the radiation. The lamps are housed in enclosures; when the enclosure is opened, the current to the lamps is automatically turned off. Depending on the radiation source, a greater or lesser amount of ozone is generated. Since ozone is toxic, the UV unit must be ventilated to carry it away.

One of the limitations of UV curing is that the distance between the lamp and the coating on various parts of the object being coated must be fairly uniform. Hence, UV curing is most easily applicable to coating flat sheets or webs that can be moved under the UV lamps or cylindrical objects that can be rotated under (or in front of) the lamps.

For efficient initiation there must be absorption of radiation by the photoinitiator (or as discussed later some substance that leads to the generation of an initiator). The fraction of radiation absorbed, I_A/I_0, is related to the molar absorptivity, ε, the concentration of the photoinitiator, C, and the optical path length of the radiation in the film, X. Assuming there is no other absorber present and neglecting surface reflection, the fraction of radiation of a given wavelength absorbed can be expressed by the Eq. 32.1. Just as discussed in Section 16.2.2 in connection with color, the molar absorptivity can be concentration dependent so that the range of concentration over which the equation is valid may be limited.

$$\frac{I_A}{I_0} = 1 - 10^{-\varepsilon CX} \tag{32.1}$$

The molar absorptivity varies with wavelength, so the fraction of radiation absorbed varies with wavelength. Also the intensity of radiation from the source varies as a function of wavelength. The total number of photons absorbed per unit time depends on the combination of these factors.

When the photoinitiator absorbs a photon, it is raised to an excited state. Some reaction of this excited state leads to the generation of the initiator. But there are also other possible fates of the excited state photoinitiator. For example, it may emit energy of a longer wavelength, that is, it may fluoresce or phosphoresce. It may be quenched by some component of the coating or by oxygen. It could undergo other reactions besides those that lead to initiator generation. The efficiency of generation of initiating species is an important factor in the selection of a photoinitiator.

The rate of polymerization reactions is related to the concentration of initiating radicals or ions. It would, therefore, seem that the higher the initiator concentration, the faster the curing. As one increases from very low concentrations to somewhat higher concentrations, the rate of cure does, in fact, increase. However, for any given system, there is an optimum concentration. The rate of cure in the lower part of the film decreases above this concentration. If the concentration is high enough, such a large fraction of the radiation is absorbed in the upper few micrometers of the film that little radiation reaches the lower layers to be absorbed. Since the half life of free radicals is so short, they must be generated within a few nanometers of the depth in the film where they are needed to initiate polymerization. Although the half life of certain acid species may be substantially longer, migration is limited by diffusional constraints as polymerization and cross-linking proceed.

The optimum photoinitiator concentration is film thickness dependent: the greater the film thickness, the lower the optimum concentration. Lower concentration is favorable from the standpoint of cost, since the photoinitiator is generally an expensive component. However, comparing the cure speed of films of different film thicknesses, each containing the optimum concentration of photoinitiator for its film thickness, the time required to cure a thick film is longer than to cure a thin film. This follows since less radiation is absorbed in any volume element with increasing film thickness. The problem is further exacerbated when surface cure is oxygen inhibited, as is the case with free radical polymerization. (Oxygen inhibition is discussed in Section 32.1.1.3.) In general, one should determine the concentration of photoinitiator that is just sufficient to give the required extent of cure at the surface of the film; this concentration will give the maximum rate of cure for that film thickness of that system at the lowest cost.

The problem of achieving both surface and through cure can be ameliorated by using a photoinitiator, or mixture of photoinitiators, having two absorption maximums that have distinctly different molar absorptivities near different emission bands from the UV source. In such a case, the emission band that is highly absorbed by the photoinitiator(s) will be absorbed more strongly near the surface and have less UV available for absorption in the lower layers. This band will be most important for counteracting oxygen inhibition (see Section 32.1.1.3), but will not contribute substantially to through cure. Therefore, it is important that there be weaker absorption of a second emission band that will be absorbed more uniformly throughout the film to provide through cure.

While film thickness is one variable that affects path length, it is not the only one. If one applies the same coating over a black substrate and over a highly reflective metal substrate, it is common to find that the rate of cure over the metal substrate is close to twice as fast as over the black substrate. UV radiation that passes through the film to the black substrate is absorbed, but that reaching a smooth metal substrate is reflected and passes through the film twice: the path length is twice the film thickness so there is almost twice the opportunity for absorption.

Another factor that can affect absorption of UV by the photoinitiator is the presence of competitive absorbers or materials that scatter UV radiation. In designing vehicles for use in UV cure systems, it is usually desirable to minimize their absorption of UV in the range needed for the excitation of the photoinitiator. Pigments can absorb and/or scatter UV radiation; the effects of pigmentation are discussed in Section 32.1.3.

The extent of conversion of polymerization reactions is affected by free volume availability. As polymerization proceeds, the T_g of the binder increases. If, as is commonly the case, the T_g begins to approach the temperature at which the curing is being carried out, the rate of reactions will slow down. In most cases the reactions will cease at temperatures only a little above $T_g = T_{cure}$. Since there is heat from the radiation source and a reaction exotherm, T_{cure} is somewhat above ambient temperature.

32.1.1. Free Radical Initiated UV Cure

In these coatings, free radicals are photogenerated and initiate polymerization by adding to vinyl double bonds, primarily acrylates. There are two broad classes of photoinitiators used: those that undergo unimolecular bond cleavage, and those that undergo bimolecular hydrogen abstraction from some other molecule. Free radical initiating species are generated in both processes.

32.1.1.1. Unimolecular Photoinitiators

A wide range of unimolecular photoinitiators has been studied [2]. The first ones used on a large scale commercially were ethers of benzoin. Benzoin ethers undergo cleavage to form benzoyl and benzyl ether radicals:

$$C_6H_5{-}C(=O){-}CH(OR){-}C_6H_5 \xrightarrow{h\nu} C_6H_5{-}\dot{C}{=}O + C_6H_5{-}\dot{C}H(OR)$$

Both of these radicals can initiate the polymerization of acrylate monomers [3]. There is doubt whether the benzyl ether radical can initiate the polymerization of styrene [2]. The package stability of UV cure coatings containing such benzoin ethers tends to be limited. Apparently, this is due to the ease of abstraction of the hydrogen on the benzyl ether carbon. Any organic material contains hydroperoxides that will slowly decompose. The resulting radicals can abstract the hydrogen leading to initiation of polymerization and hence poor package stability. Package stability is substantially improved if the benzyl carbon is fully substituted.

Accordingly, the ketal, 2,2-dimethoxy-2-phenylacetophenone, is an effective photoinitiator with good package stability. Photocleavage produces benzoyl and dimethoxybenzyl radicals. The latter is a sluggish initiator; however, it can undergo further cleavage to the highly reactive methyl radical; the extent of this cleavage increases with increasing temperature.

$$Ph{-}C(=O){-}C(OMe)_2{-}Ph \xrightarrow{h\nu} Ph{-}\dot{C}{=}O + Ph{-}\dot{C}(OMe)_2$$

$$Ph{-}\dot{C}(OMe)_2 \longrightarrow Ph{-}C(=O){-}OMe + Me\cdot$$

2,2-Dialkyl-2-hydroxyacetophenones are also commercially important photoinitiators with good package stability. These photoinitiators tend to give less yellowing than phenyl substituted acetophenones, including benzoin ethers as well as the above ketal, probably because benzylic radicals are not generated by photocleavage.

$$Ph{-}C(=O){-}C(OH)(R'){-}R \xrightarrow{h\nu} Ph{-}\dot{C}{=}O + R{-}\dot{C}(OH){-}R'$$

All of these photoinitiators have the acetophenone chromophore, the absorptivity of which is enhanced by electron donating substituents on the benzoyl ring.

This is exemplified by the morpholino-substituted photoinitiator, 2-dimethylamino-2-benzyl-1-(4-morpholinophenyl)-butan-1-one, which is recommended for pigmented coatings.

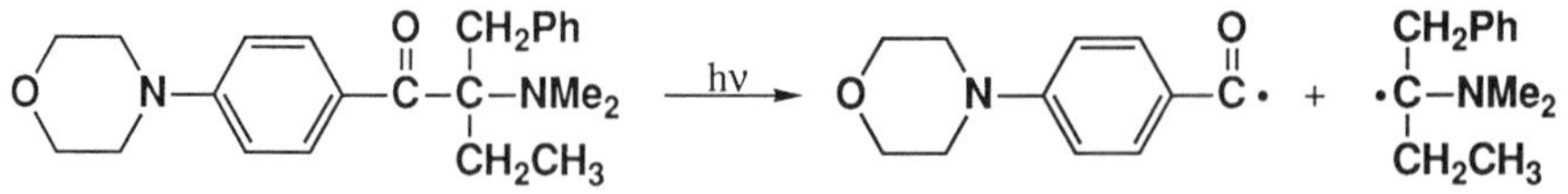

A commercially important class of unimolecular photoinitiators, which does not have the acetophenone chromophore, is acyl phosphine oxides, notably diphenyl-2,4,6-trimethylbenzoylphosphine oxide. Irradiation generates the corresponding benzoyl and phosphinoyl radicals. Acyl phosphine oxides tend to be nonyellowing, have good package stability, and have desirable absorption characteristics for both clear and pigmented coatings.

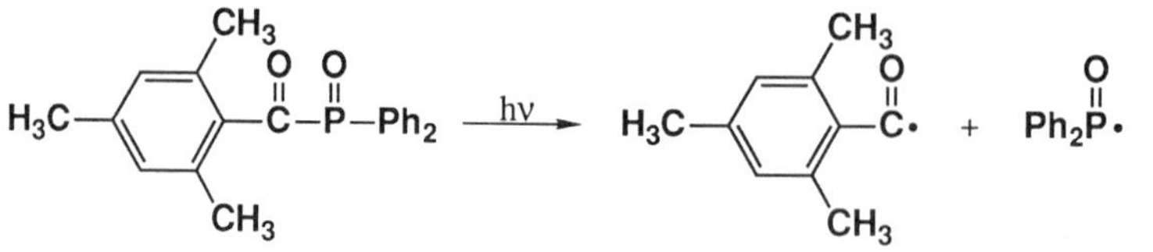

32.1.1.2. Bimolecular Initiators

Photoexcited benzophenone, and related diarylketones such as xanthone and thioxanthones, do not cleave to give free radicals but can abstract hydrogens from an appropriate hydrogen donor to yield free radicals that initiate polymerization. Thioxanthones, such as 2-isopropylthioxanthone, are used when their high absorption in the very near UV is desirable to permit absorption in the presence of pigments that absorb UV strongly in the shorter UV range.

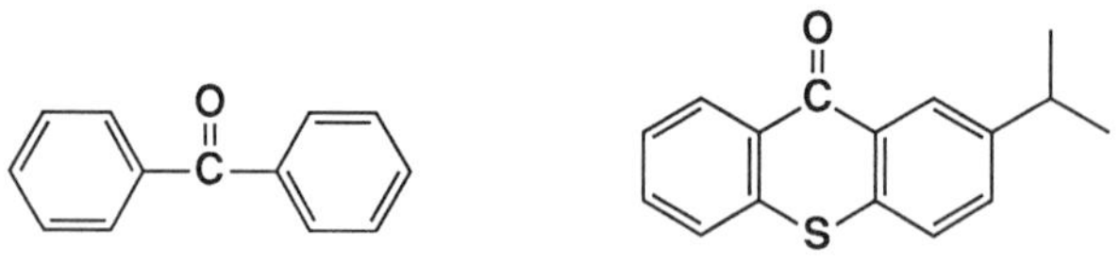

Benzophenone

2-Isopropylthioxanthone

The most widely used hydrogen donors are tertiary amines with hydrogens on α-carbon atoms such as 2-(dimethylamino)ethanol (DMAE) and, especially in printing inks, methyl *p*-(dimethylamino)benzoate. It has been shown that the accompanying ketyl free radical does not initiate polymerization.

$$Ph_2C{=}O \xrightarrow{h\nu} Ph_2C{=}O^* \xrightarrow{R_2NCH_3} Ph_2\dot{C}(OH) + R_2N{-}\dot{C}H_2$$

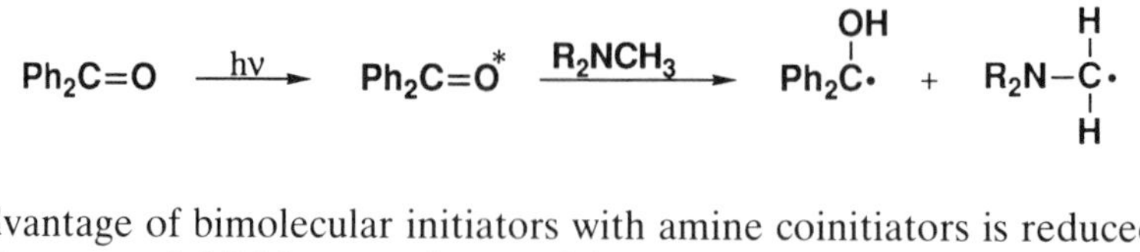

An advantage of bimolecular initiators with amine coinitiators is reduced problems with oxygen inhibition, as discussed in the next section. A disadvantage is that the excited states of these initiators are more readily quenched by oxygen as well as by vinyl monomers with lower triplet energies; hence they cannot be used as photoinitiating systems with styrene.

32.1.1.3. Oxygen Inhibition

Oxygen inhibits free radical chain-growth polymerizations. In coatings, this inhibition is particularly troublesome since coating films have such a high ratio of surface area, where oxygen exposure is high, to total volume. Oxygen reacts with the terminal free radical on a propagating molecule to form a peroxy free radical. The peroxy free radical does not readily add to another monomer molecule; thus the growth of the chain is terminated. The terminal radical from methyl methacrylate has been shown to have a rate constant for reaction with oxygen that is 10^6 times that for reaction with another monomer molecule. Furthermore, the excited states of certain photoinitiators are quenched by oxygen, thereby reducing the efficiency of generation of free radicals. Reaction with oxygen is most prevalent at the air/coating interface where the concentration of oxygen is two orders of magnitude greater than inside the film.

Several approaches are available to minimize this problem. One can carry out the curing in an inert atmosphere, but this is relatively expensive. One can incorporate paraffin wax in the coating. As the coating is applied and cured, a layer of wax comes to the surface of the coating, shielding the surface from oxygen. While effective, the residual wax detracts from the appearance of the film and makes recoating difficult. High-intensity UV sources minimize, but do not eliminate, the problem with fast curing systems. In effect, free radicals can be generated so rapidly that their high concentration can deplete the oxygen at the film surface, permitting other radicals to carry on the polymerization before further oxygen diffuses to the surface.

Free radicals on carbon atoms alpha to amines react especially rapidly with oxygen. In the case of DMAE, there are eight potentially abstractable hydrogens on each molecule, and each DMAE molecule can potentially react with eight oxygen molecules. Thus, benzophenone/amine initiating systems substantially reduce oxygen inhibition. Furthermore, one can also reduce oxygen inhibition in coatings using unimolecular photoinitiators by adding small amounts of an appropriate amine.

32.1.1.4. Vehicles for Free Radical Initiated UV Cure

In the earliest work, styrene solutions of unsaturated polyesters (see Section 13.1) were used along with benzoin ether photoinitiators. However, styrene is sufficiently volatile that a significant amount evaporates between application and curing. Furthermore, the rate of polymerization is relatively slow as compared to acrylate systems. The cost of these compositions is low, and they continue to find some application because of the cost advantage.

Most current coatings use acrylated reactants (see Section 13.2). Primarily acrylate, rather than methacrylate, esters are used since acrylates cure much more rapidly at room temperature; they are also less oxygen inhibited. Furthermore, polymerization of acrylates tends to terminate by combination, whereas methacrylate polymerization terminates largely by disproportionation (see Section 4.1.1). The extent of cross-linking as well as higher molecular weights are favored when termination of growing radicals, derived from multifunctional oligomers/monomers, occurs by combination relative to disproportionation.

Two different ways of expressing functionality of molecules with reactive groups can be potentially confusing. As the term is used in discussing numbers of functional groups, a molecule with one acrylate ester group is called monofunctional, difunctional with two, and poly- or multifunctional with two or more groups. On the other hand, each of the acrylate double bonds can actually react twice; first with a free radical to undergo addition, followed by reaction with another double bond in a chain reaction. In the latter kinetic sense, a molecule with one acrylate ester group is difunctional. We follow the common practice of referring to molecules that contain one acrylate ester group as monofunctional, two as difunctional, and so on.

In general, the vehicle consists of two types of acrylate esters: acrylate-terminated oligomers, generally di- and trifunctional, and acrylate monomers ranging from mono- to hexafunctional, most commonly mixtures of mono-, di-, and trifunctional monomers. The monomers are also called *reactive diluents*. The multifunctional oligomer contributes to a high rate of cure owing to its polyfunctionality and, in large measure, controls the properties of the final coating because of the effect of the backbone structure on such properties as abrasion resistance, flexibility, adhesion, and so forth. On the other hand, the higher molecular weight of the oligomer means that its viscosity is relatively high. The multifunctional acrylate monomers also lead to fast cure owing to polyfunctionality but with lower viscosity than the oligomer. The viscosity is lowered further by the monofunctional acrylates, although speed is generally compromised and "cured volatiles" increase.

The two principal types of oligomers are derived from epoxy- and isocyanate-functional starting materials. Low molecular weight BPA epoxy resins can be reacted with acrylic acid to yield predominantly difunctional acrylate derivatives (see Section 13.2). Epoxidized soybean oil acrylates have been used where the higher functionality and lower T_g are desirable. Epoxidized linseed oil derivatives have even higher functionality.

The urethane oligomers are obtained from isocyanate-terminated resins by reaction with 2-hydroxyethyl acrylate (see Section 13.2). Any hydroxy-terminated resin can be reacted with a diisocyanate to give an isocyanate-terminated resin. Polyesters have been the most widely used base resins but polyurethanes, acrylics, and other hydroxy-functional resins are also used. If color retention is important, aliphatic diisocyanates are used; if not, aromatic diisocyanates are used because of their lower cost (see Section 12.3.2).

Many multifunctional acrylate monomers have been used. Examples are trimethylolpropane triacrylate, pentaerythritol triacrylate, hexanediol diacrylate, tripropyleneglycol diacrylate, and others. Care must be used in handling them because many are skin irritants and some are apparently sensitizers.

A wide range of monofunctional acrylates has also been used. Those with lowest molecular weight tend to reduce viscosity most effectively, but may be too volatile. Ethylhexyl acrylate has sufficiently low volatility. 2-Hydroxyethyl acrylate has low volatility, high reactivity, and imparts low viscosities, but the toxic hazard is too great in many applications. Ethoxyethoxyethyl acrylate, isobornyl acrylate, 2-carboxyethyl acrylate, and others are also used. *N*-Vinylpyrrollidone (NVP) is an example of a nonacrylate monomer that will copolymerize with acrylate functionality in the coatings at speeds comparable to acrylate polymerization; NVP is particularly useful since the amide structure promotes adhesion to metal, but it introduces a possible toxic hazard.

Acrylated melamine–formaldehyde resins (see Section 13.2) have been reported that permit formulation of coatings that can cure by two different routes: UV curing through the acrylamide double bonds and thermal curing through residual alkoxymethylol groups [4]. If a system containing an acrylated MF resin is first UV cured and then thermally cured, it is reported that films with increased hardness, improved stain resistance, and greater durability are obtained.

Another free radical chain reaction that is adaptable to UV curing is based on a thiol-ene system. The hydrogen on sulfur of thiols (mercaptans) (RSH) is very readily abstracted by bimolecular photoinitiators as well as by free radicals. The resulting thiyl free radicals (RS·) can add not only to activated double bonds such as acrylates, but also to less reactive double bonds such a allyl derivatives that do not undergo homopolymerization.

$$\mathrm{RSH} + \mathrm{I}\cdot \longrightarrow \mathrm{RS}\cdot$$

$$\mathrm{RS}\cdot + \text{—O—CH}_2\text{—CH=CH}_2 \longrightarrow \mathrm{RS\text{-}CH_2\text{-}\dot{C}H\text{—}CH_2\text{-}O\text{—}}$$

$$\mathrm{RSH} + \mathrm{RS\text{-}CH_2\text{-}\dot{C}H\text{—}CH_2\text{-}O\text{—}} \longrightarrow \mathrm{RS}\cdot + \mathrm{RS\text{-}CH_2\text{-}CH_2\text{-}CH_2\text{-}O\text{—}}$$

If one has a polymercaptan and a polyallyl-substituted oligomer, one can have a chain cross-linking reaction. A typical polythiol is the ester of trimethylolpropane (TMP) and 2-mercaptopropionic acid. Polyallyl derivatives can be illustrated by the reaction products of isocyanate-terminated polyesters with allyl alcohol.

The thiol-ene systems are fast curing and exhibit little oxygen inhibition. Their principal limitation is odor. While pure TMP trimercaptopropionate is odorless, it is virtually impossible to make it commercially without any trace of the highly odorous mercaptopropionic acid or other low molecular weight mercaptan byproducts.

32.1.2. Cationic UV Cure

In these systems, strong acids are photogenerated and initiate cationic chain-growth polymerization. When the idea of using cationic polymerization in cross-linking coatings is mentioned, some polymer chemists assume the idea is not feasible. They think of cationic polymerization being carried out at very low temperature in the rigid absence of water in order to achieve high molecular weight. However, in coatings, the objective is not to make high molecular weight linear polymers, but to cross-link polyfunctional reactants. The reactions that terminate cationic polymerizations, in general, still give cross-links. Therefore, cationic polymerization reactions can be carried out in coating films even under humid conditions.

32.1.2.1. Cationic Photoinitiators

Commercially important cationic photoinitiators are onium salts of very strong acids [5]. Iodonium and sulfonium salts of hexafluoroantimonic and hexafluorophosphoric acids are important examples.

Irradiation of diaryliodonium and triarylsulfonium salts yields strong protic acids of the corresponding counter anions as well as radical cations, both of which can initiate cationic polymerization. The onium salts can also be utilized as photoinitiators for free radical polymerization, as well as for concurrent cationic/free radical polymerization, since free radical species are also formed in their photolysis. These reactions are shown for a triphenylsulfonium salt (anion omitted) following the primary unimolecular bond cleavage.

$$Ph_3S^+ \longrightarrow [Ph_2S\cdots Ph]^+$$

$$[Ph_2S\cdots Ph]^+ \longrightarrow Ph\text{—}C_6H_4\text{—}SPh + H^+$$

+ ortho and meta isomers

$$[Ph_2S\cdots Ph]^+ \longrightarrow Ph_2S^{\cdot +} + Ph\cdot$$

Diaryliodonium and triarylsulfonium salts absorb radiation only weakly above 350 nm, however, their spectral response can be extended into the UV-visible, as well as into the midvisible, range by the use of photosensitizers [6].

32.1.2.2. Vehicles for Cationic UV Cure

Homopolymerization of oxirane groups (see Section 11.3.5) is the major type of cationic polymerization that has been carried out commercially thus far. The reactions can be illustrated using a proton as initiator:

$$R\text{-}\overset{\overset{O}{\diagup\diagdown}}{CH\text{—}CH_2} + H^+ \rightleftharpoons R\text{-}\overset{\overset{H}{\overset{|}{O^+}}}{CH\text{—}CH_2} \xrightleftharpoons{H_2C\overset{O}{\text{—}}CH\text{—}R}$$

$$R\text{-}\underset{OH}{CH}\text{-}CH_2\text{-}\overset{+}{O}\langle{}^{CH_2}_{CH\text{—}R} \xrightleftharpoons{H_2C\overset{O}{\text{—}}CH\text{—}R} R\text{-}\underset{OH}{CH}\text{-}CH_2\text{-}O\text{-}\underset{R}{CH}\text{-}CH_2\text{-}\overset{+}{O}\langle{}^{CH_2}_{CH\text{—}R}$$

In contrast to free radicals, cations do not react with each other. Consequently, in the absence of nucleophilic anions, cationic initiated cross-linking can continue after exposure to the radiation source until the reactive cations become immobilized. Typical termination reactions are with water and alcohols. It should be noted that cross-linking occurs even if only one pair of epoxy groups reacted before reaction with water. Furthermore, the termination reaction is really a chain transfer reaction since a proton is regenerated. The absence of oxygen inhibition further distinguishes cationic from free radical polymerization.

The counterions must be very weak nucleophiles, in other words, anions of very strong acids. Hexafluoroantimonates are the most effective followed by hexafluorophosphates. Hydrochloric acid will not initiate polymerization of epoxides, rather it reacts to give a chlorhydrin (see Section 11.3.5).

Cycloaliphatic epoxides such as 3,4-epoxycyclohexylmethyl-3′,4′-epoxycyclohexane carboxylate, **1**, show the highest reactivity. This results from the added ring strain due to the fused ring system that promotes ring opening of the oxonium ion during the propagation step shown above.

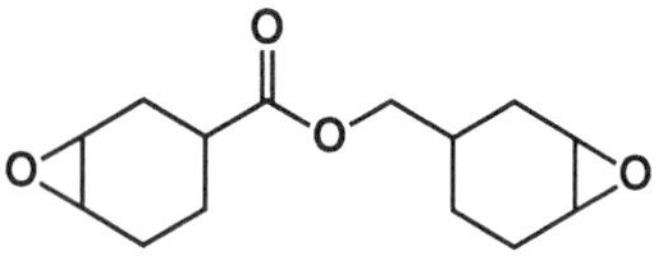

1

Bisphenol A epoxy resins react slowly at ambient temperatures. Optimum cure temperatures are about 70–80°C. Rapid curing of coatings based on BPA epoxy resins can be effected by using a combination of UV and infrared sources. In some cases, a UV curable epoxy coating is applied to one side of a coil strip and a thermally cured epoxy coating to the other. The initial partial UV cure of the UV curable coating is sufficient to permit that coated side to go over a roller without sticking during application of the thermal cure coating. Subsequent thermal cure also advances the cure of the UV curable epoxy.

Vinyl ethers and styrene are readily polymerized cationically. Cationic photopolymerization of vinyl ethers is more rapid than polymerization of epoxy-functional reactants [5]. Less photoinitiator is required; moreover, multifunctional vinyl ether monomers are said to show a very low order of toxicity. Vinyl ether monomers are interesting in their own right and may also be particularly useful as a highly reactive component in cationically cured epoxy resin coatings. An example of a divinylether monomer that is reported to have a very high rate of polymerization is the divinylether derived from the reaction product of ethylene oxide with bisphenol A [5].

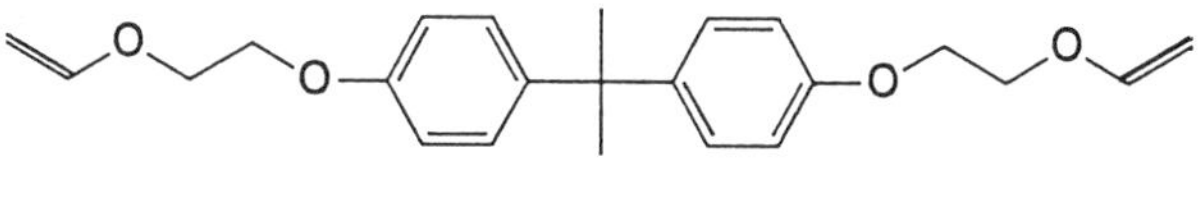

32.1.3. Effects of Pigmentation

Since many pigments absorb and/or scatter UV radiation, they generally inhibit UV curing to some degree [7]. Scattering leads to reflection back out of the film, and absorption also reduces UV availability to the photoinitiator. The effect, of course, becomes more serious as film thickness increases since the pigment plus the photoinitiator reduce the amount of UV that can reach the lower layers of the film. In the case of a strongly absorbing pigment like carbon black, the films that can be cured, even with very large amounts of photoinitiator, are limited to one or two micrometers in thickness. Hence, black printing inks can be UV cured but not black coatings.

The most widely used pigment in other classes of coatings is rutile TiO_2. However, rutile TiO_2 is poorly suited to UV cure since it absorbs some violet light and absorbs essentially all but the very nearest UV even at thin film thickness. Anatase TiO_2 is more useful since it does not absorb very near UV to a great extent; major absorption is only at wavelengths less than about 360 nm. Therefore, thicker films can be cured than with rutile. However, the hiding obtained with anatase is not as high.

Curing is favored by using UV sources with higher emission in the very near UV and by using photoinitiators, such as 2-chlorothioxanthone (CTX), with higher absorption coefficents in the very near UV wavelengths. As previously discussed, there is an optimum photoinitiator concentration that decreases as film thickness increases. Pigmented coatings are more sensitive to this effect than unpigmented coatings.

The effect of reflection by the substrate can be critical in determining the film thickness that can be cured. In practice, with rutile TiO_2 the upper limit of film thickness of white coatings that can be cured on a reflective metal surface by free radical polymerization when using an appropriate photoinitiator and light source is about 15 μm. This film thickness does not give adequate hiding for most applications. It is acceptable for some applications such as, for example, exterior can coatings.

Equations have been developed that permit calculation of the fraction of UV of each wavelength that will be absorbed by the photoinitiator in the presence of pigments that absorb and scatter UV [8]. The equations permit calculation of the total fraction of radiation absorbed as a function of wavelength but, perhaps, more importantly permit calculation of the absorption in the bottom layer of films of different film thicknesses. The data base needed and the calculations would be very extensive, so that it is not practical to make the actual calculations. However, model calculations using the equations illustrate the effect of variables and can be useful in guiding a formulator in considering selection of pigments, photoinitiators, and their concentrations.

Figure 32.2 shows graphically the results of model calculations on the effect of different pigments on UV absorption by photoinitiator in the bottom 0.1 μm of films as a function of film thickness. It must be emphasized that these calculations were based on a single wavelength value and with several simplifying assumptions (see Ref. [8] for details). For comparison, the calculated absorption of 1% benzoin ether (BE) in a pigment-free coating is shown as line 8. This is a concentration that permits curing of a wide range of types of UV cure coatings. The corresponding curve for 1.2% CTX is shown on line 7.

Figure 32.3 illustrates the effect of concentration of photoinitiator on absorption of UV by photoinitiator in the bottom 0.1μm of 15-μm films. The calculations were based on assumptions intended to illustrate the effect of 20 PVC of rutile TiO_2 in the film. The photoinitiator absorption was based on CTX. With the assumptions used, the appropriate concentration of CTX for this system would be about 0.33%.

A further problem of pigmentation in curing of films can result from a large differential of absorption of UV at the surface and at the bottom of the films. This can result in wrinkling (see Section 23.5). The surface layer cures while the bottom layer is still fluid; then when the bottom layer cures, the film shrinks, causing the top of the film to wrinkle. This effect is particularly likely to be seen with free

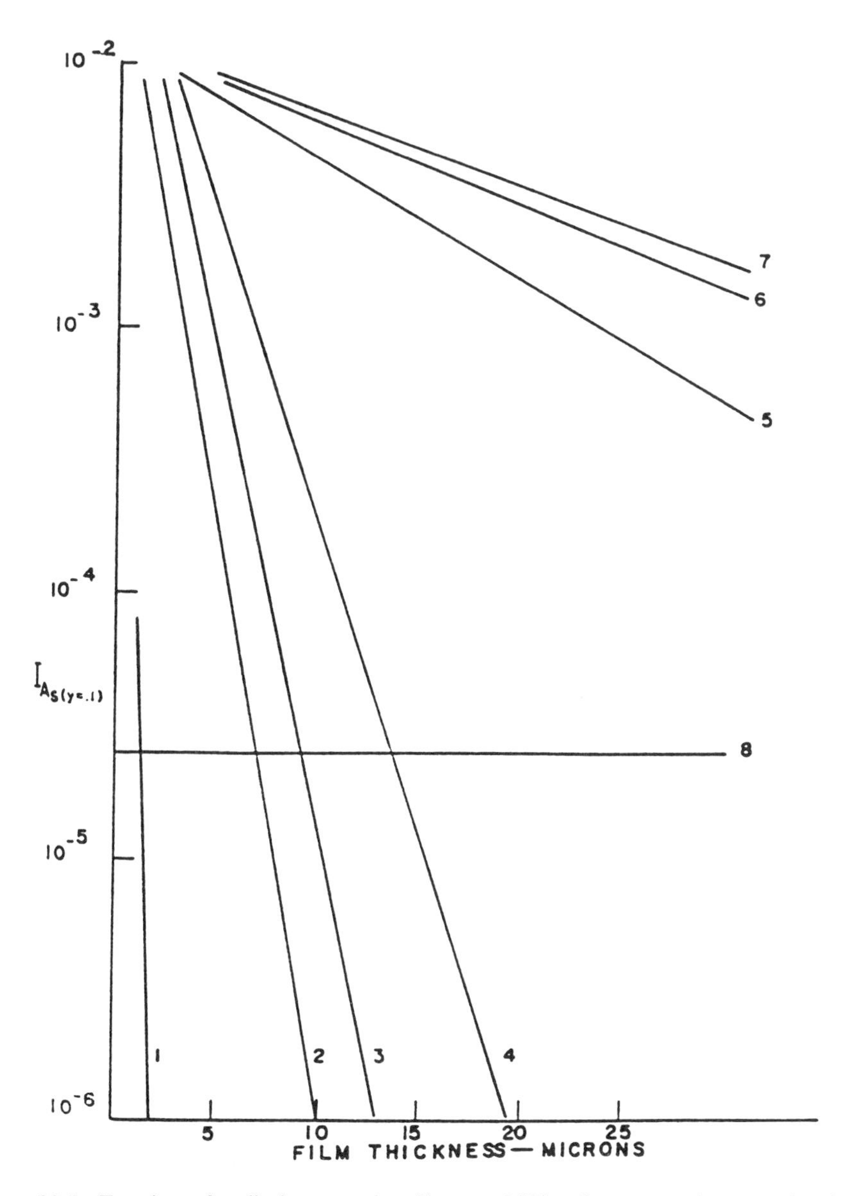

Figure 32.2. Fraction of radiation entering films on 85% reflectance substrates that is calculated to be absorbed by photoinitiator in the bottom 0.1 μm as a function of film thickness. Coatings 1 through 7 contain 1.2% CTX. Coating 8 contains 1% BE. Pigments in coatings 1 through 6 are: 1-carbon black, 2-molybdate orange, 3-lemon chrome yellow, 4-anatase TiO_2, 5-madder lake, 6-ultramarine blue. See Ref. [8] for assumptions. (From Ref. [8], with permission.)

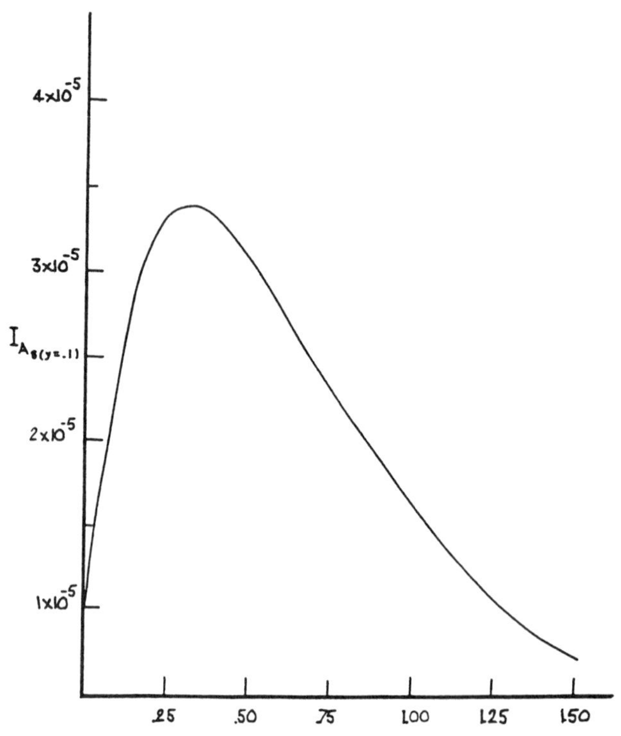

Figure 32.3. Absorption by CTX of 375 nm radiation in the bottom 0.1 μm of 15 μm films with 20 PVC over an 85% reflectance substrate as a function of CTX concentration. (From Ref. [8], with permission.)

radical cure done in an inert atmosphere or cationic cure. In both cases, there is no inhibition of the surface cure by oxygen.

Although the general tendency is to concentrate on the effects of pigmentation on cure, there can be even more serious problems as a result of pigmentation on flow properties. In almost all cases, UV cure coatings are applied without solvent. This means that the volume of pigment in the wet coating is almost as high as the PVC of the final film. Without solvent, viscosity tends to be higher than desirable for good flow in the short time between application and cure, and even at 20 PVC, there is a significant increase of viscosity due to the presence of the pigment. With the use of sufficient monomer replacing oligomer, the viscosity can be reduced to application viscosity, but this reduces the curing speed. As one tries to make higher PVC coatings, the problem becomes more difficult. As PVC approaches CPVC, viscosity of a wet coating approaches infinity. The problem is compounded by the difficulty of stabilizing the pigment dispersion against flocculation using the relatively low molecular weight vehicle system. The flow problems are just as serious for cationic systems as for free radical systems.

In many wood finish applications, transparent low gloss coatings are desired, but they are difficult to achieve with radiation cure coatings. Low gloss is attained in transparent lacquers used in finishing furniture and paneling by adding low concentrations of small particle size SiO_2 (see Section 34.1). When the lacquer is applied, the solvent evaporates, setting up convection currents in the drying lacquer film. The SiO_2 particles are carried up to the surface of the film by these convection currents. As the viscosity of the surface increases, the SiO_2 particles are trapped at the surface. The resultant high PVC at the surface gives low gloss. In UV cure coatings there is no solvent to evaporate and, therefore, no mechanism to concentrate the pigment at the surface. If the level of pigmentation is increased to the extent necessary to provide the low gloss, the viscosity is too high for application.

Some progress has been made in reducing gloss by the use of a dual cure method. A UV cure coating containing as much small particle size SiO_2 as possible from a viscosity point of view and designed so that its cure is strongly oxygen inhibited is applied and cured in air. The lower layer of the coating cures, the top layer does not. When the lower layer cures, it shrinks, exerting an unbalanced stress on the pigment particles in the upper parts of the film, forcing them toward the surface. An alternative explanation is that reactive monomer/oligomer undergoes net migration to the polymerizing regions, thereby concentrating pigment in the nonpolymerizing region. Then, the coating is cured again; this time under an inert atmosphere. Now the surface layer cures; some further shrinkage occurs, further increasing the PVC near the surface. In this way medium gloss transparent coatings can be made. Several production lines have operated using this dual cure method.

32.2. ELECTRON BEAM CURE COATINGS

Rather than using UV radiation with a photoinitiator, high energy electron beams (EB) can be used directly to polymerize acrylate coatings. In EB cure compositions, the high energy electrons cause direct excitation of the coatings resins (P) as well as their ionization into radical cations ($P^{\cdot +}$) and secondary electrons, e^-.

$$P + EB \longrightarrow P^* + P^{+\cdot} + e^-$$

$$P^{+\cdot} + e^- \longrightarrow P^*$$

$$P^* \longrightarrow I\cdot$$

The major fate of the radical cations is recombination with the relatively low-energy secondary electrons to yield additional excited state resins. The excited state resins (P^*) undergo homolytic bond cleavage to free radicals that initiate polymerization of the acrylated resins.

The high-energy electrons are generated by charging a tungsten filament at high negative potential, 150–300 keV. They are directed by magnets in a curtain through a metal "window" to the coatings to be cured. (The term electron curtain is also used.) The types of vehicles used are the same as those used in free radical initiated UV curing. Since the polymerization reactions are oxygen inhibited and there is little differential in absorption between the top and the bottom of the film, it is essential to do EB curing in inert atmospheres.

Electron beam curing by cationic polymerization is not generally feasible owing to the very short lifetimes of the cationic species. However, it has been found recently that cationic polymerization of vinyl ethers, as well as highly reactive epoxides, can be induced by EB radiation in the presence of diaryliodonium and triarylsulfonium salts. The readily reducible onium salts have been implicated in the generation of cationic initiators by oxidation of free radical species, as well as by capture of secondary electrons. Scavenging of secondary electrons by onium salts tends to lengthen the lifetime of cations and to improve the prospects for cationic polymerization [6,9].

The principal advantages of EB curing over UV curing are that no photoinitiator is needed and that pigments do not interfere with the curing. These advantages are often offset by the higher capital cost of the electron beam or curtain generating equipment, the need to use inert atmospheres, and the need for shielding to protect workers from the electron beam. The advantage of being able to cure pigmented systems is real but of relatively limited importance in coatings since flow is adversely affected by pigmentation.

32.3. APPLICATIONS OF RADIATION CURING

The major advantages of radiation curing are:

Very short cure times are possible at ambient temperature with package stable coatings. In the case of clear acrylate and thiol-ene compositions, curing times required are fractions of a second.

Since the curing can be done at ambient temperatures, the coatings are applicable to heat sensitive substrates such as paper, some plastics, and wood.

The coatings can be formulated without solvent and, therefore, lead to minimal VOC emission.

The energy requirement is minimal. The major losses in energy encountered in thermal curing systems are not experienced. The cure is done at or near to ambient temperature so no energy is required to heat the object being coated. (The IR radiation from the UV lamps does cause a modest temperature increase.) The heat loss from air flow through a baking oven required to keep the solvent concentration below the lower explosive limit is eliminated. The extent of this advantage is, of course, dependent on the cost of energy. The use of radiation curing in Europe and Japan has been greater than in the United States, at least in part because the cost of energy is higher in Europe and Japan than in the United States.

Capital cost for UV curing is low. This results primarily from the small size of the curing units as compared with ovens for baking. A UV cure coating only needs about 0.5 s exposure for curing. This means that a line being run at a speed of 60 m min^{-1} only need be exposed to UV over a distance of about 0.5 m. A UV curing unit with 4 lamps can do this in a total length of about 2 m. In contrast, a curing oven for thermal cure has to be a hundred or more meters long to provide time for curing at this speed. In addition, there has to be space for flash off of at least some of the solvent from conventional coatings before the coated object enters the oven and space for it to cool back down to handling temperature. The saving in building space can thus be very large.

There are, however, major limitations to radiation curing:

As noted earlier, it is most applicable to flat sheets or webs where the distance to the UV source or the window of the EB unit can be approximately constant.

Cylindrical, or nearly cylindrical, objects can be rotated in front of the radiation source, but irregularly shaped objects are not easy to expose uniformly.

In the case of UV curing, pigmentation can severely limit the thickness of films that can be cured. The practical limit for white coatings is about 15 μm, too thin for many applications. In the extreme case of carbon black pigmented films, the limit is about 2 μm. Black printing inks can be UV cured but black coatings cannot. While cationic systems are less limited than free radical systems, even in this case, wrinkling becomes a serious limitation much above 15 μm. Electron beam coatings do not have this limitation because there is no gradient of initiation from top to bottom of the film.

Owing to the effects of pigmentation on flow, the level of pigmentation is limited in UV and EB coatings. Furthermore, low-gloss coatings cannot be applied. These limitations can be overcome, of course, by using solvent containing UV cure coatings, but this eliminates many of the advantages.

Shrinkage during curing can lead to severe adhesion problems on metal and some plastic substrates (see Section 23.2). When polymerization results from chain addition of free radicals to double bonds, there is a substantial decrease in volume. An important factor is that the length of the carbon–carbon bonds that are formed is shorter than the intermolecular distance of the monomer units from each other. The degree of shrinkage is directly related to the number of double bonds reacted. In a typical acrylic UV cure coating, the potential shrinkage is 5–10%. Since cure occurs in less than a second, there is not time for much of the volume shrinkage to take place before the free volume in the coating is so limited that movement becomes restricted. The stress from the restricted shrinkage applies a force to offset the adhesion forces holding the coating onto a substrate so that less external force is required to remove the coating. Shrinkage can be minimized by using higher molecular weight oligomers or higher ratios of oligomer to low molecular weight reactive monomers, but these expedients increase viscosity and tend to increase the required cure time. The same limitation applies to EB cure. Curing of epoxy systems by cationic photoinitiators leads to less shrinkage. The decrease in volume resulting from the formation of polymers from monomers is partially offset by an increase in volume from the ring opening. Typical epoxy UV cure compositions shrink less than 3%.

Porous substrates such as paper and wood do not give adhesion problems since the mechanical effect of penetrating into the surface of the substrate holds the coating on the substrate. However, there can be a curling problem with thin substrates as a result of the tension from shrinkage on one side of the substrate.

The photoinitiators are only partly consumed in the curing of the films so that substantial amounts are left in the films of UV cured coatings. These photoinitiators can then accelerate photodegradation of films exposed outdoors. Furthermore, conventional UV stabilizers and antioxidants tend to reduce UV cure rates which limits their usage. These constraints limit the use of UV cure coatings to interior applications. While there has been work done using precursors, including photoinitiators, that convert to UV stabilizers on exposure, the results do not appear to be satisfactory.

Cationic coatings have the advantage that they do not generate free radicals when films are exposed outdoors so that there is not this problem, but acid is generated, leading to potential catalysis of hydrolysis. Electron beam systems have the distinct advantage that there is no photoinitiator in the system. However, even with EB, the coatings used must have excess unsaturation to cure in rapid schedules and residual unsaturation can adversely affect exterior durability.

A further factor that has limited the growth of radiation curing is the cost of the coatings. The effect of this cost increment has been to limit the use of radiation curing to applications where there is some performance advantage that cannot be achieved with conventional coatings.

A large use of UV curing is in printing inks, particularly lithographic inks. The viscosity of lithographic inks is relatively high, on the order of 5–10 Pa·s. This eliminates the flow problem with highly pigmented compositions. The films are thin, 2 μm or less, so that the interference of pigment with curing can be offset by the use of highly absorbing photoinitiators and/or high photoinitiator concentrations. Since the films are not continuous, adhesion or curling difficulties resulting from shrinkage are not a major problem. The advantage of fast cure is large—it allows the webs to be wound up quickly without offsetting on the back of the substrate. In the printing of tin plate and aluminum for cans, there is the advantage that the four colors can be printed in line by simply having a UV lamp after each printing station; operating cost can be lower than with heat cure inks.

Paper coatings are an important end use because of the high speed curing at low temperatures. Adhesion is not a problem due to the porous substrate and the coatings are high gloss unpigmented coatings.

An early commercial use of UV curing was in Europe for top coating wood furniture. A large part of European furniture is manufactured in flat sheets and assembled after coating (see Section 34.1). Furthermore, the finish on much European furniture is glossy. UV curing is ideal for such an application. The low temperature, fast cure is appropriate for wood. Early coatings were styrene–polyester, but acrylic coatings were later adopted because they cure more rapidly and the volatility of styrene is avoided. There has been relatively little use of UV cure top coats on wood furniture in the United States because a major fraction of U.S. furniture is assembled and then coated. Furthermore, most U.S. furniture has a low gloss top coat that, as discussed above, is difficult with solvent-free UV cure coatings.

There is an important furniture application in the United States, the filling of particle board. Particle board has a rough surface and before use in furniture tops it must be coated with a filler and sanded smooth to give an even surface for the application of base coat and printing (see Section 34.1). This was done with highly pigmented solvent-borne coatings applied at about 30% volume solids. However, the low solids resulted in so much shrinkage that two or three coats had to be applied in order to have sufficient film thickness to permit sanding to a smooth surface without cutting down into the substrate. With UV cure, the shrinkage is only a few percent so that an adequate layer can be applied in one coat. This together with the fast cure means that UV cure fillers are less expensive to apply. The pigments used are clays and silica that absorb or scatter only minor amounts of UV. The high level of pigmentation leads to poor leveling, but it does not matter since the surface is sanded. Furthermore, the pigmentation results in less resin being at the surface so that air inhibition is less of a problem and, again, the surface is sanded off in any case. Originally, the fillers were styrene/polyester systems, but the higher speed of acrylate system has led to their adoption in many cases despite their higher cost.

Another large-scale application whose success results from the low temperature fast curing possible with UV cure coatings is the top coating of vinyl flooring. It is especially applicable to expanded flooring since the foam pattern is heat sensitive. A major component of the binders is acrylated urethane oligomers since the resulting urethane coating exhibits the high abrasion resistance needed in a floor covering.

Radiation cure abrasion resistant clear coatings for plastics have been developed. The poor abrasion resistance of clear plastics is a distinct disadvantage in application, such as replacement for glass in uses ranging from glazing to eyeglasses. Silicone-based coatings have been used for many years to improve the abrasion resistance (see Section 24.3). These coatings require long thermal cure schedules and have high VOC levels. Radiation cure coatings based on alkoxysilane–acrylate monomers and colloidal silica cure in seconds and have very low VOC emission [10]. They are said to have excellent adhesion to polycarbonates and other plastics while giving abrasion resistance equal to or superior to the solvent-based coatings [10,11].

An application for cationic UV cure epoxy coatings is for coil coating on tin plate for the exterior side of can ends. The superior adhesion of epoxy coatings makes them preferable for this application as compared to acrylate coatings. In some cases, a UV cure coating is applied on one side of the strip and UV cured. Then a thermally cured epoxy coating is applied on the other side. When the strip is put through the oven to cure the thermally cured coating, the cross-linking of the UV cure coating is also advanced.

An interesting relatively new application of UV curing is in coating optical fibers for wave guides used in telecommunication cables. The process is carried out at high speeds in towers up to 25 m tall. The glass fiber is pulled from a heat softened blank of glass and then coated through a *die coater*, that is, a reservoir with holes on each end through which the glass is pulled. The UV lamps are aligned parallel to the fiber as it comes out of the coater with elliptical reflectors both behind the UV source and on the opposite side of the fiber from the source. The fiber is at the focal point of the pair of reflectors. In this way, the coating is given very high dosages of UV, permitting cure speeds of 20 m s^{-1} without an excessive bank of UV lamps. Acrylated urethane oligomers are used since abrasion resistance is a critical requirement. Two UV cure coatings are applied, a soft first coat and a harder outer coat.

UV curing is also the basis for important photoimaging processes utilized in the printing and electronic industries [11,12]. In these applications, UV curable coatings on selected substrates are exposed through an image bearing transparency, such as a film negative. Cross-linking occurs selectively under the transparent regions. The unexposed regions are subsequently removed, commonly by washing out with aqueous or organic solvent. The resulting image transfer into the coated substrate is a critical step in the manufacture of printing plates as well as printed circuit boards and integrated microcircuits (e.g., computer chips).

GENERAL REFERENCES

P. K. T. Oldring, Ed., *Chemistry & Technology of UV and EB Formulation for Coatings, Inks & Paints*, (Six volumes), SITA Technology, London, 1991 to 1993.

S. P. Pappas, Ed., *UV Curing: Science and Technology*, (two volumes), Technology Marketing Corp., Norwalk, CT, 1978 and 1985.

S. P. Pappas, Ed., *Radiation Curing: Science and Technology*, Plenum, New York, 1992.

C. G. Roffey, *Photopolymerization of Surface Coatings*, Wiley, New York, 1982.

REFERENCES

1. A. J. Bean in *Radiation Curing: Science and Technology*, S. P. Pappas, Ed., Plenum, New York, 1992, pp. 308–314.
2. H. J. Hageman, *Prog. Org. Coat.*, **13**, 123 (1985).
3. L. H. Carlblom and S. P. Pappas, *J. Polym. Sci.; Polym. Chem.*, **15**, 1381 (1977).
4. J. J. Gummeson, *J. Coat. Technol.*, **62** (785), 43 (1990).
5. J. V. Crivello, *J. Coat. Technol.*, **63** (793), 35 (1991).
6. S. P. Pappas, in *Radiation Curing: Science and Technology*, S. P. Pappas, Ed., Plenum, New York, 1992, pp. 1–20.
7. S. P. Pappas and Z. W. Wicks, Jr., in *UV Curing: Science and Technology*, Vol. 1, S. P. Pappas, Ed., Technology Marketing Corp., Norwalk, CT, 1978, pp. 78–95.
8. Z. W. Wicks, Jr. and W. Kuhhirt, *J. Paint Technol.*, **47** (610), 49 (1975).
9. S. C. Lapin, in *Radiation Curing: Science and Technology*, S. P. Pappas, Ed., Plenum, New York, 1992, pp. 241–271.
10. J. D. Blizzard, J. S. Tonge, and L. J. Cottington, *Proc. Water-Borne, Higher-Solids, Powder Coat. Symp.*, New Orleans, LA, 1992, p. 171.
11. L. N. Lewis and D. Katsamberis, *J. Appl. Polym. Sci.*, **42**, 1551 (1991).
12. B. M. Monroe in *Radiation Curing: Science and Technology*, S. P. Pappas, Ed., Plenum, New York, 1992, pp. 399–440.
13. D. Funhoff, H. Binder, S. Nguyen-Kim, and R. Schwalm, *Prog. Org. Coat.*, **20**, 289 (1992).

CHAPTER XXXIII

Product Coatings for Metal Substrates

Approximately 34% of the volume (1.37×10^9 L) and 37% of the value ($4.24 billion) of U. S. shipments of coatings in 1991 were of product coatings for OEM (original equipment manufacture) application [1]. In this and the following chapter on coatings for nonmetallic substrates, we discuss some of the major OEM product coating end uses as illustrations of the factors involved in selecting coatings for particular applications. The largest fraction of industrial coatings is used on metals. A multitude of products are coated; space only permits discussion of a few. We have selected four of the larger end uses: automobiles, appliances, container coatings, and coil coating. The coatings used are generally proprietary and are continually evolving. Thus, these discussions focus on the principles involved, not on specific formulations.

33.1. OEM AUTOMOTIVE COATINGS

The OEM automotive coatings are those used in painting cars and trucks on assembly lines. The coatings for repair and refinishing of cars are discussed in Section 36.3. This discussion is further limited to coatings applied to the exterior body of the automobiles. Many other coatings are used on air filters, car interiors, wheels, truck linings, and so on.

There are two major reasons for painting an automobile: appearance and corrosion protection. Initial appearance can be a critical factor in the sale of a car. The paint is the first thing a potential buyer sees. The purchaser can be attracted by good appearance of the paint; furthermore, if the paint job is poor, the purchaser may assume that the manufacturer is careless not only in applying the paint but in other aspects of manufacturing the automobile. Maintaining excellent appearance of the paint over many years of service is also critical. If the paint does not continue to look good, not only will the owner of the car be unhappy, but others may decide not to buy that make of car because they have seen that the paint does not stand up well. High-gloss coatings are used, and gloss retention is a critical requirement. Thirty or more years ago, many people waxed their cars frequently. Since the performance of the paint has been substantially improved, that is no longer necessary and most people do not want to have to wax their cars often, if at all. Corrosion protection is another critical performance requirement. The use of salt on icy highways provides an environment that can lead to rapid corrosion of steel. In the 1970s and 1980s coatings were developed that provide protection against corrosion for the lifetime of the car.

33.1.1. Metal and Metal Cleaning and Treatment

Until the 1980s, the outer shell of most automobiles was predominantly cold-rolled steel. Substantial amounts of cold-rolled steel are still used; however, some manufacturers use galvanized steel, or corrosion-resistant zinc or nickel steel alloys for the components of cars most vulnerable to corrosion. Each type of metal can present a unique set of painting requirements.

Additional complexities result from the increasing use of plastic and rubber components in car bodies. In some cases, plastic and rubber parts are completely coated before assembly; in others, they are primed and attached to the car body before application of the top coat. Many fillers and sealants are used. In some cases, they are applied to the original steel; in others they are applied after priming but before top coating; and in yet others, they are applied after top coating. Furthermore, there has been increasing use of decals and trim items that are adhesively bonded to the top coat. All these complexities impose additional requirements on the painting process and on the properties of the paints. Automobile producers want to use the lowest possible baking temperatures to avoid deformation or deterioration of polymeric components, but there is inevitably a tradeoff between paint film performance and reduced baking temperatures.

The most critical single factor affecting corrosion is the steel [2,3]. There are large variations in cold-rolled steel that can lead to greater or lesser probabilities of corrosion failures. The principal differences result from the annealing process. When steel is cold rolled, oil is used as a lubricant; the oily, rolled steel is then annealed. There are two processes of annealing. In one, coils of the steel are put in annealing ovens and heated to about 500°C to relieve stresses set up during the rolling process. In the other, called strip annealing, the steel is run as a single layer through an annealing oven. The oils can get baked into the surface of the steel in the coil annealing process. It is apparently particularly likely to happen if the coils are wrapped tightly so that the oils cannot volatilize away from the steel surface. The oils partly decompose, leaving organic carbon compounds embedded in the surface that are not removed by subsequent cleaning steps; these organic compounds make uniform conversion coating more difficult. The effect can be irregular; only one side of the coil may be affected or only one end of the coil.

The body of the car is fabricated and then is cleaned to remove all dirt and particularly oil. Solvent degreasing is commonly used (see Section 26.4). Detergent cleaning is also used; the cleaners are proprietary mixtures of trisodium phosphate, surfactants, metal chelating agents, and sodium polyphosphate. Commonly an *activator*, such as titanium phosphate, is included in the cleaner; the activator is said to condition the metal surface to react with the conversion coating that is applied next [4]. If part of the car body is galvanized steel, the pH must be kept below 9.5

since zinc dissolves under basic conditions. The body is washed to remove any residual cleaning agents.

As discussed in Section 26.4.1, the surface is next conversion coated. The major component of the conversion coating treatment is primary zinc phosphate, $Zn(H_2PO_4)_2$. This treatment results in the deposition of a layer of mixed crystals of $Zn_3(PO_4)_2 \cdot 4H_2O$ and $Zn_2Fe(PO_4)_2 \cdot 4H_2O$ [4]. The weight of the phosphate layer is about 2 g m^{-2} and the thickness about 5 μm. The conversion coating is critical to performance and must be carefully controlled. If part of the steel is galvanized, the surface of the zinc will become coated with zinc phosphate. Zinc–manganese–nickel conversion coatings have been recommended for treating zinc-coated metal for cationic electrodeposition [5].

After treatment, the metal must be thoroughly rinsed. Any residual soluble salts can accelerate corrosion, as explained in Section 27.3.2. The last rinse water contains chromic acid, which reacts with the surface to give a small amount of chromate to act as a passivating agent against corrosion until the surface is coated. The treated car body then goes into a dry-off oven. Not only is free water dried off the surface, but the tetrahydrate phosphate salts are converted to the more stable dihydrate crystals [4]. It is critical to keep the treated surface of the metal clean until the primer is applied. If, for example, someone should touch the car body with bare hands, oil and salts could be deposited on the surface, potentially leading to blisters and inferior corrosion protection.

33.1.2. Primers and Primer-Surfacers

While the quality of the steel and pretreatment are critical, all cold-rolled steel would rust rapidly without a protective coating. The primary purpose of the primer is to provide adhesion to the phosphate coating and metal. As discussed in Section 27.3.2, adhesion is the primary key to corrosion protection, particularly adhesion that will resist displacement by water. Of course, it is also critical that the primer serve as a suitable base for the top coats providing for smoothness and adhesion.

Almost all automobiles are now primed with cationic electrodeposition (E-coat) primers, as discussed in Chapter 30. The E-coat primers provide excellent adhesion and virtually completely resist displacement by water. The strong driving force of the electrophoretic application apparently aids in penetration into the phosphate crystal mesh on the steel surface. The amine groups on the binder resin provide strong interaction with the phosphate coating and the steel surface. The binders are designed to be nonsaponifiable so that even if the paint is scraped through, undercutting will be very slow and the coating will continue to protect the metal next to the gouge against corrosion.

The E-coats cover all metal surfaces, including recessed areas such as the interior of rocker panels below the doors that cannot be coated by spray application. The process provides a relatively uniform film thickness on all metal surfaces of about 25 μm of dry film. While E-coat primers provide excellent corrosion protection, there are two problems involved in their use: (1) the film thickness can be so uniform that all the surface irregularities in the metal and conversion coating are copied in the surface of the primer film, and (2) it can be difficult to achieve the required adhesion of top coat to the primer surface.

Primers have been formulated that provide greater smoothness (the term *metal filling* is often applied) by reducing the pigment content of the primer. As a result, the viscosity of the uncured primer film is somewhat lower and, when the primer car body is baked, more leveling can occur than if PVC were higher. While this aids in surface smoothness, adhesion of top coat to the smoother, glossier primer surface is more difficult.

The surface roughness and adhesion problems can be minimized by spray applying a primer–surfacer over the E-coat before top coating. The binder in the primer–surfacer can be a polyester or an epoxy ester that will provide better adhesion to the epoxy-based E-coat primer than the acrylic top coat. In some cases, a two-package (2K) polyurethane primer–surfacer is used. This primer–surfacer has a lower cross-link density than the primer, providing greater opportunity for the solvent in the top coat to penetrate into this surface than into the surface of the E-coat primer. The primer–surfacer is formulated with a higher PVC so that the surface roughness provides a greater opportunity for top coat adhesion. The PVC of primer–surfacers is commonly greater than CPVC, providing for penetration of a little of the vehicle from the top coat into the primer–surfacer film. Also, such a highly pigmented primer–surfacer can be relatively easily sanded so that local areas of excessive roughness can be sanded smooth and film thickness is sufficient to minimize the danger of sanding through to bare metal.

A different approach to improving the adhesion of top coats is to apply a so-called *sealer* to the electrodeposited primer surface. The sealer is a low solids top coat with a high content of relatively strong solvents. It is applied just before the top coat; the solvent softens the surface of the primer slightly, permitting penetration of top coat binder into the surface, providing increased adhesion.

A *chip-resistant coating* is frequently applied over the E-coat primer on the lower parts of the car body before top coating. The coating is designed to be especially resistant to impact by stones thrown up from a road against the car body. The chip-resistant coatings are organosols or polyurethanes using blocked isocyanates as cross-linkers. Reference [6] gives a discussion of the mechanical properties involved in stone chip resistance and approaches to testing.

If electrodeposition is not used, primer–surfacers with BPA epoxy-based binders are used since they provide superior adhesion and saponification resistance. To reduce VOC emissions, water-reducible primers are used (see Section 11.4.1). These primers provide equal performance to solvent-borne epoxy ester primers with lower VOC emissions, but generally not as good corrosion protection as cationic E-coat primers.

The color of automobile primers and primer–surfacers is either red or gray. There is no evident reason why a red primer, pigmented with iron oxide, should be used. It has been said that the practice is based on the fact that in the early days of corrosion resistant primers red lead was used as a corrosion inhibiting pigment. Red iron oxide is not a corrosion inhibiting pigment, but people associate red primers with corrosion protection. As a neutral color, gray is a more logical choice. The trend has been to use darker grays. This results from the problem of exterior durability of BPA epoxy-based primers. If the film thickness of the top coat is inadequate, UV radiation can reach the primer surface leading to degradation and loss of adhesion of top coat to primer. When a relatively dark gray primer is used, inadequate top coat coverage is evident from the greater effect of the dark color on the color of the top coat. UV absorber in the top coat that strongly absorbs UV in the wavelength range of 280–310 nm (see Section 25.2.1) can also help protect the primer from degradation.

33.1.3. Top Coats

The primary purpose of the top coat is to provide the desired appearance. Virtually all are high-gloss coatings. The top coats must be designed to maintain their appearance for long periods. While the primary factors in maintaining appearance are resistance of the resin binder to photoxidation and hydrolysis, durability of the pigment in the film is also a critical factor (see Chapter 25). There are many other requirements: resistance to bird droppings, acid rain, sudden thunder showers on a car that has been sitting in the hot sun, the impact of pieces of gravel striking the car, gasoline spillage, and so on.

In the United States a large majority of top coats are so-called metallic and other polychromatic colors. As discussed in Section 16.4 and 18.2.5, metallic coatings give an attractive appearance because of the change in color with angle of viewing. When viewed at an angle near the perpendicular, the color is light; and when viewed at larger angles, the color is darker. The extent of this phenomenon is called the degree of color flop. High color flop depends on three factors: (1) minimal light scattering by the coating matrix between the aluminum flakes, (2) a smooth, high-gloss surface, and (3) orientation of the aluminum flakes parallel to the surface.

Minimal light scattering by the coating matrix requires that pigment selection and dispersion provide transparent films when prepared in the absence of the aluminum flake. Also all resins and additives in the coating must be compatible so that there will be no haziness in the unpigmented film.

Surface smoothness can be affected by many factors. It has been shown that the major single factor is the roughness of the initial metal substrate/phosphate layer [7]. Other important factors are leveling after spraying and, in the case of base coat/clear coat systems, the thickness of the clear coat film.

The required orientation of aluminum flake parallel to the surface has been achieved in everyday production for decades, but it has never been entirely clear why orientation is as good as it is [8,9]. Most workers have concluded that an important factor is shrinkage of the film after application; shrinkage is accompanied by increasing viscosity of the film, most rapidly near the surface as the solvent evaporates, leading to a viscosity gradient. This viscosity gradient causes the upper edge of an aluminum platelet to be immobilized before the lower edge, and the platelet swivels toward parallel orientation as the film shrinks. In general, the lower the solids of the coating as it arrives at the substrate, the better the alignment will be. Some workers also feel that when spray droplets strike the surface they spread out, and the resulting flow forces tend to align the flake particles parallel to the surface. It has also been suggested that the spray droplets must penetrate the surface of the wet film and crash on the substrate [9]. Perhaps there are still further factors influencing orientation.

Variations in spraying can give substantial variations in flake alignment [10]. For example, changes in atomization, air flow, solvent evaporation rates, and gun to surface distance can affect orientation. Transfer efficiency is greatly improved by electrostatic spraying; however, it is more difficult to obtain good surface smoothness and metal orientation using electrostatic spray. It has been suggested that there may be some alignment of the metal flakes parallel to the lines of force in the electrostatic field that are perpendicular to the substrate. Use of automatic high-speed rotary bell spray guns further complicates the problem of obtaining even film application. Reference [10] provides a discussion of the variables involved in automated spray application of metallic coatings.

Historically, there were two main types of top coats: thermoplastic coatings, called *lacquers*, and thermosetting coatings, called *enamels*. Lacquers were originally based on nitrocellulose. Because of their far superior exterior durability, acrylic lacquers replaced nitrocellulose lacquers. Lacquers had to be applied at very low solids—10 to 12 NVV (nonvolatile volume percent)—because of the effect of the high molecular weight thermoplastic acrylic polymers on viscosity (see Section 4.1.1). Acrylic lacquers had the distinct advantage of providing excellent color flop in metallic coatings, especially when some cellulose acetobutyrate (CAB) was incorporated into the formulation. The low solids provided the needed shrinkage to orient the aluminum flakes parallel to the surface of the coating. Somewhat higher solids—around 27 NVV—was achieved with nonaqueous dispersion (NAD) lacquers (see Section 4.1.2). However, leveling was a severe problem and the VOC emissions were still much too high. Lacquers are no longer used in OEM finishing. They are still used in refinish coatings for automobiles (see Section 36.3).

All OEM top coats are now enamels since much higher solids can be achieved. Alkyd top coats, using nonoxidizing alkyds (see Section 10.5) cross-linked with melamine–formaldehyde (MF) resin are still used by one automobile manufacturer for solid color top coats but not for metallic coatings. The alkyd coatings give an appearance of greater depth and are less subject to film imperfections than acrylic coatings; but the exterior durability, especially in metallic colors, is inferior to that of high-quality acrylic coatings. By far the majority of top coats are now based on thermosetting acrylic resins sometimes blended with polyester and CAB resins. The most widely used cross-linkers are MF resins (see Chapter 6); however, there are advantages to polyisocyanate cross-linkers (see Chapter 12), which may supplant MF resins in some applications (see Section 33.1.3.2). Other cross-linking reactions, such as epoxy/carboxylic acid reactions (see Section 11.3.2), are also used.

The increasing use of plastics in the exterior construction of automobiles leads to difficulties in designing coatings. Some of the plastic moldings will distort at temperatures normally required for baking top coats. Some plastics are flexible and some have rigidities similar to steel. Coatings designed for flexible plastics are generally too soft for car exteriors and coatings designed for steel or rigid plastics are rigid and subject to cracking when used over flexible substrates. In many cases, plastic components are coated (see Section 34.3) and then installed on the car body. There would be substantial advantages in the use of a "universal coating" that could be applied to flexible and rigid plastics and to steel in one operation. Reference [11] discusses the design criteria for a universal coating.

33.1.3.1. Single Top Coat Systems

There are two major classes of top coat systems: single coat systems and base coat/clear coat systems. Single top coats were used exclusively until recent years. They are now primarily used for solid color enamels. The first thermosetting acrylic coatings were solution acrylics having hydroxy functionality and an $\overline{M}_W$ of the order of 35,000 together with Class II MF resins as cross-linkers. They were applied at about 30 NVV. The solids were increased to around 35 NVV by using thermosetting NAD acrylic resins. In order to achieve high gloss, pigmentation levels are low. In solid colors, PVCs of 8–9% are used; in the case of metallic colors, PVCs are much lower—on the order of 2–4%. Owing to the low pigment content, relatively thick films are required for hiding—on the order of 50 μm. Since the coatings are spray applied, some areas will have significantly over 50 μm of dry coating, for example, around the "A post" between the windshield opening and the front door opening.

Conventional solids acrylic enamels are less expensive than acrylic lacquers primarily because solvent content is lower and less expensive solvents can be used. The exterior durability of optimized lacquers and enamels is essentially equal. Enamels have greater resistance to swelling with gasoline—especially unleaded gasoline with its higher aromatic and oxygenated compound content. However, enamels are more difficult to repair than lacquers. The brilliance and degree of color flop that can be achieved in metallic enamels is somewhat inferior to lacquers; however, when the volume solids are sufficiently low, the metallic effect is still pleasing.

In order to reduce VOC emissions, water-reducible thermosetting acrylic resins of the type discussed in Sections 7.3 and 29.1 were formulated into water-reducible enamels using MF resins as cross-linkers. The coatings were used on a large commercial scale in several auto assembly plants. The field performance of the coatings was fully equal to that of the solvent-borne enamels of the time. The coatings were applied at 18–20 NVV. As a result of the low solids, excellent metallic color flop was obtained—the colors being applied with acrylic lacquer in other assembly plants were matched. Although the solids were low, VOC emissions were also fairly low since about 80% of the volatile material was water. The VOC emissions were approximately equivalent to those from a solvent-borne coating with 60–65 NVV, but with superior metallic colors resulting from the low solids and excellent performance properties related to the higher molecular weight of the resins.

However, there were significant problems with the use of these water-reducible coatings that resulted in comparatively high applied costs. These problems were all related to the use of water as the major volatile component. In order to avoid corrosion in the paint handling tanks and lines, more expensive stainless-steel construction was needed. Since the rate of evaporation of water depends on relative humidity, the degree of volatile loss during spraying and flash off varied with humidity. The first assembly plants had to be designed without knowledge of how critical this humidity effect would prove to be. To be on the safe side, the plants were designed so as to be able to control the humidity and temperature in the spray booths within relatively narrow limits. Since the air flow through the spray booths was very high, the cost of air conditioning all this air was very large. Experience showed that it was not necessary to control humidity so closely, but still there was significant additional cost.

The more critical problems related to the greater difficulty of controlling popping of these water-reducible top coats as compared to solvent-borne top coats. As discussed in Section 23.6, the maximum film thickness of water-reducible coatings that can be applied without popping under standard conditions is significantly lower than with solvent-borne systems. The film thickness that must be applied in painting cars is high, especially in areas such as around the "A posts" and along the "coach joint" between the rear window and the trunk opening. The coatings were applied successfully by utilizing more spraying passes and longer flash-off times; but this required more personnel and longer, more expensive spray booths. The ovens were zoned so that more water could evaporate before the top surface of the film increased in viscosity to the point that popping would occur; but this required longer ovens. Extra space in an assembly plant is at best expensive and at worst unavailable without rebuilding the plant. As a result of these problems, use of these water-reducible coatings for single top coats was replaced by higher solids solvent-borne coatings.

The highest solids at which it has been possible to formulate high performance solvent-borne top coats that still exhibit some reasonable degree of aluminum orientation to give color flop is about 45 NVV. This has also been near the upper limit of solids that provides outstanding exterior durability with thermosetting acrylic resins. As the molecular weight is reduced in order to increase the solids further, the average number of functional groups per molecule and molecular weight get so low that the fraction of molecules with single functional groups becomes too large to permit good properties with conventional free radical synthesized thermosetting acrylic resins (see Section 7.2).

As discussed in Section 23.2, as solids increase, the problems of controlling sagging increase. Automotive coatings are particularly vulnerable to sagging since the film thickness applied is large and variable. In metallic colors, even a small amount of sagging is very noticeable because it changes aluminum orientation around the sag marks. As the solids have been increased, it has become necessary to add thixotropic agents to increase the viscosity at low shear rates to minimize sagging [12]. Conventional thixotropic agents make a coating hazy due to light scattering; such haze interferes with color flop. Therefore, additives have had to be designed with refractive indexes similar to those of the acrylic binder to minimize haze. Those most widely used are acrylic microgels, highly swollen gel particles that are lightly cross-linked so that they can swell but not dissolve in the liquid coating (see Section 23.2). The swollen particles distort at high shear rates and regain their spherical shape when the shear rate decreases; hence viscosity drops as the coating is sprayed, but increases after it arrives on the surface, reducing the tendency to sag. Rheology of microgel dispersions is discussed in Ref. [12].

Relatively high solids thermosetting acrylic enamels are still used for solid color single top coat finishes. Metallic coatings are dominantly applied by the base coat/clear coat system described in the next section.

33.1.3.2. Base Coat/Clear Coat Systems

Application of a clear coat over a pigmented coating is, obviously, a way of achieving higher gloss. Historically, clear coats were not used in automotive finishes because of the high cost of an extra coating step and because the exterior durability of available clear coatings was thought to be inadequate. However, with the development of better binders and especially with the development of stabilizing systems based on combinations of HALS and UV absorbers (see Section 25.3.3), clears can have long-term exterior durability.

In the base coat/clear coat system, the total film thickness is only a little greater than that of a single coat system. The base coat contains roughly twice the PVC of a single coat system with a dry film thickness of about 20–30 μm, depending on the color. This provides approximately the same hiding as the 50 μm single coating with the lower PVC. The clear coat thickness is about 35 μm. The resulting coatings have a very high gloss and long-term gloss retention. The advantage is particularly marked for metallic coatings. The same overall VOC emission can be achieved using lower solids in the base coat and higher solids in the clear coat. The lower solids in the base coat permits better orientation of the aluminum in the coating and hence superior color flop. The higher level of pigmentation leads to a somewhat rougher surface of the base coat as compared to a single coat, but this is covered up by the clear coating.

While it would be possible to bake the car after application of the base coat, then apply the clear, and again bake the car, it is, obviously, more economical to apply the clear over the wet base coat and bake the overall system just once. Part of the solvent must evaporate from the base coat film before the clear coat is sprayed over it; otherwise, the force of the spray of the clear coat would distort the base coat. On the other hand, if too much solvent has evaporated, intercoat adhesion may be adversely affected. Therefore, the solvent must be permitted to flash off until the distortion does not occur, but then the top coat should be applied without further solvent evaporation. The flash-off time required is about two minutes, but varies with spray booth ventilation conditions as well as coatings formulations. In some systems, in order to permit application of the clear after two minutes, it was necessary to incorporate a small amount of wax or zinc stearate in the base coat formulation [14]. The additive apparently orients at the surface and minimizes distortion of the layer. Care must be exercised since excess additive could interfere with intercoat adhesion. Incorporation of some CAB in the base coat formulation has also been shown to help [14].

Several combinations of coating types are possible: high solids solvent-borne base coat and clear coat; lower solids base coat with an extra high solids clear coat; water-borne base coat and clear coat; water-borne base coat with solvent-borne clear coat; powder coatings are a further possibility as discussed later.

In most cases, binders in liquid base coats are thermosetting acrylic resins with MF cross-linkers. It has been reported that hydroxy-functional urethane modified polyesters are also useful in base coats [14]. There is a trend toward use of water-borne base coats with high solids solvent-borne clear top coats. By mid-1991 close to a million light trucks had been coated with water-borne base coats in Canada, and limited production or major trials were underway in about 10 assembly plants in the United States and Europe [15]. The lower film thickness of the base coat and the flash-off time required before applying the clear top coat reduce the popping problem that plagued water-borne single coats [16]. Control of sagging during application and distortion of the surface on application of the clear coat requires that the water-borne base coat be shear thinning [17].

The low solids (15–20 NVV) of water-borne base coats facilitates aluminum orientation. Opaque colors can be applied at 25–35 NVV. Water-borne base coats are formulated for application at relative humidities (RH) of about 55%, since it is less expensive to increase RH than to decrease it. The VOC of water-borne base coats at 16 NVV varies from about half to about equal that of high solids base coats at 41 NVV. Regulations often permit VOC savings in the base coat to be traded for extra solvent in the clear coat; the corresponding utilization of higher molecular weight resins in a high solids clear coat significantly reduces the risk of failure on exterior exposure.

Published papers on water-borne base coats mention MF cross-linkers with water-reducible acrylics, water-reducible polyester/polyurethanes, and acrylic latexes [15–18]. The acrylic latex approach appears to be gaining favor as experience is gained in designing the complex combinations of rheological characteristics needed for application and flake orientation [8,18].

The high solids clear coats are predominately low molecular weight acrylic resins cross-linked with MF resins. As discussed in Section 7.2, volume solids obtainable is limited to well below 50 NVV by the necessity that essentially all of the resin molecules must have at least two functional groups. Considerable research has been aimed at blending low molecular weight polyester resins with acrylic resins to increase solids without reducing the molecular weight of acrylic resins below the level that will provide adequate film properties.

The MF resins remain the dominant cross-linkers for high solids clear coatings. Mainly because of concern about acid etch resistance (see Sections 6.3 and 25.5), there have been extensive efforts to develop alternatives. Polyisocyanate 2K coatings are quite widely used in Europe and to a more limited degree in North America. They are said to afford superior acid etch resistance than many MF cross-linked coatings and may provide somewhat better exterior durability. As discussed in Section 28.2.2, new polyisocyanates and coreactants are being designed which may permit lower VOC coatings. The 2K urethane coatings can be cured at lower temperatures than MF cross-linked coatings. However, weighing against the use of 2K coatings is the need for dual spray systems, concern about toxicity, and cost. Another approach to incorporating urethane groups in the final coating is to use hydroxy-functional urethanes. For example, a low molecular weight hydroxy-functional urethane derived from the reaction of a triisocyanate and a diol such as neopentyl glycol has been reported to give good clear coat properties when cross-linked with MF resin [19]. Clear coats based on epoxy-functional acrylic resins made using glycidyl methacrylate as a comonomer (see Section 11.1.2) have reached commercial use because of their excellent acid etch resistance but may be somewhat deficient in abrasion resistance.

Low surface tension may reduce the probability of acid etching. Acrylic resins modified with unsaturated silane coupling agents are being used in clear coats. Highly fluorinated resins (see Section 13.4.3) cross-linked with MF resins or polyisocyanates are being used as additives or main binders but they are expensive.

Lowest emissions could be achieved if powder coatings (see Chapter 31) could be used for exterior automotive coatings. Powder automobile top coats have been in a development stage since the late 1960s. In Japan in the early 1980s powder coatings were used as nonmetallic single top coats. Their major limitations are the need for a separate spray facility for each color (or very expensive cleanup between color changes) and the very poor color flop of metallic powder coatings. Since these limitations do not apply to clear coatings, clear powder coatings are a possible future development, but the increasing use of plastic components will further complicate development problems.

33.1.4 Factory Repair Procedures

During the assembly of an automobile after painting, it is quite common for the paint to be damaged, resulting in a need for repair. Once the glass, upholstery, tires, and the like have been installed, the car can no longer be baked at the temperature of 120°C or more required for cross-linking MF-based top coats. Either the whole car can be baked at 80°C or the area repaired can be heated somewhat above 80°C with infrared lamps.

When cars were painted with acrylic lacquer, repair was relatively simple since the thermoplastic systems stay soluble in the solvent in the repair lacquer. However, enamels are more difficult to repair. Achieving adhesion to the surface of the cross-linked coating is more difficult and, when properly done, the whole panel in which the damage occurred has to be refinished. The top coat(s) is removed, any bare metal is primed, and special repair top coat(s) is applied to the whole panel. Since the coating cannot be baked at as high a temperature as the finish on the whole car, additional strong acid catalyst must be added to coatings having MF resins to allow curing at lower temperature. The excess catalyst remains in the film and can lead to more rapid hydrolysis. The gloss retention of such repairs is good, but not as good as that of the original coating.

Urethane 2K repair coatings are being used increasingly since they cure at relatively low temperatures without as great a loss in long-term durability as can occur with MF cross-linked coatings. Alternatively, it is possible to repair enamels with lacquers.

33.2. APPLIANCE COATINGS

Another large market for OEM coatings is for finishing appliances. Major markets are coatings for washing machines, driers, refrigerators, air conditioners, and ranges. There is a wider range of performance requirements than for automobiles. In some cases, single coats are used, but a major part of the market is for primer/top coat systems. In applications such as washing machines and air conditioners where corrosion protection is a critical performance requirement, cationic E-coat primers (see Section 30.2) are used on the highest quality products.

Nonelectrodeposition primers are frequently applied by flow coating (see Section 22.4.). In order to minimize VOC, water-reducible epoxy ester based primers are particularly appropriate. The solids of solvent-borne primers tend to be limited by the effect of the high pigment loading on viscosity, as discussed in Section 28.1.3.

Cationic E-coat epoxy coatings are used in some applications as single coats. Although epoxy coatings chalk badly on exterior exposure, the drum of a dryer or the interior of an air conditioner does not get exposed outdoors, but they need a relatively high degree of corrosion protection. The uniform coverage of edges permits the use of 12-μm E-coats to replace 50-μm solution epoxy coatings on air conditioners with substantial reduction in manpower required for application while maintaining the necessary performance [20].

White E-coats have been applied to appliances. In the case of anionic E-coats, as discussed in Section 30.1, the discoloration resulting from iron salt formation and the loss of conversion phosphate coating have limited the applications. Cationic top coats based on acrylic resins (see Section 30.2) avoid these two problems but do not provide the corrosion protection that a cationic primer based on epoxy resins provides. This limits the use of single E-coats to applications where corrosion protection is not a major requirement.

Thermoset acrylic coatings are generally used over primers. For one-coat compositions, polyesters are more commonly used because they tend to exhibit better adhesion to treated steel or aluminum than acrylics. The most commonly used cross-linkers are amino resins. In the case of washers and dishwashers or other applications where detergent resistance is critical, the amino resins are benzoguanamine–formaldehyde resins (see Section 6.4.1) because of their greater resistance to alkaline detergents. In other applications, conventional MF resins are used because of the lower cost. In end uses such as hot water heaters, where performance requirements are not severe, lower cost semioxidizing alkyd/MF based coatings may well be the most appropriate.

The use of powder coatings as top coats in appliance applications has been growing rapidly (see Chapter 31). The long runs of single colors make appliances a natural application for powder coatings. A limitation is the greater difficulty in achieving good leveling with powder coatings as compared to liquid coatings. While some orange peel can be desirable to conceal metal irregularities, powder coatings tend to have too much orange peel. However, in some cases, particularly in Europe, consumers are used to the orange peel that is typical of porcelain enamels and the finish from powder coatings is readily accepted. The low VOC emissions, low fire risk (with proper precautions in handling powders), low energy requirements, and reuse of overspray powder are strong economic and environmental reasons for using powder coatings.

For applications such as washers where corrosion protection is essential, an epoxy powder coating primer can be used or a powder coating can be applied over a cationic E-coat primer. If corrosion protection is not a major requirement, one-coat powder coatings are used. Hydroxy-functional polyester or acrylic resin are used as binders with a blocked isocyanate or a tetramethoxymethylglycoluril (see Section 6.4.3) as cross-linker. Alternatively carboxylic acid-functional resins with triglycidyl isocyanurate (see Section 11.1.2) or tetra(hydroxyethyl)adipamide (see Section 13.4.1) as cross-linker are used.

Another approach to reducing VOC emissions at the appliance manufacturer's factory is the use of coil coated metal. The solvent emissions occur at the coil coating factory where they can readily be burned to provide part of the fuel for curing the coatings (see Section 33.4).

33.3. CONTAINER COATINGS

Container coatings were classified as *metal decorating* coatings historically since a major portion of the business was in coating flat sheets, then lithographically printing on that coating, followed by a *finishing varnish* (clear top coat) to protect the print. There were many uses for such sheets—metal boxes, waste paper baskets, bottle caps and crowns, and most importantly, cans. Over time, plastics have replaced coated metal in applications other than bottle caps, crowns, and cans. The field is now usually referred to as container or can coatings.

The major portion of the cans coated are for either food or beverage containers. As a result, one of the key requirements is that there be no possibility of introducing toxic compounds into the foods or beverages [20]. In the United States, all can linings must be acceptable to the Food and Drug Administration (FDA) or, in the case of meat products, the Department of Agriculture. Contrary to what many people seem to believe, the FDA does not approve coatings; it lists acceptable ingredients. Rules are published in the Code of Federal Regulations (CFR), Title 21, Part 175. The most important of the rules is 21CFR175.300 which deals with resinous and polymeric coatings. In most cases, if all of the components of the coating have been used in earlier can coatings, a new coating will be acceptable. In some cases, it is necessary to prove that no material is extracted into any food or beverage that will be packed in the can. However, new raw materials must pass extensive tests.

Toxicity considerations predominantly affect interior can coatings, but in some cases there are also restrictions against possible contamination from exterior can coatings. This is particularly so when metal sheets coated on both sides are stacked, in which case the exterior coating on one sheet is in direct contact with the interior coating on the next sheet in the stack. In this configuration, migration of low molecular weight components between coatings is possible.

A further critical factor unique to can coatings is their potential effect on the flavor of the food or beverage packed in the container. Again, while the flavor requirements are particularly important in the case of interior coatings, care must also be exercised with coatings for exterior application. Flavor changes can result from either extraction of some contaminant from the coating or failure to isolate the food or beverage from the metal of the can. A common potential problem is residual solvent in the coatings. In order to assure that all residual solvent is driven out of the coatings, high baking temperatures are used. These high temperatures, obviously, help drive other possible volatile flavor detractors out of the coatings. Flavor can be affected by minute amounts of substances. The only way to evaluate effects of coatings on flavor is by making test packs of the food or beverage in the container and later tasting the food or beverage. As a result, major suppliers of can coatings maintain flavor panels of people trained to taste and, particularly, to use consistent words to describe flavors.

In the case of beer cans, a major aspect of the flavor problem is to eliminate contact between the beer and the can. Apparently metals catalyze flavor changes in beer. For this reason, the final interior coating is spray applied after formation of the can to avoid a potential problem from breaks in the can lining resulting from stresses during can forming operations. Linings are spray applied to the interior of soft drink cans not only to protect flavor, but also to protect the can; the acid present in most soft drinks could eat through the metal can unless the coating acts as a barrier. An interesting side light on history is that in the early days of packing pineapple and grapefruit juice in tin cans, coatings for lining these cans that would resist the high acidity of these foods were not available. They were packed in tin cans with heavy tin plate linings to protect the steel bodies. The tin affects the flavor of the pineapple products and grapefruit juice. Even though organic coatings became available that could be used with these products, most are still packed in heavily tinned cans. Apparently consumers are used to and prefer canned pineapple products and grapefruit juice that taste "like tin."

There are two major classes of cans—three-piece and two-piece cans. In a three-piece can, one piece is the body, a flat sheet formed into a cylinder and sealed by soldering or welding, or with an organic adhesive. Soldering was formerly widely used but, in general, soldered cans are no longer used for foods or beverages because of concern about lead toxicity. The other two pieces are the can ends that are punched out from flat metal and formed separately. A gasket is formed in place in the lip of each end so that when the can is assembled it will not leak.

In the case of three-piece cans, either flat metal sheets are coated or continuous strips of metal are coil coated. Blanks for the can bodies are stamped out of the coated metal, formed, and sealed into the cylindrical body. In many soldered or welded cans, the side seam is sprayed with a fast drying coating called a *side striper* to cover the exposed metal resulting from the heat of soldering or welding. Most side stripers are now powder coatings. They are applied while the metal is still hot, resulting in fusion of the powder. The interior of the can body is sprayed with the final lining. The ends are made separately by stamping out of coated sheets or coil coated metal, forming, and equipping with a formed-in-place rubber gasket. One end is put on by the can manufacturer and the other end by the food or beverage packer after the can has been filled.

Coatings on sheets are generally applied by direct roll coating (see Section 22.5). In some cases, transfer rollers with sections cut out are used so that the coating is only applied in selected areas. For example, the edges that will be soldered or welded at the sides of the body of the can are commonly not coated. On coming out of the coater, the sheets are fed onto wickets attached to a conveyor. This permits the sheets to go through the baking oven in an almost vertical position, reducing the necessary length of the oven substantially. The ovens are equipped with high-velocity hot air units that direct the hot air over the coating once it has been heated enough so that most of the solvent has evaporated. This substantially increases the efficiency of heat transfer and, therefore, reduces the baking time that is necessary. Baking schedules of the order of 5 min at 175°C are typical. Inks are applied by offset lithography. The ink is transferred from the lithographic plate to a rubber blanket and is in turn offset from the rubber blanket to the sheet.

There are two processes for making two-piece cans. Drawn and wall-ironed (DWI) cans are formed by drawing a cup from a flat blank and then ironing the walls to thinner thickness and greater depth. In draw-redraw (DRD) cans, a coated blank is formed into a shallow cup and then drawn two further times to achieve the desired height and configuration and shape of the bottom of the can. The other piece of a two-piece can is an end similar to the ends of the three-piece can. (In the United States almost all beverage cans are two-piece cans, but three-piece ones are still made in some other countries.)

Examples of DRD two-piece cans are shallow cans for tuna fish and taller cans for vegetables, soups, and pet foods. Flat sheets are coated and then the can is drawn and formed. The single end is formed separately. In some cases, such as shoe polish and auto wax cans, printed sheets are formed into two-piece cans. The design must be *distortion printed*; that is, the print must be designed so that it looks right after the distortion resulting from fabricating the can.

In general terms, it costs less to coat and print flat sheets before forming than to coat and print the formed can. However, the ability of the coating to withstand this degree of formation depends on the depth of draw. It is not just a matter of how deep the can is but also how wide it is. Forming a large diameter can involves less distortion than a narrow can; a deep can involves more distortion than a shallow can. Bottle caps and crowns are coated and distortion printed on sheets and then post formed.

Beverage cans are predominantly DWI cans; uncoated metal is drawn and formed, then the exterior coatings and inks are applied to the individual cans, and the lining is sprayed into the interior of the can. When cans are to be coated and printed after forming, coating is applied by rotating the can against the transfer roll of a small coater and inks are transferred from a litho plate to a soft rubber roller that in turn transfers the inks to the can surface. Baking is done on short time cycles at high temperature. Currently lines are running at a rate of over a thousand cans per minute. It is estimated that the coating reaches a peak temperature of about 205°C for only about 1 s. Only partial cross-linking occurs in the short time. The cross-linking is completed when the can is baked again after the interior lining is applied.

Interior coatings are applied by spray. A small spray gun is automatically inserted into the spinning can, it sprays, and is pulled out of the can. In order to assure loss of solvent, the final part of the curing is done with air being directed into the can bodies. Typically the cure schedule is about 2 min at 200°C. Final forming of the top of the can to shape it to fit the end is done after the coating and printing are completed.

There are three major types of metal used in cans: tin plate, steel, and aluminum. The choice of metal depends on the end use. Many food cans are three-piece cans made from tin-plated steel. A large fraction of pet food cans are made from treated steel—so-called black plate—since it is less expensive and there is not the need for the highly reflective tin coating. By far the largest fraction of beverage cans are two-piece aluminum cans. Pressure from the carbonation of either soft drinks or beer keeps the thin walled aluminum can sufficiently rigid. In the case of food cans which do not have significant interior pressure, aluminum does not compete with steel because thicker walls are required to achieve adequate rigidity. There has been some discussion of using aluminum in food packing by putting a small piece of dry ice into the pack before sealing. Vaporization of the CO_2 provides the pressure necessary for rigidity. However, people associate slightly bulging food cans with spoilage so consumer acceptance has been slow.

33.3.1. Interior Can Linings

The composition of interior coatings for a can depends on the food or beverage to be packed in the can. In most cases, if the food is cooked, it is cooked in the can. A common cooking cycle is 60 min at 121°C. Most beer is pasteurized in the can at somewhat lower temperature. In both cases, the interior and exterior coatings must maintain their adhesion and integrity through the cooking or pasteurizing process.

Most common vegetables and fruits are packed in cans with an interior coating called an *R enamel*. Historically R enamel was a phenolic/drying oil varnish. Now it is more common for R enamel to be a phenolic resin modified with some drying oil derivatives during its synthesis or an epoxy/phenolic coating. For the packing of vegetables, such as corn, that give off hydrogen sulfide during cooking, fine particle size ZnO pigment is dispersed in the coating, and it is called a *C enamel*. The ZnO reacts with H_2S to form white ZnS. This prevents or conceals formation of unsightly black tin sulfide from reaction of tin oxides with H_2S.

Fish cans are generally lined with phenol/formaldehyde resole resin coatings (see Section 11.8.1). In order to obtain the needed formability to permit drawing of two-piece fish cans, the cross-link density is reduced by using a mixture of *p*-cresol and phenol in making the resole phenolic resin. Poly(vinyl butyral) is commonly incorporated in the formulation to promote adhesion and act as a plasticizer. Thin film thickness is used; this helps permit forming without film rupture. The phenol/formaldehyde resins are sufficiently highly cross-linked to avoid swelling and softening by oils from the fish (and, in some cases additional oil in which the fish is packed) during processing and storage. Aluminum flake pigment is sometimes incorporated in the coating to minimize permeability through the coating. In linings of cans for meats, such as ham, one of the key requirements is that the coating permit the meat to slide out of the can easily after the top is removed. This requires incorporation of a release agent, such as petroleum wax or silicone, in the can lining.

Ripe olives are an example of a special product requiring a specifically developed can lining. In order to avoid any risk of botulism, ripe olives must be cooked in the can at 120°C for 2 hr. Especially in view of the high oil content, the development of a lining that would withstand this processing in the early days of the metal decorating business was not easy. In 1904 a tung oil/hard resin varnish with high drier content was developed that would withstand the processing. At least as late as 1978, essentially this same varnish was being used in ripe olive cans. This is not to say that one could not develop an alternative suitable lining, but the development work required to find such a product and to be assured of its efficacy and safety has to the best of our knowledge not yet been done. Properly, the interior can lining business for food cans is a very conservative one; people stay with what has proven to work. The value of the lining in a ripe olive can is a trivial fraction of the value of the ripe olives, and furthermore one must be very sure that there would be no risk of botulism with a change in process.

An increasing fraction of food cans are DRD two-piece cans. The draw-redraw process requires that coatings possess a greater degree of ductility than the coatings used for interiors of three-piece food cans require. Most commonly, vinyl organosol coatings lightly cross-linked with phenolic resins or MF resins have been used since the vinyl chloride polymer binder shows ductility below T_g (see Section 24.2) [21]. In some European countries there is concern about toxic emissions from factories recycling vinyl chloride copolymer coated cans [22]. This has led to work on other more distensible coatings. For example, epoxy resins have been designed that will take considerably deeper draws than the conventional BPA epoxy coatings [23].

Very large volumes of coatings are used in lining beverage cans. Historically, the spray interior coatings were solvent-borne vinyl chloride copolymer or epoxy/melamine coatings. The volume solids of the soluble PVC copolymer coatings at application viscosity were very low, about 12% NVV. The high cost of solvent and the high level of VOC emissions has forced a change. A water-borne coating based on a graft copolymer is now widely used in lining both beer and soft drink cans [24,25]. Styrene/ethyl acrylate/acrylic acid side chains are grafted onto a BPA epoxy resin. The resin is "solubilized" with dimethylaminoethanol (DMAE) in glycol ether solvent and reduced with water (see Section 11.5). In order to lower the cost of some soft drink can linings, it has been possible to use a special latex binder with small amounts of the graft copolymer described above and epoxy phosphate as adhesion promoters.

33.3.2. Exterior Can Coatings

Most food cans are not coated on the outside of the can; rather paper labels are used. Paper labels would not stand up on beverage cans since they may be put in ice chests or be refrigerated so that water condenses on the outside of the can when they are brought out into humid air. Therefore, beverage cans are coated and printed on the can exterior itself. This process also permits manufacture of an attractive can; printed cans are used for some other products besides beverages in spite of the higher cost compared to paper labels.

The general procedure used on the exterior of sheets for the bodies of three-piece cans is to apply a *base coat*, often called an *enamel*, print on the base coat up to four colors by offset lithography, and finally top coat with what is called a *finishing varnish*. The most common color for base coats is white but a wide variety of colors is used for specific products. Color stability and lack of color change on baking are critical requirements. In order to minimize VOC emissions, water-borne acrylic coatings cross-linked with MF resins are widely used vehicles. The lithographic printing inks are either baking or UV cure inks. The vehicles for the baking inks are long oil alkyd resins with some MF resin cross-linker while the vehicles for the UV cure inks are blends of acrylated epoxidized soy (or linseed) oil with acrylated epoxy resin and acrylate reactive diluents (see Chapter 32).

The finishing varnishes are acrylic/MF binder compositions. In order to decrease friction in forming machines and conveyors, the finishing varnish commonly contains a small amount of petroleum wax or fluorinated surfactant to reduce surface tension so the cans will have a low coefficient of friction. In the case of baking inks, two inks are applied, the sheets are baked, the other two inks are applied, the finishing varnish is applied, and the sheets baked again. This is called *wet-ink varnishing* since the varnish is coated on the printed surface before the last two inks are cured. With UV inks, each ink is at least partially cured by passing under a UV lamp before the next ink is applied. Commonly, finishing varnishes are not used over UV cure inks.

The coatings on the exterior of two-piece cans are most commonly water-borne acrylics. The film thickness of white coatings on either three- or two-piece cans was formerly 15 μm. This was not enough to provide complete hiding, but the coating was definitely white. In order to reduce cost, film thicknesses have been decreasing to as low as 8 or 9 μm in some cases. In still more cases, white coatings have been replaced with white printing inks where the thickness is on the order of 2–3 μm. The resultant whites are very gray. On some cans, a transparent yellow brown coating is used to give a brass or gold color over the relatively shiny metal surface.

UV coatings are sometimes used on beverage cans (see Chapter 32). Exterior finishing varnishes are acrylated resin coatings cross-linked by photogeneration of free radicals. Coatings for the exterior side of can ends are epoxy resins cross-linked by photogeneration of acid. The UV cure end coating is applied to the side of the sheets that will be on the outside of the cans and passed under the UV lamps, giving a partially cured film; the other side, which will be on the inside of the can, is coated with a FDA listed epoxy/phenolic thermal cure coating. When the epoxy/phenolic is cured in an oven, the cure of the UV epoxy coating is advanced to completion by the photogenerated acid still present in the films. This thermal complement to the UV cure also enhances adhesion of the coating.

33.4. COIL COATING

Steel and aluminum are manufactured in long strips which are rolled into *coils*. In many cases, the coils are cut, formed, and fabricated before they are coated, as in the production of automobiles, for example. In other cases, it is possible to coat metal coils and later fabricate the final product from the precoated metal. When precoated metal can be used, there can be substantial advantages (see Section 33.4.1), and coil coating has grown into a major industrial process [26].

Coil coating started in 1935 as a process for coating Venetian blind slats. The strips of metal were about 5 cm wide and the line was run at a rate of about 10 m min^{-1}. Modern coil lines can coat metal up to 2.5 m wide at rates as high as 300 m min^{-1}. Most lines run at a rate of around 125 m min^{-1} with metal that is up to 1.8 m wide. A schematic drawing of a coil line is shown in Figure 33.1.

The metal is shipped from the steel or aluminum mill in coils weighing up to 25,000 kg (55,000 lb) that are 0.6–1.8 m (2–6 ft) wide and 600–1800 m (2000–6000 ft) long. As shown in Figure 33.1, the first step in the process is precleaning; brushes and sanding remove any physical contaminants. The strip then goes to the entry accumulators. The rollers of the accumulators move apart to accumulate a significant length of coil so that when one coil is about to run out, the next coil can be stitched on while the accumulator rolls move together supplying strip to the line without interruption of the process. After the stitching is done, the new coil is fed to the coating line. As the process continues, the accumulator rolls gradually separate to store strip for the next change of coils.

The strip is carried next through the metal treatment area. Detergent washing and rinsing are followed by phosphate conversion coating for steel (chromate conversion coating for aluminum), rinsing, and final chromic acid rinse. All the cleaning and conversion coating procedures must be designed to work at high speeds. Since the metal is moving at 100 m min^{-1} or more, the total time for cleaning and treatment is a minute or less. Next, the strip is carried through a dry-off oven and finally to coaters and a baking oven. In Figure 33.1, a laminator is also shown, although laminating of film in line with coating is not common. It is more usual to have two or three coating stations, each followed by a baking oven. This permits coating both a primer and a top coat on one side of the strip and another coat on the back side. It is fairly common to coat, cure, then print one or more colors.

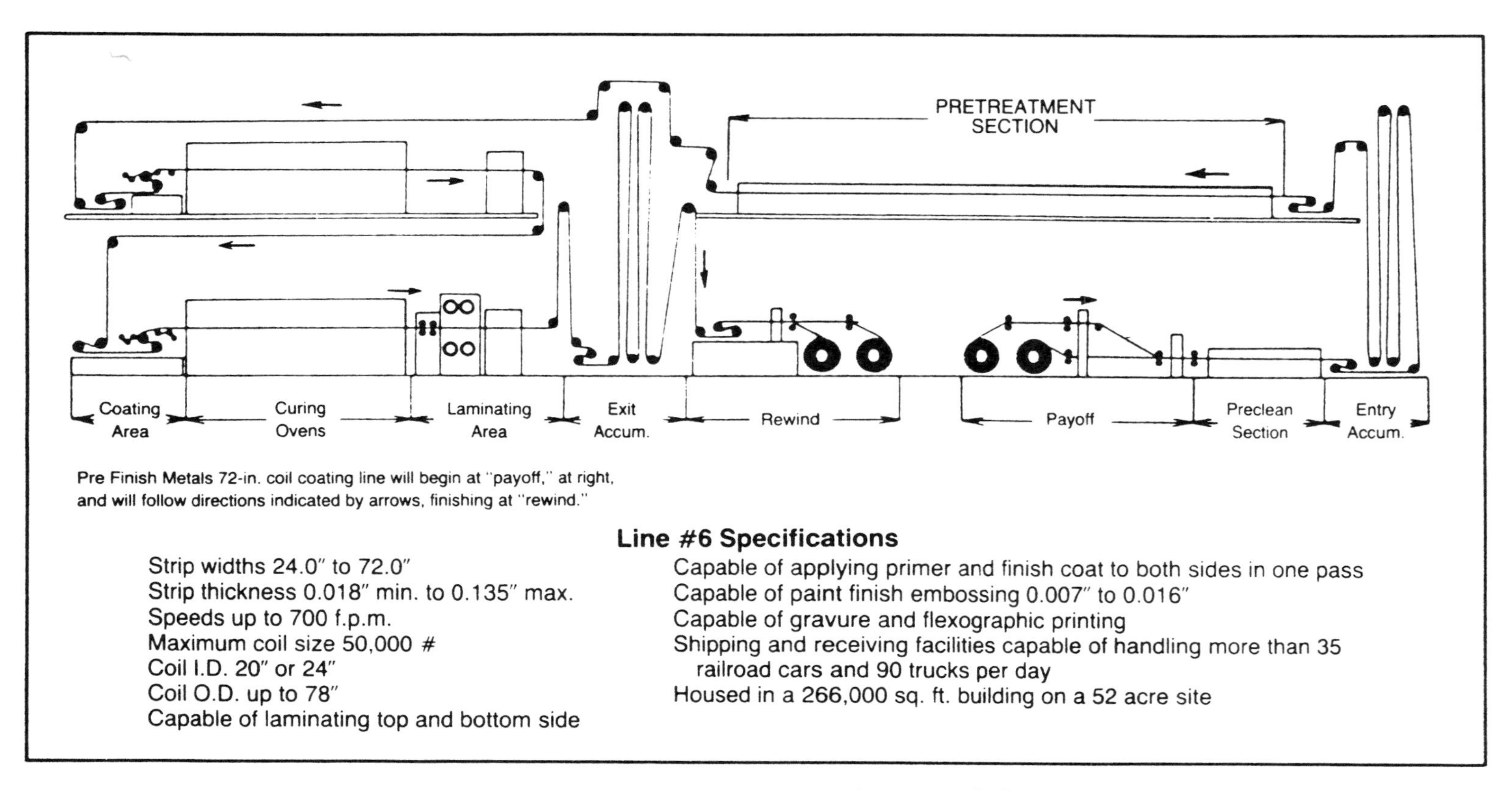

Figure 33.1. Schematic drawing of a coil coating line. (From Ref. [26], with permission.)

Most commonly coatings are applied by reverse roll coating, but sometimes direct roll coating is used for thinner films where greater precision is needed (see Section 22.5). Reverse roll coating is used for thicker films where close control of thickness is not as critical, but where the flow advantages resulting from the wiping action in contrast to the film splitting action in direct roll coating are important. In order to achieve even film thickness across a strip 2 m wide, the rolls have to be crowned, that is, designed so their diameter is greater in the middle than on the edges since the pressure involved tends to bow the rolls to a small degree. To avoid damage to the rollers, the line is programmed so that very shortly before a stitched section joining two coils is going to pass through the coating nip between the rolls, the rolls are automatically separated slightly and then almost immediately return to normal operating pressure.

At the high speeds at which the lines are run, even with long ovens and looping the strip back and forth through the oven, the dwell time in the ovens is less than a minute. In some cases, the dwell time is a low as 15 s, but more commonly in the range of 30–45 s. After a short initial period, hot air is directed over the surface of the coatings at high velocity. The air temperature can be as high as 375°C. The temperature reached by the coating on the metal is the critical temperature for curing the coating. This temperature cannot be directly measured but is closely related to the metal temperature which can be measured. The temperature considered most important is the *peak metal temperature*; this can be as high as 260°C. After the coating(s) is baked, the strip passes through the exit accumulator to the rewind. The exit accumulator stores coated strip during removal of a coated coil. In some lines, the strip passes over chilling rolls to reduce the temperature before the strip is rewound into a coil. The pressure in the center of the rewound roll is very high. The viscosity of the coating on the metal that is being rewound must be very high and/or it must be adequately cross-linked to avoid blocking.

The exhaust air from the hoods over the coaters and particularly from the oven contains solvents. On most lines, the exhaust air streams are used as part of the air that is used to burn the gas to heat the ovens. In this way, part of the residual heat from the oven exhaust is recycled and the solvent is burned. Burning the solvent essentially eliminates the VOC emission problem and the fuel value of the solvent is recovered. As a result, there has been less incentive to change to waterborne or high solids coatings in coil coating applications than in other applications. There are still pressures to reduce solvent content because the fuel value of the solvents is low compared to the cost of the solvents and because some lines are not equipped to burn solvent.

Coatings on aluminum are frequently single coat; but on steel, primer/top coat systems are most widely used. Binders for primers are almost always based on BPA epoxy resins; epoxy esters (see Section 11.4) and epoxy/MF resins (see Section 11.3) are examples.

Many types of coating binders are used for top coats. Oxidizing alkyds with MF resin are the lowest cost compositions. They are widely used on the reverse side of the coated strip as a *backer*. This coating is commonly unpigmented and contains a small amount of incompatible wax. The purpose of the backer is to avoid metal marking of the top surface coating by rubbing against a bare metal reverse side in the coil. Alkyd/MF coatings are also used as top surface coatings where corrosion resistance requirements and/or exterior durability requirements are modest. Polyester/MF binders are widely used, especially as single coats; exterior durability and corrosion protection are generally superior to that obtained with alkyd coatings. Polyester/blocked isocyanate coatings have been used to a degree for applications where abrasion resistance is particularly important. Close temperature control in the ovens is especially important with urethane coatings since urethanes discolor and decompose relatively rapidly at the baking temperatures involved in coil coating. Thermoset acrylic coatings are sometimes used, usually over primers.

For greater exterior durability, one can use silicone-modified polyesters and silicone-modified acrylic resins (see Section 13.3.1). For example, one might use 50% silicone-modified acrylic resin with a small amount of MF resin as a supplemental cross-linker as the binder for color top coats for high performance residential siding. In the same quality line, the white might well be just an acrylic/MF coating. After many years exposure, the white might start to chalk slightly, but this would not adversely affect the appearance. However, even a small amount of chalking of a color coating makes an easily seen change of color due to the change in surface reflection. Such changes are particularly serious in an exterior siding application since the exposure varies depending on the location on the building. The resulting nonuniform chalking of the color coating sticks out like a sore thumb. For the highest exterior durability, fluorinated resin coatings are used (see Sections 4.2.3 and 13.4.3). In some cases, such coatings show no indication of change after exposure outdoors for more than 20 years.

Organosol coatings (commonly misnamed plastisol coatings) are used to a significant degree in coil coating. As discussed in Section 4.2, the vehicle in these coatings is a dispersion of vinyl chloride copolymer in plasticizer and solvent. Such coatings provide reasonable exterior durability with excellent fabrication properties. Organosol coatings have the advantage that they need not be cross-linked and hence can be run with as little as 15 s dwell time in the oven. Solution vinyl resins are still used to a degree in coil coating metal for can ends for beverage containers.

Increasing use is being made of latex vehicles for coil coating. They have the distinct advantage of high molecular weight so that mechanical properties can be achieved without need for much, if any, cross-linking. High-gloss coatings cannot be made and there can be greater problems of leveling than with solvent-borne coatings. The leveling problems are minimized by using reverse roll coating and by using associative thickeners to control viscosity. Associative thickeners minimize latex particle flocculation relative to conventional water-soluble polymeric thickeners such as hydroxyethyl cellulose (see Sections 23.1 and 35.3).

33.4.1. Advantages and Limitations of Coil Coating

There are important advantages of coil coating that have promoted the growth of the business to a major component of the industrial coating field. For long production runs the cost is very low compared to coating preformed metal. The rate of application of coating is much faster and the labor cost is much lower. Coating utilization is essentially 100%. Oven designs are such that energy usage in curing is much more efficient. In general, floor space requirement is much less, so capital cost for buildings is lower. Since solvent is restricted to the immediate area of a roll coater, fire hazards and toxicity hazards are reduced as compared to spray application. Since in most cases solvent is incinerated, VOC emissions are generally very low.

Film thickness of the applied coatings is more uniform than is generally obtained in coating preformed products. Since the coatings are applied on uniform thickness metal, curing of all parts of the coating tends to be more uniform than in curing coatings on fabricated products.

In many applications, the performance of the coatings applied and baked on the coil provide superior performance. This difference is particularly evident in comparing high-quality precoated exterior siding as compared to house paint. A large part of this difference results from the difference between performance of baked coatings as compared to air dry coatings.

The manufacturer using precoated metal gains some substantial advantages. The fire hazards associated with coatings application are eliminated, and insurance costs drop. The manufacturer does not have to deal with the VOC emission problems associated with applying coatings. There is no waste disposal problem with sludge from spray booths. There may be a substantial saving in floor space.

On the other hand, there are limitations to the coil coating process. It is economical only for fairly long runs of the same color of the same quality of coating. The cost of changing color of the coating is high because the coater must be shut down for the cleanup and rethreading of the line. The capital cost of the line is very high; therefore, lost production time in shutdowns is expensive. In a line of coated metal with several colors, the inventory cost can be high since inventory of several colors must be kept rather than just one inventory of uncoated coils. If a stylist changes colors, the obsolete inventory cost can be high or, said the other way, the flexibility of changing colors is more limited than when the assembled product is painted.

A major challenge to the coil coating supplier is color matching. Generally, very close color matches are needed. Hiding is less than complete and the color of the metal or primer affects the color of the coating. Color can be affected by the high temperature baking schedule. It is not possible to duplicate a curing schedule of 30 s in an oven with high-velocity 375°C air in the laboratory. In effect, the color matcher must learn to compare color differences that can be expected in the laboratory with what will happen on a particular commercial coil coating line and then do the color matching so as to take the difference into consideration.

The coated metal must be able to withstand fabrication into the final product without film rupture. This means that there will have to be acceptance of somewhat softer films than could be specified if the product were coated after fabrication. When the coated metal is die cut to make the eventual product, bare edges of metal are exposed. Welding of coated metal can be a problem. Generally, coatings can be designed that will not interfere with the welding process; however, the appearance of the coated surface will be destroyed at and near the welded area. In some applications, bare edges are not a problem and the product is not welded. In other cases, it is critical for the designer of the finished product to ensure that the cut edges do not show and are not in locations where they will be particularly subject to corrosion, and that welding, if any, will be done in areas that will not show.

Examples of large applications for coil coated metal are: siding for residential use, original siding for mobile homes, Venetian blinds, rain gutters and downspouts, fluorescent light reflectors, appliance cabinets, can ends, and can bodies for fruits and vegetables.

GENERAL REFERENCES

B. N. McBane, *Automotive Coatings*, Federation of Societies for Coatings Technology, Blue Bell, PA, 1987.

J. E. Gaske, *Coil Coating*, Federation of Societies for Coatings Technology, Blue Bell, PA, 1987.

REFERENCES

1. M. S. Reisch, *Chem. Eng. News*, **70** (41), 36 (1992).
2. J. J. Wojtkowiak and H. S. Bender, *J. Coat. Technol.*, **50** (642), 86 (1978).
3. S. Maeda, *J. Coat. Technol.*, **55** (707), 43 (1983).
4. B. N. McBane, *Automotive Coatings*, Federation of Societies for Coatings Technology, Blue Bell, PA, 1987.
5. C. K. Schoff, *J. Coat. Technol.*, **62** (789), 115 (1990)
6. E. Ladstadter and W. Gessner, *Proc. XIIth Intl. Conf. Org. Coat. Sci. Tech.*, Athens, 1986, p. 203.
7. C. D. Cheever and P.-A. P. Ngo, *J. Coat. Technol.*, **61** (770), 65 (1989).
8. D. C. van Beelen, R. Buter, C. W. Metzger, and J. W. Th. Lichtenbelt, *Proc. XVth Intl. Cong. Org. Coat. Sci. Technol.*, Athens, 1989, p. 39.
9. K. Tachi, C. Okuda, and S. Suzuki, *J. Coat. Technol.*, **62** (782), 43 (1990).
10. G. T. Weaks, *Proc. ESD/ASM Adv. Coat. Technol. Conf. 1991*, 201 (1991).
11. R. A. Ryntz, *SAE Technical Paper Series 880597*, Society of Automotive Engineers, Warrendale, PA, 1988.
12. D. R. Bauer, L. M. Briggs, and R. A. Dickie, *Ind. Eng. Chem. Prod. Res. Dev.*, **21**, 686 (1985).
13. M. S. Wolfe, *Prog. Org. Coat.*, **20**, 487 (1992).
14. M. Broder, P. I. Kordemenos, and D. M. Thomson, *J. Coat. Technol.*, **60** (766), 27 (1988).
15. C. B. Fox, *Proc. ESD/ASM Adv. Coat. Technol. Conf., 1991*, 161 (1991).
16. A. J. Backhouse, *J. Coat. Technol.*, **54** (693), 83 (1982).
17. Z. Vachlas, *J. Oil Colour Chem. Assoc.*, **72**, 139 (1989).
18. I. Wagstaff, *Proc. ESD/ASM Adv. Coat. Technol. Conf. 1991* 43 (1991).
19. J. L. Gardon, *J. Coat. Technol.*, **65** (819), 24 (1993).
20. T. J. Miranda, *J. Coat. Technol.*, **60** (760), 47 (1988).
21. P. J. Palackdharry, *Polym. Mater. Sci. Eng.*, **65**, 277 (1991).
22. M. Hickling, *Polym. Mater. Sci. Eng.*, **65**, 285 (1991).
23. R. A. Dubois and P. S. Sheih, *J. Coat. Technol.*, **64** (808), 51 (1992).
24. J. T. K. Woo, V. Ting, J. Evans, R. Marcinko, G. Carlson, and C. Ortiz, *J. Coat. Technol.*, **54** (689), 41 (1982).
25. J. T. K. Woo and R. R. Eley, *Proc. Water-Borne Higher-Solids Coat. Symp.*, New Orleans, 1986, p. 432.
26. J. E. Gaske, *Coil Coatings*, Federation of Societies for Coatings Technology, Blue Bell, PA, 1987.

CHAPTER XXXIV

Product Coatings for Nonmetallic Substrates

Many products made from wood and plastics are coated in factories; examples are wood furniture and paneling, hardboard paneling and siding, book covers, plastic auto body parts, and computer cases. Do-it-yourselfers also paint wood and plastic products at home, but the products they use are generally different and are discussed in Chapter 36. While many plastic articles are manufactured with the appropriate color and do not need coatings, many others must be coated for decorative or functional purposes or both. As the use of plastics expands, the field of coatings for plastics expands correspondingly. In this discussion we include elastomers (rubbers) in the broad term plastics.

Some important types of coatings for cellulosic and plastic substrates are not classified by the U. S. Department of Commerce as coatings and are, therefore, outside the scope of this book (see Section 1.1). Examples are coatings applied to paper during the paper making process (on-machine coatings), which are classified as chemicals used by the paper industry, and most coatings applied after paper is manufactured (off-machine coatings), which are supplied by the ink industry. Other examples are coatings on textile fibers, plastic packaging film, and photographic film. Many of the principles set forth in this book can be applied to such coatings.

34.1. COATINGS FOR WOOD FURNITURE

There are many styles of furniture, and furniture coatings are very much affected by styling. Furniture styling and manufacture are quite different in the United States than in most other countries in the world; most of our discussion deals with the U. S. market.

A large fraction of wood furniture in the United States is styled to resemble antiques, especially French, English, and Spanish antique furniture styles. Styling is generally initiated in the high end of the market, the expensive wood furniture that involves a great deal of hand labor and artistry in its manufacture. The bulk of the market is furniture made to look as much like this expensive furniture as possible, but using manufacturing and finishing techniques that permit lower prices.

Even though the volume of the high end of the furniture market is small, a representative finishing process is described in some detail since a major part of furniture is finished to resemble high style furniture.

Fine quality furniture is made from a combination of woods. The tops and sides are made from five-ply plywood, the legs and rails are made from solid wood, and carved wood decorative pieces are commonly attached to the furniture. The center ply of the plywood can be wood, but more generally is chipboard made by pressing wood chips and an adhesive binder into sheets and curing the sheets. Chipboard is desirable core material because it is more dimensionally stable than wood. A ply of wood veneer is then laid up with glue on each side of the core. A further layer of veneer is then laid up on each side cross-grain to the first layer, and the whole is cured under pressure and heat. The top or face veneer, that is, the side seen on the furniture, is usually from some kind of hardwood and is selected for the beauty of its grain pattern.

Grain patterns are affected by the kind of tree from which the wood is obtained and by the way the veneer is cut from a log. The top of a table is generally too large for the face veneer to be from only one piece. It has to be laid up from several pieces. In the highest quality furniture, these pieces are carefully selected and put together to give a particularly beautiful pattern. Adhesive is needed not only between the layers in the plywood but also to connect the edges of the individual pieces of veneer in the top face. Each table top is unique because the grain patterns in the separate pieces of veneer are different. In many cases, elaborate patterns are laid up. Many kinds of wood are used: pecan, walnut, and oak are among the most common; rosewood is an example of a rarer, more expensive wood that is popular because of esthetic grain patterns. All of these hardwoods have quite open pore structures and prominent grain patterns. In some styles, mahogany, which has much straighter grains and shallower pores, has been popular. Woods that have little pore structure, like maple and birch, have a more limited market than those with bold patterns.

In U. S. furniture manufacturing, the various components are cut for a run of the same style and sets of furniture. The furniture is assembled before it is finished. The various components are of different colors; if the final color of the furniture is to be lighter than any part of any of the component pieces, the wood is bleached with a solution of 30% hydrogen peroxide in methyl alcohol; the bleaching solution is activated with sodium or potassium hydroxide. Methyl alcohol solutions rather than aqueous solutions are used because methyl alcohol does not raise the grain of the wood as much as water does. Water raises the grain because the cellulose of the wood absorbs large amounts of water; the swelling is uneven, varying with the density across the grain. The part that swells the most expands most so that its surface is higher than the adjoining wood.

Glue Sizing. In former times, a dilute solution of hide glue was sprayed on the surface. Now, low solids poly(vinyl alcohol) solutions are used. After the size has dried, the little wood fibrils that stick up from the surface have been stiffened so the surface can be smoothed by sanding. Glue sizing minimizes concentration of stain at the fibrils which would give an unattractive appearance. With some grades of wood, sanding without glue sizing gives satisfactory results.

Staining. The wood is colored by using a solution of acid dyestuffs in methyl alcohol with a small amount of slow evaporating polar solvent. Dyes, not pigments, are

Table 34.1. Dark Brown Filler

Materials	Weight Ratio	Volume Ratio
Long oil linseed alkyd	123.0	14.64
Linseed oil	127.8	16.47
Limed rosin and driers	159.0	20.97
Amorphous silica	901.6	40.85
Talc	81.9	3.54
Mineral spirits	23.0	3.53
Burnt umber	37.8	1.2
Raw umber	37.8	1.29
Bone black	30.0	1.41
	1521.9	103.9

used because they do not conceal the grain of the wood. Since methyl alcohol is the dominant solvent, they are *nongrain raising* stains (NGR stains). The stain is selected and applied to give the overall base color to the furniture that the stylist has selected. Spraying the stain requires skill to obtain an even coloration.

Wash coating. A wash coat is a low solids, about 10 wt% (NVW), low viscosity lacquer. A thin layer of this lacquer is applied with two purposes. The first is to minimize any displacement of the stain and to stiffen wood fibers for sanding. The second purpose is to prepare the surface for the next step, filling.

Filling. The purpose of filling is to color the pores of the wood so as to emphasize the grain pattern and to fill the pores to near the same level as the rest of the wood. The pores must be filled with the colored *filler*, but leaving no filler on the surfaces between the pores. Usually the color of the filler is a shade of dark brown and the stains are lighter yellowish or reddish browns, but for special effects one can use white filler with black stain or other color combinations. A formulation for a dark brown filler is given in Table 34.1 [1]. As is common in industrial practice, the formulation is given in both weight and volume units, usually in the United States the units are pounds to make a 100-gal batch. In some cases, as in this one, the total volume is somewhat more than 100, the amount is such that the yield would be expected to be 100 gal after production losses in the coating factory.

The color pigments used are natural earth pigments with a high loading of inert pigments. The PVC of this dry filler is over 50%. The binder can be a varnish of linseed oil and hard resin (limed rosin in this case) or a blend of linseed oil and long oil alkyd. This formulation has only a small amount of mineral spirits, but the filler is reduced with further mineral spirits and/or VM&P naphtha before application. The whole piece of furniture is sprayed with a liberal coat of the filler, the filler is "padded" into the pores by rubbing vigorously with a pad of cloth, then excess filler is wiped off. If the wood has been properly wash coated and the filler is wiped evenly when the degree of solvent flash off is right, the pores of the wood are filled and no filler is left between the pores.

Adequate drying of the filler is critical. Residual aliphatic hydrocarbon solvents can result in shrinkage, graying, and loss of adhesion of succeeding costs of finish.

Sanding sealer. The purpose of the sanding sealer coat is to seal the stains and filler, and to prepare the surface for the application of a top coat. An essential

Table 34.2. Lacquer Sanding Sealer

Materials	Weight Ratio	Volume Ratio
Nitrocellulose—RS 1/4 sec (IPA)	69.2	6.46
Nitrocellulose—RS 1/2 sec (IPA)	23.1	2.15
Blown soybean oil	64.6	7.87
Modified rosin ester	46.1	4.88
Zinc stearate	12.3	1.34
MEK	124.5	18.55
MIBK	72.1	10.81
Isopropyl alcohol	104.3	15.45
Toluene	123.3	17.00
Xylene	112.2	15.48
	751.7	100.00

requirement is that the sealer be easy to sand smooth without "gumming" the sandpaper. A formulation for a typical sanding sealer is given in Table 34.2 [1]. Sanding sealers contain 3–7% of zinc stearate based on the lacquer solids to aid sanding. The principal binder is nitrocellulose (see Section 4.3.1). The IPA notation in Table 34.2 indicates that the nitrocellulose is shipped wet with isopropyl alcohol (70% NC, 30% IPA). Dry nitrocellulose is a much more serious fire hazard than if it is shipped wet with solvent. The RS grade of nitrocellulose does not dissolve in isopropyl alcohol, but does dissolve in the mixture of solvents including the isopropyl alcohol (see Section 14.2). The 1/4 sec and 1/2 sec refer to the viscosity of standard solutions of nitrocelluloses. The weights (and volumes) listed for nitrocellulose include the isopropyl alcohol.

Blown soybean oil (see Section 9.3.1) is a plasticizer for the nitrocellulose. The modified rosin ester is a maleated rosin esterified with a polyol; it is a "hard resin" which facilitates sanding. The weight solids of the formulation is given as 25%. The volume solids is approximately 20%. After diluting with lacquer thinner to spray viscosity, the sealer is sprayed on the furniture, and, after drying, is sanded smooth.

Shading and padding stains. Wood is not naturally as uniform in color as the overall staining has made it. By spraying different shades of stains to limited selected areas, the color can be varied and the grain highlighted. Shading is a highly skilled art; in recent years use of shading stains has been replaced in large measure by padding stains. Padding stains are made with similar dyes but have some binder and somewhat slower evaporating solvents; they are applied by hand with a rag moistened with the stain and less application skill is required.

Top coat. The furniture is then sprayed with the first coat of top coat. Table 34.3 gives the formulation of a gloss rubbing and polishing lacquer top coat [1]. Nitrocellulose is used as the primary binder in almost all high-style furniture lacquers for three major reasons. First, the appearance of furniture finished with nitrocellulose lacquers is outstanding; the lacquers provide an appearance of depth, fullness, and clarity of the grain pattern that has not been matched by any other

Table 34.3. Lacquer Top Coat

Materials	Weight Ratio	Volume Ratio
Nitrocellulose—RS 1/2 sec (IPA)	51.7	4.84
Nitrocellulose—RS 1/4 sec (IPA)	51.7	4.84
Short oil cocoa nut alkyd (60 NVW)	71.3	8.3
Modified rosin ester	26.7	2.91
Dioctyl phthalate	36.1	4.41
Butyl acetate	175.8	23.95
Isobutyl alcohol	24.8	3.71
Isopropyl alcohol	25.3	3.71
MEK	75.2	11.21
Xylene	233.5	32.2
	772.3	100.00

coating. The reasons for the appearance advantage are not known, but an experienced furniture finisher can spot the difference from across a room. Second, lacquers dry quickly so that they can be rubbed in a short time after application and then packed and loaded into a truck for shipment without *printing*. The term printing in this context means surface imperfections resulting from wrapping material denting the surface of a coating that is too soft. Third, lacquers are thermoplastic and permanently soluble so that they are easily repaired in case of damage during shipment. Note that half the nitrocellulose is 1/4 sec and half is 1/2 sec grade. The mechanical properties of the final film are better with higher molecular weight nitrocellulose. In the highest quality lacquers made some years ago, it was common to use 1-sec nitrocellulose.

The short oil coconut alkyd (see Section 10.5) is used as a plasticizer along with the dioctyl phthalate. The balance of nitrocellulose, plasticizer, and hard resin is critical. If the coating is too soft, it will be difficult to rub; if it is too hard, the lacquer will be subject to cracking as the wood expands and contracts with changes in moisture content and exposure to rapid temperature changes. If the lacquer is applied over a blond style (i.e., very pale wood color), UV absorber (see Section 25.2.1) is added to the formulation to reduce yellowing of the wood. Citric acid is also commonly added to chelate with any iron salts that produce reddish colors with phenolic compounds naturally present in wood.

Distressing. Distressing uses a variety of techniques designed to give an antique appearance to the furniture. Stylists are not trying to make fake antiques to cheat the public; rather the furniture is styled to have an antique appearance. In the old days, quill pens with black India ink were used, resulting in drops of ink falling on table tops. Now little drops of black pigment glazing stain are applied. Sometimes a colonial forefather was a little tipsy when he sat down to write and put the feather end of the quill pen into the inkwell, and then set it down on the table top. Sometimes those colonials walked on their tables with hob nailed boots or banged the table top with their pewter beer mugs. It is startling to walk by the finishing line in a fine furniture factory and see someone swinging a chain at the tops of those carefully prepared pieces of furniture. Another common thing to see is a little black pigmented stain rubbed into inside corners. Over the years in antique furniture, dirt accumulated in such corners, so the new furniture is made permanently dirty in the corners. Proper distressing requires artistic skill; excess distressing looks gross, but some distressing in the right places adds character to the furniture.

Finally another coat of top coat is applied and the lacquer is dried, usually in a force dry oven at 40–60°C. Then the lacquer surface is *rubbed*, first by sanding with fine sandpaper and a lubricant, and then rubbing with cloth and rubbing compound. The result is a soft appearing, low gloss finish.

While solubility of nitrocellulose lacquer finishes is an advantage in that it permits easy repair, it can also be a disadvantage. If solvents, like nail polish thinners, are spilled on a lacquer surface, the finish will be marred. A way of solving this problem and still obtaining the beauty of a nitrocellulose finish is to add a polyisocyanate cross-linker to the lacquer just before application. The isocyanate cross-links with the hydroxyl groups on the nitrocellulose. All solvents must be urethane grade, that is, contain no more than traces of alcohols and water, and the nitrocellulose must be wet with xylene rather than with isopropyl alcohol.

Wood furniture and finishes produced by this overall process are beautiful but expensive. A much larger part of the market is for furniture designed to look as closely as possible to this high-style furniture, but which can be manufactured and finished at lower costs. There are many approaches with varying degrees of cost savings; an intermediate cost approach is described.

Instead of plywood tops and sides made with expensive hardwood veneers that must be painstakingly finished, one can use printed tops and sides with solid wood legs and frames. *Particle board,* that is, pressed wood chips and particles with a resin binder, is first filled with a UV cure filler (see Section 32.3). Note that this is a very different filler than referred to above for filling the pores in wood. Particle board filler is applied over the entire surface to reduce surface roughness. A UV cure filler consists of an unsaturated polyester/styrene (or acrylated) vehicle pigmented with inert pigments that absorb little UV radiation. The filler is applied to the particle board with a roller coater having a brush roller. Shrinkage occurs during UV cure, but the shrinkage is much less than results from solvent evaporation from a filler with a high volatile content. As a result, only one coat of UV curing filler is needed. The high level of inert pigment minimizes the problem of oxygen inhibition. After curing, the surface of the filled board is sanded smooth with automated sanding machines.

The filled, sanded panel is next coated with a *base coat,* a highly pigmented nitrocellulose lacquer whose color will be the overall underlying color of the piece of furniture—corresponding to the color of the stain in the high-style furniture. The stain used on the solid wood parts of this furniture approximately matches the color of the base coat. The base coat is then ready for offset printing.

The gravure printing cylinders for offset printing are made from photographs taken of carefully selected hardwood top veneer plywood that was stained, filled, shaded, and distressed by highly skilled finishers. One cylinder prints the darkest tones of the original finished wood, a second cylinder prints the medium depth tones, and the third cylinder prints the lightest tones. The printing ink colors for the three cylinders are selected so that they approximate three depths of color on the original. The inks are pigment dispersions in plasticizer so that they adhere readily to the base coat and to the subsequent lacquer coats. A light coat of lacquer is sprayed over the panels to protect the print.

The frame and legs of the furniture are assembled and finished up to the stage of applying top coat, followed by assembly of the printed tops and sides on to the furniture. Then the whole piece of furniture is coated with semigloss nitrocellulose lacquer top coat. Note that semigloss lacquer is used, not gloss lacquer. The attempt is to make a finish with a *hand-rubbed effect* without extensive hand rubbing. The semigloss effect is obtained by pigmenting with a low level of fine particle size SiO_2, such that the PVC of the final film averages 2–4%. When the solvent evaporates after application, the convection currents resulting from the solvent loss carry the pigment to the surface of the film where it is "trapped" in the viscous surface layer. As a result, the PVC of the top layer of the film is high enough so that low gloss can be obtained, while the overall PVC is low enough that only a minor degree of light scattering occurs. Similar low-gloss lacquers are also used on all but the very most expensive real wood furniture since the cost of rubbing is so high.

A disadvantage of the furniture finishing systems discussed so far is that they require use of low solids coatings with very high VOC emissions. The solvent is expensive and the low solids requires multiple coats or at least more passes with a spray gun so that the application cost is also high. Hot spray (see Section 22.2.4) is widely used, especially in California. A temperature of 65°C permits increasing solids from about 20 NVW to about 28 NVW. This gives a significant reduction in VOC emissions, but still far short of probable future requirements.

A new approach is the use of supercritical carbon dioxide as a solvent [2] (see Sections 15.7 and 22.2.5). Use of CO_2 as a solvent can significantly reduce the VOC of a lacquer. The system is being studied extensively and as of 1992 a few production units are in use. Further work will be required to determine whether the reduction of VOC will be sufficient to meet future VOC regulations. Preliminary studies indicate that solvent requirements may be reduced by up to 70%.

The VOC emissions from furniture factories have been high. The regulatory agencies have been understanding of the technological problems of reducing VOC emissions, but the pressure to reduce VOC is increasing.

There have been efforts for many years to replace nitrocellulose lacquers to achieve higher solids. The greatest fraction of the work has been concentrated on top coats. The most successful have been alkyd/urea-formaldehyde (UF) top coats. Being thermosetting systems, they are frequently called *conversion varnishes* or *catalyzed finishes*. Table 34.4 gives the formulation of a gloss top coat of this class [1]. The poly(dimethylsiloxane) is in the formulation to minimize orange peel (see Section 23.1) since these thermosetting coatings are generally not rubbed. In actual use, flatting pigment dispersion would be added because there is generally more demand for low-gloss wood furniture finishes in the United States. Just before use, one adds about 5% of *p*-toluenesulfonic acid catalyst based on UF resin solids. The top coated furniture is force dried at 65–70°C for 20–25 min. The NVW of this formulation is 38.5%, close to double that of lacquers. Hot spray can be used to further increase the solids, but pot life must be carefully watched, and, as the solids increase, clear low-gloss coatings become more difficult to make. The depth

Table 34.4. Gloss Alkyd/Urea Top Coat

Materials	Weight Ratio	Volume Ratio
Butylated UF resin (50% in xylene/ isobutyl alcohol)	221.4	26.33
Short tall oil alkyd (50% in xylene)	406.2	48.32
Xylene	134.3	18.51
Butyl alcohol	32.2	4.76
High flash naphtha	12.7	1.71
Poly(dimethylsiloxane) (1% in toluene)	2.8	0.37
	809.6	100.00

of appearance is not as good as that obtained with a nitrocellulose lacquer, but is still presentable. The coating is solvent resistant and more heat and gouge resistant than lacquer coatings. Repair is more difficult but less frequently needed. Conversion varnishes have been most widely used on furniture for commercial use and for kitchen cabinets. Somewhat higher solids and faster cure can be obtained at a higher cost with two package polyurethane coatings.

In Europe, and to a more limited degree in the United States, UV cure top coats are used (see Section 32.3). These have the advantage that they have little or no VOC emission. They are applied by roller coating on separate tops that have already been finished up to the top coat. The binders were originally wax-containing styrene/polyester compositions, but have shifted to acrylated resins that are much faster curing. The coatings are gloss or high semigloss, and are, of course, solvent resistant and have excellent mechanical properties. Use in the United States is limited since most furniture here uses low semigloss coatings and is finished after assembly.

Considerable R&D effort has been expended on water-borne finishes for wood furniture, but, thus far, commercial use is minor. Application of a water-borne coating directly on wood leads to excessive grain raising, which limits their use to applications where there is already a solvent-borne sealer on the wood.

Substantial reductions of solvent usage may be possible by emulsifying nitrocellulose lacquers into water [3]. Lacquers with VOCs of 300–420 g L^{-1}, excluding water, can be made compared with VOCs of the order of 750 g L^{-1} for conventional NC lacquers. Solids of the NC solution internal phase of the emulsion can be maximized by using only true solvents, esters, and ketones. Short oil alkyds using ester solvents (other than the small amount of xylene needed for azeotroping water in producing the alkyd) are selected as plasticizers. Water wet NC is used rather than the conventional isopropyl alcohol wet NC. A surfactant is incorporated in the lacquer as an emulsifying agent; sodium alkylphosphate surfactants are recommended.

The appearance of wood finished with nitrocellulose lacquer emulsion is reported to approach that obtained with conventional NC lacquers and to be superior to the appearance of water-borne acrylic coated wood [3]. The advantage of easy strippability for repair is retained. However, the dry time and early print resistance are not nearly as good as with conventional lacquer.

Water-reducible acrylic resin (see Section 7.3) based coatings have been used. At a sufficiently high molecular weight, they can be used as thermoplastic coatings. Thermosetting resins can be cross-linked with polyaziridine cross-linkers (see Section 13.4.5). Such cross-linkers have been used in other compositions that are applied by roller coating, but the top coat on wood furniture is spray applied, and there is concern about the toxic hazard of aziridine cross-linkers in spray application. Methylated urea-formaldehyde resins can be used as cross-linkers under force dry conditions. Thermosetting latex top coats with UF cross-linkers are being actively explored.

The above approaches are combined in nitrocellulose/acrylic latex-based lacquers. Furniture sealer and top coat formulations with VOCs in the 240–400 g L^{-1} range have been described [4]. Film physical and appearance properties are reported to be superior to acrylic latex counterparts.

Limits of 275 g L^{-1} of VOC for wood furniture are scheduled to take effect in California in 1994. As VOC regulations become more and more stringent, it seems probable that a larger fraction of furniture tops will be high-pressure laminates with wood grain prints. Laminates have been used for many years on commercial and institutional furniture; now their use in household furniture is increasing.

34.2. PANEL AND SIDING FINISHING

Large quantities of coatings are applied to plywood and hardboard sheets to make precoated interior paneling and exterior siding. A small part of the market is coatings for finishing expensive high-quality hardwood plywood used for paneling offices for executives who feel the need for luxurious surroundings. This paneling is finished in essentially the same way used for the highest end of the wood furniture market. A wide variety of other products constitutes the bulk of the market.

One class of wall paneling is made using three-ply plywood where the top veneer is a relatively featureless, low-cost hardwood like luan. The panel is dried to remove surface water, grooved to make a plank effect, and then sanded. A dark-colored lacquer is applied to the grooves by pin point automatic spray guns and the panel is dried. A lacquer sealer colored with dyes is applied. The panel is sanded again and then printed with two or three prints to give the appearance of walnut, rosewood, or some other attractive wood with prominent grain patterns. Finally, a low-gloss top coat is applied. The coatings have generally been nitrocellulose lacquers. While this business was first developed in the United States, in large measure the production has shifted overseas, initially to South Korea and more recently to the Philippines. The predominant source of the wood is the Philippines, so finished product rather than raw material can be shipped; also air pollution regulations in South Korea and the Philippines are less stringent.

A related, but in many ways quite different, application is "door skins." In the manufacture of interior doors, a common approach is to assemble the door and then laminate the top veneer (the *skin*) to the particle board surface. The skins are veneer, laid up and adhered to kraft paper. The veneer is commonly birch or luan, sometimes printed with a hardwood grain pattern. Commonly, the coating on door skins is gloss or high semigloss. Whereas low gloss is preferred on wall paneling so as to avoid glare from the reflection of lights, higher gloss is preferred on doors because it is easier to clean. Since the surface area of doors is relatively small, glare is not a serious problem. This combination of factors has led to the widespread use of UV cure coatings for door skins.

Another major part of the interior paneling market is filled by precoated *hardboard*. Hardboard is made by mixing wood fibers and shreds and curing in hydraulic presses. The lignin in the wood acts as a binder to hold the wood fibers together, sometimes the lignin is supplemented with phenolic/oil resins. Depending on the pressure applied, the density of hardboard can be varied. Furthermore, if the surface of the steel platen against which the upper face of the hardboard is formed has a pattern, the negative of that pattern is embossed into the surface of the hardboard. A wide variety of hardboards is made. Using smooth surface hardboard, one can apply a basecoat, prints, and a semigloss top coat to make 4- by 8-ft panels that look like any kind of wood imaginable. Smooth surface hardboard is usually coated by curtain coating (see Section 22.6) since this gives smooth films.

Tongue and groove paneling can be simulated by routing out grooves and painting them a dark brown color. The paneling is finished by applying base coat that does not flow appreciably into the grooves, followed by printing, and top coating. A pattern of *ticks* (short line dents scattered over the surface as in hardwoods) can be embossed. When the board is finished, it not only has the grain of the wood, it also has the little dips of a tick pattern, as well as the grooves to give the effect of planked paneling.

One can simulate wood with holes, like *pecky cypress*. The hardboard is embossed with the hole pattern copied from particularly striking real pecky cypress paneling. The *hole coat*, generally a dark brown, is applied using roller brushes to assure that the coating completely covers surfaces inside the holes. Of course, the rest of the board is coated too. Then a base coat is applied by *precision coating*. A precision coater is a roll coater in which the application roll is a gravure printing roll with the whole surface uniformly covered with cells so that the entire upper surface of the panel is coated, but no coating is applied in the holes. Generally, the color is a relatively light brown to provide a contrasting background color of wood. Then grain pattern prints are applied, followed by a semigloss top coat. Of course, one is not restricted to pecky cyprus or wormy chestnut; brick, stone, travertine marble, and so on can be simulated.

Another class of interior hardboard paneling is designed primarily for bathrooms. The coatings in this case are high-gloss pigmented coatings sometimes without prints. Sometimes, joints are embossed in to make the panels look like tiles. Sometimes prints, such as marble, are used with a clear gloss top coat.

Since, in contrast to plywood, hardboard can withstand high temperatures, baking coatings with all their advantages in performance properties are used. Since hardboard does not undergo grain raising like wood, water-reducible finishes are used on a large scale. Although the average film thickness of a base coat is relatively high, control of popping is not a major problem since the coating thicknesses applied by roll coating and curtain coating are much more uniform than when coatings are applied by spray (see Section 23.6). When water-reducible coatings are used, they are generally acrylic/MF coatings. When high solids, solvent-borne coatings are applied, they are generally polyester/MF coatings.

High-density hardboard is also widely used for exterior siding. The largest volume of siding is factory primed with a primer that is designed to have at least 6 months exterior durability before being top coated with exterior house paint (see Section 35.1). Acrylic/MF binders are used in the primers, generally water-borne in view of VOC regulations.

A challenge in formulating these primers is to be sure that the coating cross-links sufficiently rapidly that the coated boards coming out of the oven can be stacked without blocking, but that the paint to be applied to them in the field will have good adhesion. The combination requires careful control of cross-link density. The coatings are low gloss because high loading with inert pigment reduces cost, and enhances adhesion of paint to the surface. In some cases, the primer is designed with a PVC slightly higher than CPVC to assure good paint adhesion.

The largest volume is used in siding for tract houses. The primed siding is erected on to the house, but painting is postponed until the house is sold, so the buyer can pick the paint color. An extensive study of exterior durability of different hard-boards with different combinations of coatings has been published [5]. Best results were obtained when preprimed board was field coated with an alkyd primer and acrylic latex top coats.

Petrolatum and other oily substances are present in some hardboards. If the paint applied on the siding is porous, over time some substances can migrate to the upper surface of the paint leading to discoloration or sticky areas that will pick up dirt to give a blotchy appearance. The problem can be minimized by using only paints with PVC $<$ CPVC.

Fully prefinished exterior siding is manufactured on a smaller scale than the preprimed siding. In this case, the primed board is top coated in the factory with a baking acrylic enamel, or, for greater exterior durability, a silicone-modified acrylic resin (see Section 13.3.2) based coating. The durability is superior to field-applied paint, but flexibility in color selection for tract homes is sacrificed. Fully prefinished siding is more commonly used on commercial buildings.

34.3. COATING OF PLASTICS

Coating of plastic products has become a major application for coating materials. Design of the coatings is complicated by the wide variety of polymers used in plastics and the wide range of approaches that are possible in decorating the surfaces. Reference [6] has several useful chapters discussing various approaches.

In a large fraction of their applications, plastics are not coated. Colored pigments can be incorporated in the molding compound so that the whole part is colored. When coatings are required there are two broad alternatives: (1) *in-mold coating,* in which the inside of a mold is coated and then the plastic material put into the mold—when the part is taken out of the mold the outer surface is the coating that was applied to the inside of the mold; or (2) *post-mold* coating, in which the molded part is coated.

34.3.1. In-Mold Coating

In-mold coating is widely used in making glass reinforced styrene–polyester molded products such as boats and shower stalls. The coatings in this case are called gel coats (see Section 13.1). Gel coats are pigmented styrene–polyester resins which are cured along with the resin in the molding compound so that the coating is chemically bound to the main body of the plastic. The resins for gel coats are specially designed so as to be good vehicles for pigment dispersion. In many cases, more expensive isophthalic, neopentyl glycol resins are used since they provide superior hydrolytic stability and exterior durability. The main molding resin can be the less expensive phthalic anhydride, propylene glycol-based resins.

Rigid urethane foam parts to replace wood carvings for furniture are in-mold coated. The inside of the mold is sprayed with a lacquer which is color matched to the base coat of the furniture (see Section 34.1).

In some automotive parts that are to be assembled onto the steel body of the car and painted with the same top coat, the interior of the mold can be coated with a coating that will be the primer to provide both adhesion to the plastic and intercoat adhesion with the top coat. Sometimes such a primer is made with conductive pigments, such as acetylene carbon black, so that electrostatic spray will be effective over the plastic parts.

A wide variety of types of coating materials are used for in-mold coating, depending upon the plastic. To assure good adhesion to the plastic it is usually desirable to use compositions related to the composition of the plastic, for example, the use of gel coats with glass-reinforced plastic parts. In the case of polyurethane RIM (reactive injection molding) parts, in-mold coatings with free hydroxyl groups permit reaction with isocyanate groups in the molding compound. Many early in-mold coatings were solvent-borne coatings, frequently low solids lacquers. These coatings have relatively high VOC contents and are being replaced with higher solids coatings, water-borne coatings, and powder coatings.

A further alternative approach is to use coating films as a laminate for in-mold application of coatings [7]. The coating is reverse roll coated on to a heat-resistant, smooth, high-gloss polyester film. A size coat is applied to the surface of the coating and then a film of plastic of the same polymer (or a compatible polymer) as the plastic part to be coated is laminated on the film. The resulting coated film is then vacuum formed into a mold followed by a sheet of the plastic. The part is removed from the mold and the polyester film is stripped off, leaving the coating on the surface of the plastic substrate. Multiple layers of coating can be applied. For example, for finishing plastic parts for use with base coat/clear coat cars, the polyester film would be coated first with the clear coat followed by a coat of base coat and then laminated to the plastic film and molded.

34.3.2. Post-Mold Coating

Coatings are also applied to plastic articles after they are fabricated. Many of the same types of coatings used for metals can be adapted for coating plastics, and many of the selection criteria are the same as for metals. However, there are important differences. Most of all, it is frequently more difficult to achieve good adhesion to plastics than to metals. Adhesion, discussed in Section 26.5, is usually the central problem in coating plastics. In addition in some cases, the coatings must be more flexible than required for most metal coatings. This follows from the easier deformability of many plastics and all elastomers, and the coatings must be at least as easily deformed as the substrate. As a rough guideline, the elongation-at-break (see Section 24.1) of the coating should be greater than that of the substrate.

To obtain good adhesion to any plastic, a first prerequisite is that the surface of the plastic be clean. Machining oils, sanding dust, finger prints, and so on must be removed. Many plastic parts have residual mold release on the surface. If a mold release must be used in making the plastic article, it is desirable to use a water-soluble soap such as zinc stearate as a mold release, since it is the easiest to remove. Wax mold releases are more difficult to remove, and silicone or fluorocarbon mold releases should not be used for plastic parts that are to be painted. Surface contaminants and mold release are removed by spray washing in three stages. First a detergent wash is used, followed by fresh water rinsing and finally by a deionized water rinse. At the end of the washer, droplets are removed by compressed air jets and the part goes into a drying oven before coating.

After the surface is clean, the surface tension may vary widely depending on the type of plastic. In general, thermoset plastics and thermoplastics with polar structural groups, such as nylons, have relatively high surface tensions, although much lower than metals. Often they can be coated without further surface treatment. The surface of the plastic should be checked to be sure that drops of coating (or solvents with the same surface tension as the coating to be applied) will spread spontaneously on the surface of the cleaned plastic rather than bead up.

However, the surface tension of some less polar plastics, particularly polyolefins, is lower than those of most coatings and the polarity of the surfaces is very low. Some surface treatment is generally required to increase the surface tension and to provide polar groups that can hydrogen bond with components of the coating to maintain adhesion (see Section 26.5). This is an important situation because polyolefins are the least expensive and the most widely used plastics.

Oxidation of a polyolefin surface can provide such polar groups. There are several approaches to oxidation treatment. Oxidizing agents such as chromic acid/sulfuric acid baths are very effective and have been used for many years. However, disposal of chromate containing wastes is being more and more tightly controlled. In order to avoid chromium, sodium hypochlorite with a detergent has been recommended for treating the surface of polyolefin polymers to improve adhesion [8]. The surface can be oxidized by flame treatment, directing an oxidizing flame from propane or butane onto the surface of the plastic. Care is required to insure that all surfaces to be coated are adequately treated. Flame treatment is widely used in Europe for surface treating plastic parts for automobiles. Another approach to oxidation of the surface is by corona discharge. The plastic parts are passed through a cloud of ionized air generated, as described in Section 26.5, by electrodes with many wire ends at high voltage. The ions oxidize the surface of the polyolefin. For irregularly shaped parts, the corona discharge can be carried out in a chamber. Still another approach is to spray a solution of benzophenone on the surface of a polyolefin and then expose the part to UV radiation. This leads to generation of free radicals that initiate autoxidation (see Section 25.1) of the surface. This process is particularly effective with rubber-modified polyolefins.

In some cases, it is possible to promote adhesion to untreated polyolefins by spraying with a thin *tie coat* of chlorinated polyolefin. The chlorinated polymer presumably dissolves partly in the surface of the polyolefin, resulting in an adherent surface to which a broader range of coatings can adhere. Comparisons of results with a variety of treatments and tie coats have been published [9]. Sometimes the tie coats are pigmented so that the sprayer can judge whether all areas have been tie coated. Conductive tie coats permit use of electrostatic spray in application of later coats. To reduce VOC, higher solids and water-borne tie coats have been developed. While reasonable adhesion can be achieved and tie coats are used on polyolefin plastic components for automobile parts in the United States, the adhesion is generally superior with coatings over flame treated plastics as practiced more extensively in Europe.

It is preferable from an adhesion point of view for the plastic part to have a somewhat roughened surface. Plastics that are relatively highly filled with pigments have rougher surfaces to which it is easier to adhere. Alternatively, the surface of the mold can be designed so as to impart some roughness to the molded pieces.

Adhesion can be promoted by baking at temperatures above the glass transition temperature of the plastic. Unfortunately, in many cases heating above T_g leads to heat distortion of the molded part. This is one of the advantages of in-mold coating—adhesion is promoted by the heat involved in molding the plastic, avoiding the problem of heat distortion that can occur when baking post-coated plastic moldings.

Adhesion can also be promoted by using solvents in the coating that are soluble in the plastic. This reduces the T_g of the plastic at the surface, permitting penetration of binder from the primer into the surface of the substrate. Very volatile strong solvents must be avoided on high T_g thermoplastics such as polystyrene and poly(methyl methacrylate); they cause *crazing*, that is, the development of a network of microcracks in the surface of the plastic. A possible explanation for the cause of crazing is that the solvent penetrates into the plastic; then, if it is very volatile, it evaporates rapidly from the surface while there is still solvent left just below the surface. When the solvent diffuses out from below the surface, there is a reduction in volume leading to stresses sufficient to crack the surface layer of the plastic. Another potential problem with using penetrating solvents is the possibility of solvent popping caused by release of solvents from the plastic after cross-linking of the coating is well advanced. In a study [10] of factors affecting popping on sheet molding compound parts (molded fiber glass reinforced polyester plastics), it was found that the problem can be minimized by using primers with the lowest possible solvent permeability. Permeability can be reduced by using highly cross-linked or partially crystalline binders in the primer.

Electrostatic spraying of plastics is difficult since the plastics are usually insulators and, therefore, cannot be adequately grounded so as to give the charge differential needed to attract the charged spray droplets to the surface. When the object being sprayed has both plastic and metal components even when a conductive primer has been applied to the plastic, there can be nonuniform deposition of coatings since the metal can distort the electrostatic field near the plastic–metal interfaces.

Surface resistivity of plastics and time for charges on their surfaces to dissipate can be measured and correlated with the feasibility of electrostatic spraying [11]. The charge dissipation time is affected not only by the composition of the plastic but also by the humidity in the spray booth. As mentioned, sometimes conductive tie coats or primers can be applied to the plastic. The effect of a range of conductive primers on the charge dissipation time has been studied [11]. It has also been shown that the effect of surface charges on the plastic substrate in the electrostatic spraying process can be substantially reduced by placing a continuous grounded metal backing in contact with the plastic substrate [12].

The lower curing temperatures possible with two-package polyurethane coatings make them particularly appropriate for many applications to minimize the possibility of heat distortion of the plastic product [13]. The higher abrasion resistance and flexibility that are available with urethane coatings are frequently also significant advantages. While acrylic resins can be cross-linked with polyisocyanates, significantly higher solids are possible using hydroxy-functional polyesters. An example of a polyester resin is prepared from adipic acid, isophthalic acid, neopentyl glycol, and trimethylolpropane in a mol ratio of 1:1:2.53:0.19 with $\overline{M}_n$ of 730 and $\bar{f}_n$ of 2.09 hydroxyl groups per molecule [13]. Coatings cross-linked with various trifunctional aliphatic isocyanates at a 1.1:1 NCO:OH ratio had a good balance of properties, including good impact resistance at -29°C, for coating a variety of plastics. Polar plastics, such as ABS, can be coated directly while less polar plastics, such as polyolefins, require an adhesion promoting primer.

A further approach to the challenge of applying cross-linked coatings to plastics subject to heat distortion is the use of radiation curing (see Chapter 32). The temperatures involved in radiation curing are only a little above ambient temperature. For example, UV cure gloss top coats are applied to plastic flooring. The low temperature curing is particularly important in coating flooring which has a self-embossed foam decorative layer as the upper surface; elevated temperatures could collapse the foam.

In many applications, the function of the coating is decorative, such as for color, but there are many examples of functional coatings for plastics. The UV cure top coat applied to plastic flooring mentioned above has the functions of increasing wear-life, stain resistance, and gloss retention. Coatings are applied to polyethylene tanks to reduce permeability. Magnetic coatings are applied to tapes and sheets to make recording tapes and floppy disks for computers.

Transparent plastics can serve as replacements for glass in applications ranging from window glazing to eyeglasses, but they are less abrasion resistant than glass (see Section 24.3). Surface coatings to improve abrasion resistance have been developed based on alkoxysilanes and colloidal silica [14]. The coatings provide excellent abrasion resistance to plastics such as polycarbonate, but have relatively high VOC levels and long curing times. Radiation cure coatings have been developed with low VOC that cure in a few seconds (see Section 32.3).

REFERENCES

1. W. A. Krause (Aqualon), personal communication, 1988.
2. K. L. Hoy and M. D. Donohue, *Polym. Prepr.*, **31** (1), 679 (1990).
3. C. M. Winchester, *J. Coat. Technol.*, **63** (803), 47 (1991).
4. H. F. Haag, *J. Coat. Technol.*, **64** (814), 19 (1992).
5. W. Bailey, S. Bussjaeger, N. F. Dispensa, G. Early, M. Froese, R. Haines, A. Moser, L. J. Murphy, and M. A. Trigg, *J. Coat. Technol.*, **62** (789), 133 (1990).
6. D. Satas, Ed. *Plastic Finishing and Decoration,* Van Nostrand Reinhold, New York, 1986.
7. C. H. Fridley, *Proc. Soc. Manufacturing Engineers, Finishing '91 Conference,* FC91–374, 1991.
8. H. F. Haag, U. S. Patent 5,053,256 (1991).
9. R. A. Ryntz, *J. Coat. Technol.*, **63** (799), 63 (1991).

10. R. A. Ryntz, W. R. Jones, and A. Czarenecki, *J. Coat. Technol.,* **64** (807), 29 (1992).
11. D. P. Garner and A. A. Elmoursi, *J. Coat. Technol.,* **63** (803), 33 (1991).
12. A. A. Elmoursi and D. P. Garner, *J. Coat. Technol.,* **64** (805), 39 (1992).
13. S. H. Shoemaker, *J. Coat. Technol.,* **62** (787), 49 (1990).
14. J. D. Blizzard, J. S. Tonge, and L. J. Cottington, *Proc. Water-Borne, Higher-Solids Powder Coat. Symp.,* New Orleans, LA, 1992, p. 171.

CHAPTER XXXV

Architectural Coatings

In 1991 the U. S. shipments of architectural coatings were about 2×10^9 L with a value of $4.65 billion [1]. This was about 50% of the volume and about 41% of the value of all U. S. coatings shipments. Commonly, architectural coatings are called trade sales paints. There are two overlapping markets for these paints: contractors and *do-it-yourselfers,* consumers who paint their own homes and furniture. The contractor is generally particularly concerned with cost of application, which is greater than the cost of the paint. For example, contractors want, if at all possible, paints that cover in one coat. On the other hand, the do-it-yourselfer is more likely to be concerned about cost of the paint and about things like ease of cleanup, odor, and the range of colors available. In the United States, paint is sold through several distribution systems: large merchandisers, hardware stores and lumber stores, and paint stores owned or franchised by paint manufacturing companies. Some paint is sold directly to large contractors by paint manufacturers.

There are three, sometimes overlapping, classes of paint manufacturing companies: (1) large companies that carry out extensive national advertising campaigns to promote their trade name paints; (2) small companies which sell only locally, primarily to local contractors but commonly with one or more factory stores for sales to individuals; and (3) companies which sell private label paints to large merchandisers or hardware chains. The large national companies usually make the same lines of paint for sale all over the country. Local and regional paint companies frequently design their paints to be most suitable for the climate and style trends in their market area; this can sometimes be a significant advantage.

While some paints are manufactured as colored paints, the majority are manufactured and distributed as white paints. The manufacturing company supplies an extensive line of color chips with the tinting colors and the formulations necessary to tint white paints to match the color chips. This makes it possible for the paint store to have thousands of colors of paints available for customers with only a limited inventory requirement. However, it places significant limitations on changes in paint formulations. Only in unusual cases can a paint company afford to make a change in formulation that would make the color chips, tinting pastes, and/or formulations that are out in the paint stores obsolete. In most cases, new formulations must permit continuing use of the same color matching tools.

Large companies usually make three lines of products of at least the major items: good, better, and best. The good grade is usually designed for the individual whose principal criterion for selecting paint is price per unit volume. Such paint is usually adequate for the application with fair coverage. The best paint is usually designed to give the greatest life and to have as good coverage as can be designed into that class of paint. The better is, obviously, a compromise between the two.

There is a great multitude of products manufactured and sold in the architectural coatings market. We restrict our discussion to the three largest classes of paints sold in this market: exterior house paint, interior flat wall paint, and gloss enamels. Space limitations do not allow inclusion of other smaller but still significant product lines such as stains, varnishes (see Section 9.3.2 and 12.4.1), floor paints, and many specialty products.

35.1. EXTERIOR HOUSE PAINT

Exterior house paint is defined as paint applied to exterior wall surfaces. Paint for trim such as around windows, doors, and on shutters is generally gloss enamel, as discussed in Section 35.3. The majority of exterior house paint sold in the United States is latex paint. The exterior durability of latex paint is superior to that of any air dry, solvent-borne paint for field application to the exterior of houses. Latexes are made by emulsion polymerization (see Chapter 5) and undergo film formation by coalescence (see Section 3.3).

There are two situations where solvent-borne house paint should be used: if outside painting must be done when the temperature is less than 4–5°C, since latex paints do not coalesce properly at such low temperatures; or if paint must be applied over a chalky surface, that is, a surface with a layer of poorly adherent pigment and eroded old paint, since most latex paints do not adhere adequately to such surfaces. Generally a solvent-borne primer is recommended for application over chalky surfaces.

Latex paints perform better than oil or alkyd paints on wood siding. When oil and alkyd paints were widely used on wood house siding, paint failures by blistering were common. Water can get into the siding from the back side. Since the films from oil and alkyd paints have fairly low water vapor permeability, when water reaches the back side of the paint film and starts volatilizing in the heat of the sun, blisters are blown. Latex paints are not cross-linked and have T_g values below summer temperatures and, therefore, have relatively high moisture vapor permeabilities. When water gets to the back of the latex paint film, the water vapor can pass through the film; blistering is unusual.

Further, latex paints are much more resistant to grain cracking than oil or alkyd base paints. Wood expands and contracts as the moisture content increases and decreases. A coating applied with an acrylic latex binder maintains its extensibility after many years of exterior exposure and can expand and contract with the wood. Highly unsaturated oil and alkyd coating films continue to cross-link with exterior aging, thereby becoming less and less extensible and, hence, likely to crack as the wood expands and contracts. Oil and alkyd paints also commonly fail by chalking after exposure. High-quality latex paints are less likely to fail by chalking or, at least, have much longer lifetimes before chalking occurs to a serious extent.

Leveling of latex paints is generally not as good as that of solvent-borne paints. This is not viewed as too serious a problem in exterior house paints, but it is with gloss paints discussed in Section 35.3. This disadvantage of latex paints is probably more than offset in exterior house paints by the accompanying characteristic of superior sag resistance of latex paints compared to solvent-borne paints.

Since they continue to be thermoplastic, latex paints are more prone to retain dirt on their surface than are oil or alkyd paints. While undesirable, this is not a major objection in the United States. However, in countries where lignite and soft coal are widely used for heating and cooking, white latex paint turns unsightly, blotchy gray after only a few months exposure; an extreme example is the city of Shenyang in North East China.

As discussed in Section 3.3, film formation during the application of latex paints occurs by coalescence of the polymer particles. Coalescence requires that the temperature during film formation be significantly above the T_g of the particles. If the paint is to be applied at low temperature, the T_g must be correspondingly lower. However, this means that when the temperature is high during the summer, the film will be soft and susceptible to dirt pickup even when the dirt content in the air is not excessive. To minimize this problem, latex paints are formulated with coalescing solvents. These solvents reduce the T_g of the polymer particles, permitting firm formation at lower temperatures, but the coalescing solvent slowly evaporates out of the film after application so that the T_g increases over time. Latex paints are generally formulated in the United States so that they can be applied at temperatures as low as 4–5°C. If it is necessary to paint when the temperature is still lower, solvent-borne paints must be used.

The condition of the surface of wood can be an important factor in the performance of paint. Contamination with dirt and oil can seriously interfere with adhesion. When wood is exposed outdoors, the surface degrades. As a result, after painting there can be failure that looks like adhesive failure, but is actually cohesive failure of the wood. It is said that even a 3-day exposure of freshly cut wood to weather can adversely affect the apparent adhesion [2]. It is recommended that all joints and cracks be caulked and that the wood be treated first with a water-repellent preservative and then be primed [2]. On woods like pine, a latex paint can be used for both the primer and top coat. However, some woods like redwood and cedar contain water-soluble materials that will extract into a latex paint leading to reddish brown stains. The extractives are apparently naturally occurring phenolic compounds. Special *stain-blocking* latex primers are available for use over redwood and cedar. In the early days of latex paints, these primers were formulated with a somewhat soluble lead pigment which formed insoluble salts with the phenolic compounds. In order to avoid lead compounds in the formulation, other approaches for insolubilizing the phenolics have been found. One such approach is to incorporate in the formulation a cationic ion-exchange latex.

Many houses are sided with preprimed hardboard siding; see Section 34.2 for a discussion of painting this siding.

As mentioned earlier, another potentially serious problem in the application of latex house paints can be poor adhesion to chalky surfaces. The surface of chalky paint is covered with a layer of loosely held pigment and paint particles. When latex paint is applied over such a surface, the continuous phase of the paint penetrates among the particles down to the substrate, but latex particles are large compared to the interstices among the chalk particles and do not penetrate among them to any significant degree. Thus, when the water leaves the film and coalescence occurs, the paint film is resting on top of the chalk particles with nothing binding the chalk particles together or to the substrate. As a result, adhesion is poor. In the case of oil or alkyd paints, the continuous phase penetrates in between the chalk particles and as the film forms by solvent evaporation and cross-linking, the binder from the paint surrounds the chalk particles and penetrates through them down to the substrate surface, thereby minimizing the adhesion problem.

The adhesion of latex paints can be improved by careful cleaning of the surface to be painted to remove all chalk. However, it is difficult to do this over a whole house. Another approach is to replace about 15% of the latex polymer binder in a latex paint formulation with long oil alkyd resin or synthetic drying oil. The alkyd or oil is emulsified into the latex paint. After application, when the water evaporates, the emulsion breaks and some of the alkyd or oil can penetrate between the chalk particles providing improved adhesion. In order for the paint to have good storage life, the alkyd or oil should be as hydrolytically stable as possible. Of course, the presence of the drying oil or alkyd reduces the exterior durability of the paint as compared to pure latex paints. While there is no doubt that the adhesion of such paints over chalky surfaces is better than without the drying oil, many workers in the field prefer to prime a house that shows chalking with an oil-based primer and then apply a latex top coat over that primer. As the use of oil-based paints and "cement paints" has decreased, the problem of adhesion over chalky surfaces is becoming less common.

Table 35.1 gives a formulation recommended by a latex manufacturer for a "high-quality" exterior white house paint [3]. This formulation is not given to recommend this formulation over other formulations or these raw materials over other raw materials; rather it provides a framework for discussion of the rationale for the many components of a latex paint. The ingredients are listed in order of addition during making of the paint. Note that, as is generally the case, the formulation is given in both volume and weight terms. The total of the volume is approximately 100. This is done since quantities manufactured are generally multiples of 100 units of volume and also one can compare percentage volumes of ingredients. The weight figures correspond to the amounts of volume in units of pounds and gallons; thus the total weight gives an estimate of the weight per gallon.

Natrosol 250 MHR is hydroxyethylcellulose (HEC), which is in the formulation for two reasons: to increase the viscosity of the external phase of the paint during production and application, and to control the viscosity of the final paint. Viscosity of the paint as a function of shear rate particularly affects ease of brushing, film thickness applied, leveling, sagging, and settling. The viscosity of the external phase controls the rate of penetration of the external phase into a porous substrate such as wood. If penetration is rapid, the viscosity of the paint above the porous surface increases rapidly resulting in poorer leveling. The HEC is added at this stage since it must be dissolved and since the viscosity of the solution provides the viscosity necessary for pigment dispersion.

The ethylene glycol and the propylene glycol (further down the list of ingredients) are in the formulation for two reasons. First, they act as antifreeze to stabilize the paint against coagulation during freezing and thawing. The expansion of water when it freezes exerts substantial pressure on the latex particles and can push them

Table 35.1. Exterior White House Paint

Materials	Weight	Volume
Natrosol 250 MHR (Aqualon)	3.0	0.26
Ethylene glycol	25.0	2.65
Water	120.0	14.40
Tamol 960 (Rohm & Haas)	7.1	0.67
KTPP	1.5	0.07
Triton CF-10 (Union Carbide)	2.5	0.28
Colloid 643 (Rhone–Poulenc)	1.0	0.13
Propylene glycol	34.0	3.94
Ti-Pure R-902 (duPont)	225.0	6.57
AZO-11 (Asarco Inc)	25.0	0.54
Minex 4 (Unimin Specialty Minerals)	142.5	6.55
Icecap K (Burgess Pigment Co)	50.0	2.33
Attagel (Engelhard Corp)	5.0	0.25
Disperse for 10–15 min with a high-speed impeller at 1200–1500 m min^{-1} and let down at slower speed as follows:		
Rhoplex AC-64 (60.5%) (Rohm & Haas)	320.5	36.21
Colloid 643 (Rhone–Poulenc)	3.0	0.39
Texanol (Eastman)	9.7	1.22
Skane M-8 (Rohm & Haas)	1.0	0.12
NH_4OH (28%)	2.0	0.27
Water	65.0	7.80
2.5% Natrosol 250 MHR (Aqualon)	125.0	15.15
	1167.8	99.80

Formulation Constants

PVC	43.9%
Volume solids	37.0%
VOC (g L^{-1} excluding water)	93
pH	9.5
Initial viscosity	90–95 KU
Equilibrated viscosity—unsheared	95–100 KU
sheared	90–95 KU
ICI viscosity	0.1–0.12 Pa·s

together with enough force to overcome the repulsion by the stabilizing layer, thereby resulting in coagulation. The ethylene glycol lowers the freezing point of the water. Even if the temperature gets low enough to freeze the solution, it freezes to a slush so there is less pressure exerted on the latex particles.

The second reason for incorporating the glycols is to control the rate of drying of the paint to permit *wet lapping* without disruption of the edge of the paint film. As water evaporates from a latex paint film, the viscosity increases rapidly as a result of the increase in the volume fraction of internal phase. When paint is applied with a brush or roller, the edge of the wet paint film is painted over (lapped) so as to assure that no substrate is left unpainted or with only a thin layer of paint. In the case of oil paint, there is no problem with wet lapping. However, with a latex paint, by the time the lapping occurs the viscosity of the applied paint has increased enough so that the film is semisolid, but since only limited coalescence has occurred, the film is weak. The pressure of the brush or the roller can break up the film, resulting in irregular chunks of paint film along the lapped edge. This can be minimized by slowing down the rate of evaporation of the continuous phase by incorporation of the slow evaporating glycols.

The next ingredient is water. Hard water should be avoided since it can reduce the stability of dispersions; commonly, deionized water is used. The water provides the volume of liquid required for the pigment dispersion. Obviously, the amount of water is also a major factor affecting the solids of the final paint.

The next three materials, Tamol 960, KTPP, and Triton CF-10, are surfactants. Their function is to stabilize the pigment dispersion while not interfering with the stability of the latex particle dispersion or the stability of tinting color dispersions that may be added to the paint. Tamol 960 is an anionic dispersing agent. KTPP is potassium tripolyphosphate, which is an effective dispersion stabilizer, as is testified to by the wide use of polyphosphates in laundry detergents. KTPP is also desirable since it does not promote foaming. Note that the potassium salt is used, not the sodium salt. Sodium tripolyphosphate is likely to cause *frosting*. The salt is leached out of the paint film by water, and the solution evaporates on the surface, leaving a layer of crystals of the sodium tripolyphosphate. KTPP is less likely to cause this problem. Triton CF-10, a nonionic surfactant, helps stabilize the pigment dispersion and also is effective in reducing the surface tension of the latex paint so that it will wet out low surface tension substrates. It has been proposed that dynamic surface tension may be more important in wetting the substrate than equilibrium surface tension [4].

The combination of dispersing agents used in a formulation is generally empirically determined by trial and error. Careful records should be kept of the results of using different combinations with particular latexes, pigments, and surfactants, especially records of failures. This information serves to guide a formulator in selecting dispersion agent combinations for future formulations.

Colloid 643 is an antifoam agent (see Section 23.7). The minimum amount necessary for controlling foam should be used; excess antifoam agent can result in crawling when paint is applied. Selection of antifoams is empirical; the manufacturers of proprietary antifoams offer test kits with small samples of the various antifoams in their line to paint manufacturers. The formulator tries the samples to find which is most effective with the particular formulation.

As noted above, propylene glycol is part of the glycol mixture to improve freeze-thaw stability and control wet lapping. A mixture of the two glycols may have been used to help control the rate of drying: propylene glycol evaporates somewhat more rapidly than ethylene glycol.

Ti-Pure R-902 is a rutile TiO_2 pigment. While it was not stated in the formulation constants given with the formulation, the PVC of TiO_2 in the formulation is approximately 18%. As discussed in Section 16.3, this is about the optimum level of TiO_2 considering the relationship between cost and hiding. Calculation of the PVC of TiO_2 in addition to the total PVC in a white paint formulation is strongly recommended as a check on the level of TiO_2 in the final formulation. About 18% is usually an appropriate level, depending on the actual TiO_2 content of the pigment. In this formulation, dry powder TiO_2 is called for. In most cases, large manufacturers of latex paints no longer use dry TiO_2. TiO_2 "slurries," that is, dispersions of TiO_2 in water supplied by the pigment manufacturer (see Section 20.4), provide more economical material handling.

AZO-11 is zinc oxide. Zinc oxide is a white pigment and contributes somewhat to the hiding, but its scattering efficiency is poor because the refractive index difference between it and the binder is relatively small. Zinc oxide is in the formulation primarily as a reasonably effective fungicide. The level of zinc oxide used is often higher than that given in this formulation. The lower level may reflect the use of an additional fungicide, Skane M-8, in the formulation. The need for fungicides and bactericides is discussed later. Zinc oxide can cause catastrophic viscosity increases during storage of some latex paints [5]. The effect of the ZnO is very system dependent; Ref. [5] discusses the variables and possible formulation approaches to minimize the instability problems. Even though one may not be concerned with the use of zinc oxide in a paint, this paper is recommended since it discusses the dependence of latex paint properties on interactions of several variables.

Minex 4 is a sodium potassium aluminosilicate inert pigment, and Icecap K is an aluminum silicate inert pigment. These inert pigments have refractive indexes sufficiently similar to that of the binder in the paint that they contribute little directly to hiding. However, use of some fine particle size inert pigment is generally felt to increase the hiding efficiency of the TiO_2. This controversial subject is discussed further in Section 35.2.

While in this formulation all of the inert pigments are clays, a wide variety of types of inert pigments can be used. Particle size, cost per unit volume, and color are major selection criteria. Calcium carbonates are inexpensive and are sometimes used as inert pigments in exterior latex paints, but there is a potential problem. Latex paint films have quite high water permeability; water and carbon dioxide, which are in equilibrium with carbonic acid, can permeate into the film. Calcium carbonate dissolves in carbonic acid to give a calcium bicarbonate solution that can diffuse out of the film. On the surface, the water evaporates, leaving a deposit of calcium bicarbonate which then reverts to insoluble calcium carbonate. The white deposit, called *frosting*, on the surface of the film is undesirable, especially with colored paints.

The gloss of this, and most, latex exterior house paints is quite low. Gloss is controlled by the PVC/CPVC ratio. Exterior durability as well as many other properties are also affected by PVC. Very importantly, cost is strongly affected by PVC. The least expensive components of the dry film of latex paints are the inert pigments. Therefore, it is desirable to maximize CPVC, since the PVC of the paint at the same PVC/CPVC ratio increases with CPVC. With a given latex, a major factor controlling CPVC of latex paints is the particle size distribution of the pigments. It is common to have multiple inert pigments so as to maximize breadth of particle size distribution which increases CPVC and permits higher pigment loading at the same PVC/CPVC (see Sections 21.2 and 21.4).

Attagel is attapulgite clay, which is in the formulation to increase the degree of shear thinning and thixotropy. The clay particles swell substantially in water. The swollen clay particles can be distorted by shear, leading to reduction in viscosity. The particles return to their unsheared shape when shear stops, leading to relatively rapid recovery of viscosity. This flow behavior gives a paint in which the pigment does not settle out on standing and does not sag appreciably on application. Leveling tends to be inhibited since recovery of viscosity after shearing takes place quite rapidly, but leveling is not a factor of major importance in exterior house paint. Note that the attapulgite is added last to the mill base. Since, as discussed in Section 20.4.1, high-speed impeller dispersers are not very effective in dispersing shear thinning systems, the other pigments should be thoroughly dispersed before incorporating the attapulgite clay.

After the pigment is dispersed, the high-speed impeller is slowed down and the latex, Rhoplex AC-64, is added. Latexes can coagulate at the shear rates involved when the impeller is running at high speed. Rhoplex AC-64 is an all-acrylic latex designed for exterior durability. Its composition is proprietary; like most commercial latexes, about the only information the manufacturer publishes on composition is the solids of the latex (60.5% in this case). Based on the relatively high solids of this latex, it is reasonable to guess that the latex has either a bimodal or perhaps a broad particle size distribution. This would be appropriate because such latexes tend to give higher CPVCs and hence permit higher pigment loading. It is reasonable to guess that the latex is a copolymer of methyl methacrylate (MMA) with ethyl or butyl acrylate and has a T_g of 10–15°C. Such copolymers have excellent exterior durability. Latexes with part of the MMA replaced by styrene have lower costs and reasonably satisfactory exterior durability. High vinyl acetate, vinyl acetate/acrylate copolymers are less expensive. While their hydrolytic stability is inadequate for exterior use in climates with high humidities and rainfall, they can be appropriate for exterior use in arid climates.

Colloid 643, the antifoam additive, is added a second time so as to minimize the amount needed. Texanol is a coalescing agent; it is a mixture containing the monoisobutyrate ester of 2,2,4-trimethylpentane-1,3-diol. Note the high degree of steric hindrance of ester groups of Texanol which provides necessary hydrolytic stability to the coalescing solvent.

Skane M-8 is a fungicide which supplements the ZnO listed earlier. Fungicides are needed to minimize mildew growth on the paint film after it is applied. Surprisingly, this formulation does not include a bactericide. Bactericides are needed to control bacterial growth in the can of paint. There are three adverse effects of bacterial growth: a putrid odor can develop; the metabolic processes release gases that can build up enough pressure to blow off the can lid; and, since the bacteria can digest cellulose derivatives like HEC, the viscosity of the paint can drop dramatically. The most widely used biocides in exterior paints have been aromatic mercury compounds like phenylmercuriacetate (PMA), which is a very broad spectrum fungicide and bactericide. Organomercury compounds react with hydrogen sulfide to give black mercuric sulfide. At least one large paint company does not use organomercury compounds in their exterior paint for this reason. Aromatic organomercury compounds are toxic, but not as toxic as aliphatic organomercury compounds. There are stringent controls on mercury levels in waste water discharge so extra care must be taken in plant cleanup procedures when they are used. While still permitted in exterior paints, the use of organomercury compounds in interior paints in the United States was prohibited in August 1990.

Many other compounds are used as fungicides and bactericides. Testing is difficult because fungal and bacterial growth are very dependent on ambient conditions and particularly because many fungicides and bactericides are effective against only a limited number of organisms. Diluted paints are more likely to support bacterial growth than undiluted paints; a test method based on inoculating diluted latex paint with several bacteria and fungi has been reported to be more reliable [6]. Examples of biocides tested are: substituted 1-aza-3,7-dioxabicyclo(3.3.0)octanes, substituted 3,5,7-triaza-1-azonia adamantane chloride, 1,2-benzthiazolone, and methylchloro/methylisothiazolone. Reference [7] provides a brief review of other biocides for latex paints.

In the case of fungicides, there is the further problem that in order to be effective they must be somewhat soluble in water. However, if they are water-soluble they can be leached out of the film when it is exposed outdoors. An interesting approach to this problem is an investigation of the use of polymers with pendant fungicide groups that are time-released by hydrolysis [8].

A critical aspect of manufacturing latex paint is that housekeeping in the factory must be the equivalent of that in a food-processing factory. The best control of bacterial growth is to avoid as far as possible the contamination of the paint with bacteria. Furthermore, bactericides kill bacteria, but they do not deactivate enzymes that have been produced by bacteria. If the factory has places where bacteria can grow and the products can get into a paint tank, manufacturers have found to their chagrin that they have introduced enzymes into their paint. The enzymes split the HEC molecules and the viscosity drops even though there is adequate bactericide in the paint.

The ammonium hydroxide is added to adjust the pH of the paint. In this case, the pH of the final paint is said to be 9.5. The high pH assures the stability of the anionic dispersing agents. While all latex paints are packed in lined tin cans, it is possible to have breaks in the lining, and the high pH minimizes the possibility of corrosion of the cans. In many cases, the latex polymer is made with some acrylic acid as a comonomer; the viscosity of paints made with such a latex is very pH dependent (see Section 5.2).

The last two items in the formulation are water and a 2.5% solution of HEC. The amount of these two determines the solids of the final paint. There will be some variation from batch to batch of the viscosity with any paint formulation. The ratio of the water and the HEC solution can be varied so that the viscosity and the solids of the paint will both come out at the standard level.

The PVC of the paint is 43.9%; it is a relatively low gloss paint, as are most exterior latex paints. Commonly the gloss is lower with latex paints than with oil paints. The lower PVC reduces cost while still providing excellent exterior durability. The volume solids of the paint is 37%, substantially lower than exterior oil or alkyd-based paints. As a result, the coverage per unit volume of latex paints is lower than that of oil-based paints, but, on the other hand, the price per unit volume of latex paints is lower. The grams of VOC per liter of paint excluding water calculates to be 93, substantially lower than that of alkyd and oil-based house paints.

The "initial viscosity" refers to the viscosity taken with a "Stormer Viscometer" immediately after mixing the paint. Note that in spite of the widely recognized shortcomings of the Stormer Viscometer discussed in Section 19.3.7, in this case (and in virtually every paint lab in the United States), it is the viscometer that is used. The "equilibrated viscosity" signifies the viscosity after the thixotropic structure has time to build up again. Note that the unsheared equilibrated viscosity, that determined without stirring the paint before measuring the viscosity, is 5 KU units higher than the initial or the "sheared" viscosity. This shows that this paint is somewhat thixotropic. The "ICI" number refers to the viscosity run on an ICI Viscometer that measures viscosity at a high shear rate similar to that experienced in application by brush or roller. It is important to have the proper high shear viscosity because this is a major factor controlling the film thickness of the paint that is applied. The high shear viscosity of 0.1–0.12 Pa·s is on the low side for a latex paint.

35.2. INTERIOR FLAT WALL PAINT

The largest volume of trade sales paint is interior flat wall paint; essentially all of it is latex paint. Since exterior durability is not needed, lower-cost vinyl acetate copolymer latexes (see Section 5.3) are the principal binders. Since the T_g of vinyl acetate homopolymer is about 32°C, some comonomer must be used to reduce the T_g; butyl acrylate is commonly used. For interior applications where there is need for greater water resistance, acrylic latexes are used.

The major advantages of latex paints as compared to the older oil-based flat wall paints are:

Fast drying. If desired, two coats can be applied to the walls of a room during a day, the furniture moved back, and the room used that night.

Low odor. The odor of mineral spirits and byproducts from oxidation of drying oils in solvent-borne paints is quite unpleasant for days after the walls are painted. The minor odor from latex paints dissipates quickly.

Ease of cleanup. Spills, dripped spots, brushes, and rollers can be easily cleaned with soapy water in the case of latex paints; solvent is required with oil paints. On the other hand, the cleaning up must be done promptly because once the latex has coalesced, cleaning is more difficult than with oil paints.

Low VOC emissions. Latex paints were widely adopted before there was concern about VOC emissions; but if a solvent-borne coating were to be developed, VOC emissions would have to be at least as low as those of latex paints to be acceptable. There is also the advantage of substantially reduced fire hazard with water-borne paints. Not only is solvent-borne paint flammable, but also rags wet with oil-based paints in a confined space can undergo *spontaneous combustion,* that is, ignite as a result of the heat generated by autoxidation.

Flat wall paint is usually stocked as white paint, and tinting colors are added to make a color picked by the customer from a large array of color cards. This means that it is critical to maintain equal white tinting strength in quality control. Furthermore, any new formulation has to have the same white tinting strength as the formulation being replaced or else the color cards and formulations in all the dealers stores will have to be replaced.

In each quality line, two, or sometimes three, white base paints are included. One is to be used alone as a white or tinted to make pastel color paints. At the other extreme, a deep tone base that contains little TiO_2 is used for tinting to deep colors that could not be made if the regular base paint were used. Frequently, a third base is in the line which is used for intermediate depths of shade. Intermediate depths of shade could be matched using the regular base white, but the cost would be excessive because more tinting color would be needed to match the colors; hiding would be greater than needed.

Users are sometimes confused by the changes in color of latex paint as it dries. The color of the dry film of a latex paint is darker than the color of the wet paint. In the wet paint, the interfaces between the polymer particles (n = approximately 1.5), water (n = 1.33), TiO_2 (n = 2.76), and inert pigments (n = approximately 1.6) scatter the light to a much greater degree than when the paint is dry. The dry paint has fewer interfaces, since the latex particles coalesce, and smaller refractive index differences since the pigment particles are in a polymer matrix instead of the combination of polymer particles and water. Since the light scattering decreases as water evaporates and the latex particles coalesce, the color gets darker. It also follows that hiding of the wet paint is greater than hiding of the dry paint.

When painting ceilings, one is particularly anxious to get hiding in one coat since painting over one's head and moving the ladder is more of an effort than in painting walls. The problem is particularly challenging since ceiling paints are commonly pain white to give the greatest diffuse light reflection. Since there are no color pigments in a base white paint to absorb light, the hiding by white paints is poorer than that of any color paint made from it. The problem is compounded by the decrease in hiding when a latex paint film dries. The user thinks he or she has applied enough paint to hide the old marks on the ceiling, but comes back an hour or so later and finds that the marks show through the dry paint. Special ceiling paints are made that minimize this problem by pigmenting so that the PVC is greater than CPVC. Dry paint films with PVC above CPVC have voids of air with $n = 1$ adding to the light scattering by having new interfaces between air and polymer as well as air and pigment. Formulations can be adjusted so that wet hiding and dry hiding are approximately equal. Of course, the films do not have as high mechanical strength as films of paint with PVC < CPVC, and the resistance to staining is much poorer. But neither mechanical strength nor stain resistance is an important criterion for ceiling paints; hiding is what counts most.

Table 35.2 gives the formulation of an interior flat tint base wall paint [9]; the materials are listed in the order of addition. Kathon LX 1.5% is a biocide whose active ingredients are 5-chloro-2-methyl-4-isothiazolin-3-one and 2-methyl-4-isothiazolin-3-one. As noted earlier, since August 1990 organomercury biocides are not permitted in interior paints in the United States. Tamol 731 is an anionic surfactant and Tritons N-101 and N-57 are nonionic surfactants composed of nonylphenolethoxylates with an average of 9 and 5 ethylene oxide units per molecule, respectively. Colloid 643 is an antifoam. The propylene glycol is present for freeze-thaw stability and wet-lapping. Ti-Pure R-900 is a rutile TiO_2; the PVC of the TiO_2 in the formulation is approximately 18.4% as is appropriate for the most cost efficient hiding. Optiwhite is a calcined kaolin clay inert pigment; it has internal voids which can not wet out and hence may contribute somewhat to hiding. Its particle size (expressed as equivalent spherical diameter) ranges from 0.5 to 20 μm, with 80% by weight between 0.6 and 6 μm. Snowflake White is a ground natural calcium carbonate with a particle size distribution from about 0.3 μm to 30 μm, with 80% by weight between 1 and 17 μm. The broad range of particle sizes of the combination of pigments permits incorporation of a high amount of low cost inert pigment while still staying (presumably) below CPVC. The PVC of this formulation is stated to be 59.6%. In some other flat wall formulations, a small

Table 35.2. Interior Flat Tint Base Wall Paint

Materials	Weight Ratio	Volume Fraction
Water	119.9	14.39
Kathon LX 1.5% (Rohm & Haas)	1.8	0.22
Tamol 731 (25%) (Rohm & Haas)	6.0	0.65
Triton N-101 (Union Carbide)	2.0	0.23
Triton N-57 (Union Carbide)	2.0	0.24
Colloid 643 (Rhone-Poulenc)	2.0	0.26
Propylene glycol	45.0	5.20
Ti-Pure R-900 (DuPont)	175.0	5.25
Optiwhite (Burgess Pigment)	118.0	6.43
Snowflake White (ECC America)	118.0	5.24
Disperse using high speed impeller, then add:		
Rovace 9100 (55%) (Rohm & Haas)	230.0	25.37
Dowanol DPnB (Dow)	13.0	1.72
Colloid 643 (Rhone-Poulenc)	1.0	0.13
Natrosol 250 MR (2.5%) (Aqualon)	288.0	34.54
Aqueous Ammonia (28%)	1.0	0.13
Total	1122.7	100.00

Formulation Constants

PVC	59.6%
Volume solids	29.9%
Weight solids	47.7%
Viscosity, initial	97 KU
ICI	80 mPa·s
pH	9.5
VOC	179 (g L^{-1} excluding water)

fraction of relatively large particle size SiO_2 is incorporated in the formulation. This inert pigment improves burnishing resistance, that is, it reduces the tendency of the flat wall paint film to polish to a glossy area where it is frequently rubbed, such as around light switches. As discussed in Section 24.3, the large size particles make bumps in the film that reduce contact area and hence reduce burnishing.

Rovace 9100 is a 55% solids vinyl acetate/acrylic copolymer latex with a narrow particle size distribution averaging 0.3 μm. Unless high scrub resistance is needed, vinyl acetate copolymer latexes are used in flat wall paints because of their lower cost. Dowanol DPnB is dipropylene glycol *n*-butyl ether; it is a coalescing agent and improves freeze-thaw stability. Colloid 643 is additional antifoam. Natrosol 250 MR is HEC. The 2.5% solution is made up in advance; it is good practice to include some bactericide in the solution to avoid bacterial growth, although there is no indication in this formulation that bactericide was included. The aqueous ammonia serves to adjust the pH which can affect the viscosity of the paint and minimize the chances of corrosion of steel that might get exposed in the can.

The VOC of this paint is well within regulations in effect in 1992. However, pressure for reductions can be expected. In 1992 latex paints of undisclosed composition were introduced which have no detectable VOC in tests sensitive to 12 g

L^{-1} (see Section 29.2). They are currently available only in stock colors since current tinting colors contain some solvent. Comparisons of performance properties with conventional latex paints have not yet been published.

The most expensive major component on a volume basis of any white flat paint is the TiO_2. Several approaches are used to minimize the TiO_2 content required at a given level of hiding. Where applicable, as in the case of ceiling paints, PVCs above CPVC are used. The efficiency of hiding by TiO_2 is affected by the choice of inert pigments used with it. While there is some controversy, most workers accept that inert pigment with particle sizes less than that of the TiO_2, called "spacer" inerts, increase the efficiency of the TiO_2 [10]. A mathematical model has been developed that can be used to improve TiO_2 pigment spacing by predicting the optimum inert pigment size distribution and concentration for a given formulation [11]. The results of application of this model also predict less TiO_2 crowding with small size inert pigments.

Another approach is to use a pigment (Spindrift, Dulux Australia Ltd.) that consists of resin particles which have within them air pockets that in turn contain TiO_2 particles [12]. The larger refractive index differences and the increased number of interfaces give substantial increases in light scattering and hence hiding efficiency.

Still another approach to greater hiding at lower cost is the use of a high T_g latex, such as polystyrene, as a pigment [13]. When the latex binder coalesces, the high T_g latex does not coalesce but the particles remain separate as with any other pigment. Including the dry volume of the high T_g latex particles as part of the pigment volume, paints can be formulated with a PVC greater than CPVC without making the surface of the film porous. These paints provide equal hiding at lower TiO_2 content while still retaining good enamel holdout and stain resistance. No convincing explanation of why an intact surface film forms in such paints has been published, but the method is said to be used on a large commercial scale.

Yet another approach to minimizing the TiO_2 requirement is the use of special high T_g latexes that contain air voids within the particles as pigments, such as Ropaque (Rohm and Haas Co.) [14,15] (see Section 18.1.1). Again in calculating PVC, one includes the volume of the high T_g particles since they do not coalesce with the latex binder particles during film formation. Even at PVCs below CPVC, the air voids in the latex pigment particles provide hiding which permits a reduction of the TiO_2 level in the formulation. Such paints can have stain and scrub resistance equal to other latex paints with PVC less than CPVC.

A large fraction of latex flat wall paints are applied by roller. During roller application, some latex paints *spatter* to a major degree (see Section 22.1.3). Paints with high extensional viscosity are likely to spatter severely during roller application [16]. Extensional viscosity increases when high molecular weight water-soluble polymers with very flexible backbones are used as thickeners in a latex paint [17]. Spattering can be minimized using low molecular weight water-soluble thickeners with rigid segments in the polymer backbone such as low molecular weight HEC. Since the selling price of HEC is the same for various molecular weights and more low molecular weight water-soluble polymer is needed to reach the same shear viscosity, paint cost increases when low molecular weight HEC is used. It is common to compromise by using an intermediate molecular weight or a mixture of high and low molecular weight grades of HEC.

While in many applications, the increase in external phase viscosity resulting from the use of water-soluble polymeric thickeners is desirable, there is an application where it is undesirable, namely concrete block walls. Concrete block surfaces contain both holes that are large compared to the pigment and latex particles and pores that are small compared to them. When a solvent-borne paint is applied to concrete block, the coverage is extremely low since so much paint penetrates into the holes. When regular latex paint is applied, the coverage is better. The lower viscosity of the external (continuous) phase relative to solvent-borne paint leads to more penetration of continuous phase into the small pores of the block so that the viscosity of the remaining paint increases rapidly and there is less penetration into the holes. The coverage with latex paint can be further improved by omitting water-soluble polymer from the formulation. This modification further reduces the viscosity of the external phase, which can penetrate more rapidly into the small pores and hence give even more rapid buildup of viscosity of the remaining paint. The effect is enhanced by the requirement that the volume fraction of internal phase in the modified paint must be significantly higher than when water-soluble polymer is present in order to have equal starting viscosity. With increasing volume fraction internal phase, the viscosity increases more rapidly as the continuous phase is drained off in the small pores so that penetration into the holes is substantially reduced. A further advantage of latex paints, especially acrylic, styrene/acrylic, or styrene/butadiene latexes, over oil or alkyd paints is that they are not subject to saponification by the basicity of the cement in the concrete block.

When latex paint films form, surfactants sometimes bloom out of the film collecting on the surface. Usually this is not noticeable; however, if water condenses on the paint surface, this surfactant on the surface can dissolve, then when the water evaporates, surfactant concentrates in the last remaining water droplets leaving brown spots of surfactant on the surface of the film. The problem can be minimized by avoiding dark-colored surfactants. However, surfactant can leave whitish spots on dark color paints. It has been shown that nonionic surfactants are more compatible with latex polymers and less likely to come to the paint surface than anionic surfactants [18].

35.3. GLOSS ENAMELS

In contrast to the situation with exterior house paints and, especially, with flat wall paints, a significant fraction of gloss enamels sold in the United States are alkyd paints rather than latex paints. The term enamel connotes a hard, glossy surface analogous to porcelain enamel. The terms enamel and gloss paint are commonly used interchangeably. The term enamel has frequently been applied to thermoset coatings, but in this case the term has carried over from thermoset alkyd enamels and thermoplastic latex paints are also called enamels. Gloss enamels are used both inside and outside for trim around windows and doors, for shutters, for wood furniture, and on kitchen and bathroom walls. Professional painters have particularly favored continued use of alkyd paints while do-it-yourself painters favor latex paints.

One advantage of alkyd enamels is that they can be made with higher gloss than can latex enamels. As discussed in Section 17.3, during film formation from solution vehicle paints, a layer having very low pigment content forms at the upper surface of the coating; this is not the case with latex paints. The ratio of pigment to binder at the surface of a latex paint firm can be reduced somewhat by using a finer particle size latex, but there is still a significant difference from alkyd paint films. Another factor tending to reduce the gloss of latex paint films is the haze that results from incompatibility of surfactants (and perhaps other components) and the blooming that results when surfactants migrate to the surface of the film. These problems can be minimized by making latexes with as low monomeric surfactant content as possible and by selecting surfactants for pigment dispersion that are as compatible as possible.

Considerable work has been done in using blends of compatible water-soluble resins with latexes to achieve high gloss. In floor wax applications, morpholine salts of styrene/acrylic acid copolymers have been used. When the film forms, the solution resin concentrates in the surface layer. When the film dries the morpholine evaporates, leaving the free carboxylic acid groups. The film has reasonable water resistance but, as is an advantage for floor wax, can be removed by mopping with ammonia water. However, this level of resistance is not adequate for paint films. There have been some proprietary resins marketed that give gloss advantage with relatively smaller loss of resistance properties.

In exterior applications and to a lesser degree many interior applications, the advantage of high initial gloss alkyd enamels is more than offset by the better gloss retention and resistance to cracking exhibited by latex gloss paints. Depending on the location, alkyd enamels exposed outdoors lose so much gloss in a year or two that the coating looks like a flat paint instead of a gloss paint. While a latex paint starts out with a lower gloss it retains most of its initial gloss for several years. In interior applications, the superior color retention and greater resistance to cracking are important advantages for latex gloss paints.

The principal limitation of gloss latex paints is not, however, their lower gloss. Rather, the major problem is the difficulty of getting adequate hiding in one coat. Since the cost of application for painting contractors is higher than the cost of the paint, contractors particularly prefer alkyd paints that commonly hide in one coat (over a primed surface). There are several factors involved in the difference of hiding between alkyd and latex gloss paints.

The volume solids (NVV) of latex gloss paints is substantially lower than the NVV of alkyd paints. A typical NVV for a latex gloss paint is about 33%, whereas the NVV of alkyd gloss paints can be up to 66% or higher. Accordingly to obtain the same dry film thickness, one has to apply twice as thick a wet film of the latex paint as compared to the alkyd paint. Wet film thickness applied is primarily controlled by the high shear rate viscosity of the paint. Thus, a latex paint should be formulated to have a higher viscosity at high shear rate than an alkyd paint for maximum hiding. In practice, as is discussed later, the high shear viscosity of latex paints has usually been formulated to be lower than that of alkyd paints further compounding the problem.

Wet film thickness can be controlled to a degree by the extent to which the painter brushes out the paint. The painter tries to judge how much he should brush out by how well the wet paint is covering. In the case of alkyd paints, there is relatively little difference between the wet and dry hiding powers of the paint. However, the wet hiding power of latex paints is substantially greater than their dry hiding power. This increases the difficulty of judging how far to brush out a latex paint.

Another factor affecting hiding by gloss latex paints is their comparatively poor leveling. Assume that a uniform dry film thickness of say 50 μm of a paint provides just satisfactory hiding for some application. If the film has streaks of thinner film, say 35 μm, and thicker film, say 65 μm, the hiding of the uneven film will be poor; and it is likely to appear inferior to that of a uniform 35-μm dry film of the same paint. Since the 35- and 65-μm areas are immediately next to each other, the contrast emphasizes the poor hiding. Furthermore, not only is the hiding adversely affected, but also the contrast in colors resulting from the poor hiding in the valleys contrasted to the ridges emphasizes the poor leveling.

The poor leveling of latex paint films has several causes. In alkyd paints for brush application, the solvent is slow evaporating mineral spirits, whereas in latex paints the water evaporates more rapidly unless the relative humidity is high. As a result the viscosity of the latex paint increases more rapidly after application under most conditions. Also, the volume fraction of internal phase in latex paints is higher than in alkyd paints since both the latex particles and the pigment particles are dispersed phases. Therefore, even with the same loss of volatile material, there is a greater increase in the viscosity of the latex paint.

It has been demonstrated by Overdiep, as discussed in Section 23.1, that leveling of solvent-borne, brush-applied paints is promoted by surface tension differential driven flows in the wet paint film [19]. When solvent evaporation starts, the fraction of solvent lost from the valleys of the brushmarks is greater than from the thicker ridges. Since the surface tension of the solvent is lower than that of the alkyd, the surface tension of the more concentrated alkyd solution in the valleys becomes higher than that in the ridges. The resulting differential in surface tension causes the paint to flow from the ridges into the valleys to minimize overall surface tension, thus promoting leveling.

Unfortunately, there have been no published studies of the driving force for leveling of latex paints. In fact, except for a recent encyclopedia article by one of the authors [20], there does not seem to have been consideration given in the literature to the driving force for leveling in the case of latex paints. The surface tension of the water phase of latex paints is controlled primarily by the surfactants in the paint, which suggests that there probably is little change in surface tension as water evaporates. If this hypothesis is right, the driving force for leveling of latex paints is the relatively small force of surface tension driven leveling rather than the larger force of surface tension differential driven leveling as in the case of alkyd paints. Another factor that may be important is dynamic surface tension. It has been shown that with some surfactants, equilibrium surface tension is more rapidly reached than with others [4]. The need for research is evident.

Probably, however, the major factor affecting the leveling, and, therefore, hiding, of gloss latex paints has been their rheological properties. As they have been formulated for many years with water-soluble thickener polymers like HEC, latex paints have exhibited a much greater degree of shear thinning than alkyd paints. This led to latex paints having too low a viscosity at high shear rate, so that the applied film thickness tended to be too thin, and too high a viscosity at low shear rate to permit adequate leveling. The problem is especially severe since the rate of recovery of viscosity after exposure to high shear rates is generally rapid with latex paints. The use of Stormer Viscometers (see Section 19.3.7) has been at least partly responsible for the prolonged time before the problem was well defined. This so-called viscometer measures something related to viscosity in a midrange of shear rates, but gives no information about viscosity in either of the most critical regions—at high and low shear rates.

The reasons for this greater dependency of viscosity on shear rate in latex coatings have not been fully elucidated; at least two factors may be involved, flocculation of latex particles (and/or) pigment particles in the presence of HEC, and possible entanglement of the swollen high molecular weight HEC polymer chains [21].

Real progress in minimizing the problem has been made in recent years by using *associative thickeners* in formulating latex paints. Many kinds of associative thickeners are available. They all are moderately low molecular weight, water-soluble polymers with two or more long chain nonpolar hydrocarbon groups spaced along the backbone. Examples are hydrophobically modified hydroxyethylcellulose, hydrophobically modified ethoxylated polyurethanes, and hydrophobically modified polyacrylic acid salts. Use of such thickeners gives latex paints that exhibit less shear thinning, thus permitting formulation of latex paints with higher viscosity at high shear rates so that thicker wet films are applied, together with somewhat lower viscosity at low shear rates, so that leveling is improved at the same time [21,22]. (The thick wet film in itself helps promote leveling too since, as discussed in Section 23.1, the rate of leveling depends strongly on wet film thickness.)

Reynolds [21] has reviewed the possible mechanisms of thickening by associative thickeners and factors involved in their use in the formulation of latex paints. He emphasizes that greater care and skill is required in formulating with associative thickeners than with conventional water-soluble thickeners. The results obtained can be very sensitive to the combination of particular latex and thickener and to the amounts and types of surfactant present in the formulation.

It has been shown that formulations with associative thickeners not only level better but also give somewhat higher gloss; however, more difficulty is experienced in controlling sagging [23]. Sag control is still better than alkyd paints, but leveling and gloss are still not equal to most alkyd paints. Higher gloss as well as better rheological properties have been reported by use of a combination of associative thickeners [24]. It is said that the combination reduces flocculation of the TiO_2. Another factor in the improved gloss may be that associative thickeners are effective with small particle size latexes, which give the highest gloss [21]. It has also been shown that paints with associative thickeners spatter less when applied by roller than those thickened with HEC and that most associative thickeners resist viscosity loss by bacterial action [21].

Another shortcoming of latex paints that is particularly evident in gloss formulations is the time required to develop final film properties. Part of the problem is that users are, in a sense, fooled by the drying properties of latex paints. They dry to touch more rapidly than alkyd paints and can be handled sooner. However, they require a longer time to reach their ultimate properties. For example, even though latex paints dry more rapidly than alkyd paints, longer time is necessary to develop block resistance required to prevent windows and doors from sticking and to permit putting heavy objects on a newly painted shelf. The initial coalescence of latex particles is rapid, but full coalescence is limited by the availability of free volume. Since the (T-T_g) must be small, free volume is small. The situation is helped by using coalescing solvents. However, the loss of these solvents is controlled by diffusion rate, which is also limited by (T-T_g). It has been reported that latex particles which have inner layers of relatively high T_g with a gradient down to a relatively low T_g on the outer shell of the particles, can provide film formation at low temperatures but yet relatively quickly achieve blocking resistance [25]. It has been recommended that high T_g latexes be used with larger amounts of carefully selected coalescing agents [26]. There is need for further progress, but as this problem is unlikely to ever be completely solved, there is need for better education of consumers so they are aware of the limitations of latex paints.

Another problem of gloss latex paints that can be critical is adhesion to old gloss paint surfaces when water is applied to the new dry paint film. After wetting with water, some latex paint films can be peeled off the old paint surface in sheets. Such a film is said to exhibit poor *wet adhesion*. There is always a problem of achieving adequate adhesion when repainting old gloss paint surfaces even with alkyd paints, but the problem with latex paints is more severe. It is essential to wash any grease off the surface and to roughen the surface by sanding; but even with such surface preparation, many latex paints do not show good wet adhesion. Wet adhesion does improve as the system ages, but for several weeks or even months there can be a serious problem.

Latexes have been developed by several manufacturers that minimize the problem of wet adhesion. The compositions are proprietary. A possible approach is to use an amine-substituted acrylic comonomer such as 2-(dimethylamino)ethyl methacrylate in making the latex. Amine-substituted polymers apparently decrease the ease with which the film is displaced from the interface by water.

Progress has been made in developing gloss latex paints with adequate hiding by one coat, but further efforts lie ahead. The whole problem is made more difficult by the lack of adequate laboratory tests to measure gloss (see Section 17.4) or absolute hiding (see Section 16.2.4). The lack of adequate test procedures is particularly troublesome because many of the companies supplying raw materials to the paint industry, many of the people establishing specifications and regulations, and sadly even some people working as paint formulators, are not aware that the laboratory tests are inadequate.

Gloss alkyd paints are widely used, but there is considerable pressure to reduce the VOC emissions. There is disagreement as to what level of VOC can be achieved with alkyd gloss paints while retaining reasonable application and film properties. Most workers seem to agree that alkyd paints with VOCs of 275 g L^{-1} or higher can have satisfactory performance, but many doubt that VOC can be reduced below 250 g L^{-1}. The problems include application characteristics, through dry, color change, and durability.

High solids alkyds are discussed in Section 10.2. Solids can be increased some by solvent selection, especially the use of hydrogen-bond acceptor solvents to reduce intermolecular hydrogen bonding. While solids can be increased by reducing molecular weight and with narrower molecular weight distributions, both of these approaches lead to poorer film properties and durability if carried very far. A key obstacle to developing good alkyd paints having a VOC of 250 g L^{-1} is the need to use low T_g resins to reduce viscosity. As discussed in Section 10.1, higher T_g

alkyds initially form dry films by lacquer dry, whereas paints based on lower T_g alkyds remain tacky after solvent has evaporated. Extra cross-linking functionality required to achieve adequate film properties can be built into an alkyd by increasing the proportion of highly unsaturated fatty acids; however, high linoleic acid containing fatty acids are not available at reasonable cost, and increasing the proportion of linolenic acid causes yellowing of films on aging. Use of high functionality polyols in making alkyds permits some increase in the number of unsaturated fatty acids per molecule without major increase in molecular weight. Drying can be accelerated by increasing the proportion of driers, but this expedient can lead to formation of a highly cross-linked surface layer that may slow diffusion of oxygen to the interior causing slow through cure. Use of driers containing a high proportion of zirconium salts is a partial solution to this problem, but they are expensive and tend to promote yellowing. Furthermore, while increasing drier concentration can accelerate cross-linking, the larger amount of drier also accelerates film degradation.

A promising approach to increasing solids is the use of reactive diluents (see Section 10.2). Such additives are low molecular weight materials designed to reduce viscosity almost as efficiently as solvent, but to coreact with the oxidizing alkyd during film formation. This permits the reduction in volatile solvent content while still maintaining film properties. An example is dicyclopentenyloxyethyl methacrylate (see Section 10.2). This reactive diluent has both an acrylate double bond and an activated allylic position. In the presence of driers, it can coreact with an alkyd such as a long oil linseed alkyd. Using this reactive diluent with specially designed alkyds is reported to permit formulation of gloss alkyd paints with a VOC of 155 g L^{-1}. The properties of the films are said to approach those of an alkyd paint with a VOC of 350 g L^{-1} [27]. Condensation products of mixtures of drying oil acid amides and acrylamide with (hexaalkoxymethyl)melamine are another type of reactive diluent [28]. While the use of reactive diluents is attractive, the cost and performance characteristics still require further improvement.

Another approach to solvent-borne gloss paints is the use of nonaqueous dispersion (NAD) binders [29]. In contrast to the NAD acrylics used in automotive coatings, the T_g of NAD polymers for architectural paints can be low (5–15°C). Vinyl-terminated, low molecular weight poly(12-hydroxystearic acid) is used as the steric stabilizer. Grafting ethyl acrylate on to the stabilizer leads to a stable NAD in aliphatic hydrocarbon solvent. A small fraction of a polyfunctional comonomer, such as diallyl phthalate, is used to give a low degree of cross-linking capacity to the particles. The cross-linking is said to improve the mechanical properties, durability, and resistance to blistering of coatings made with the NAD. The amount of cross-linking must be kept low so that coalescence of the film can be essentially complete. Flow properties of paints generally require using a blend with some alkyd. The work reported thus far has concentrated on less demanding applications such as undercoats and masonry paints but, presumably, the concept is applicable to gloss enamels. It remains to be seen how VOC levels and performance will compare with latex paints.

REFERENCES

1. M. S. Reisch, *Chem. Eng. News,* **70** (41), 36 (1992).
2. D. L. Cassens and W. C. Feist, *Finishing Wood Exteriors,* U. S. Department of Agriculture, Forest Service, Agriculture Handbook No. 647, 1986.

3. Anon., *Technical Bulletin on Rhoplex AC-64, Experimental Formula XW-64–8,* Rohm and Haas, Philadelphia, PA.
4. J. Schwarz, *J. Coat. Technol.*, **64** (812), 65 (1992).
5. I. V. Mattei, R. Martorano, and E. A. Johnson, *J. Coat. Technol.*, **63** (803), 39 (1991).
6. P. K. Cooke, U. R. Gandhi, E. S. Lashen, and E. L. Leasure, *J. Coat. Technol.,* **63** (796), 33 (1991).
7. W. B. Woods, *Paint & Coatings Industry,* **3** (5), 25 (1987).
8. C. U. Pittman, Jr., G. A. Stahl, and H. Winters, *J. Coat. Technol.,* **50** (636), 49 (1978).
9. Anon., *Technical Bulletin, Rovace 9100,* Rohm & Haas Co., Philadelphia, PA, 1991.
10. J. H. Braun, *J. Coat. Technol.,* **60** (758), 67 (1988).
11. J. Temperley, M. J. Westwood, M. R. Hornby, and L. A. Simpson, *J. Coat. Technol.,* **64** (809), 33 (1992).
12. R. W. Hislop and P. L. McGinley, *J. Coat. Technol.,* **50** (642), 69 (1978).
13. A. Ramig, Jr. and F. L. Floyd, *J. Coat. Technol.,* **51** (658), 63, 75 (1979).
14. J. J. Gambino, W. J. Gozdan, and P. M. Finegan, *Resin Review*, **25** (1), 1 (1985).
15. D. M. Fasano, *J. Coat. Technol.,* **59** (752), 109 (1987).
16. D. B. Massouda, *J. Coat Technol.,* **57** (722), 27 (1985).
17. J. E. Glass, *J. Coat Technol.,* **50** (640), 53, 61 (641) 56 (1978).
18. K. W. Evanson and M. W. Urban, *J. Appl. Polym. Sci.,* **42**, 2309 (1991).
19. W. S. Overdiep, *Prog. Org. Coat.,* **14,** 159 (1986).
20. Z. W. Wicks, Jr., "Coatings," in *Enclyclopedia of Polymer Science and Engineering*, 2nd ed., Supplemental Volume, Wiley-Interscience, New York, 1989, p. 99.
21. P. A. Reynolds, *Prog. Org. Coat.*, **20**, 393 (1992).
22. R. H. Fernando, W. F. McDonald, and J. E. Glass, *J. Oil Colour Chem. Assoc.*, **69**, 263 (1986).
23. J. E. Hall, P. Hodgson, L. Krivanek, and P. Malizia, *J. Coat Technol.*, **58** (738), 65 (1986).
24. D. J. Lundberg and J. E. Glass, *J. Coat. Technol.*, **64** (807), 53 (1992).
25. K. L. Hoy, *J. Coat. Technol.*, **51** (651), 27 (1979).
26. A. Mercurio, K. Kronberger, and J. Friel, *J. Oil Colour, Chem. Assoc.*, **65**, 227 (1982).
27. D. B. Larson and W. D. Emmons, *J. Coat. Technol.*, **55** (702), 49 (1983).
28. Anon., *Technical Bulletin, Resimene AM-300 and AM-325,* Monsanto Chemical Co., St. Louis, MO, 1986.
29. C. W. A. Bromley, *J. Coat. Technol.*, **61** (768), 39 (1989).

CHAPTER XXXVI

Special Purpose Coatings

The term special purpose coatings designates those coatings that do not lie within the classifications of product coatings or architectural coatings. In 1991 they accounted for 16% of the volume (6.4×10^8 L) and 22% of the value ($2.5 billion) of the U. S. coatings shipments [1]. While the volume is the smallest of the three classes, it is substantial, and the value per unit volume is highest. Separate profit figures are not available, but it is also probable that the operating profit percent is also highest. Again, many different end uses are involved; our discussion centers on four of the larger components of the classification: maintenance paints, marine coatings, refinish automobile coatings, and aircraft coatings.

36.1. MAINTENANCE PAINTS

The term *maintenance paints* is generally taken to mean paints for field application, including highway bridges, refineries, factories, power plants, tank farms, pipe lines, and the like; not usually included are paints for office buildings or retail stores which are classed as architectural coatings. In a large fraction of maintenance paints, the major requirement is corrosion protection. Sometimes the term *heavy duty maintenance paints* is used, implying that the paints must perform more effectively in aggressive environments than trade sales paints. Corrosion protection by coatings is discussed at length in Chapter 27; cross references are provided.

Heavy emphasis is always placed on the time interval to be expected between repaintings. The frequency of repainting is especially critical in a factory since production may have to be shut down for repainting. Although selling prices are important, the major economic demands of customers emphasize proven performance and service rather than the cost of the paint.

Since there are no laboratory tests available that will predict the field performance of maintenance coatings (see Section 27.5), potential customers want to inspect actual field use examples of a coating system that is being recommended to them. The larger oil, chemical, and construction companies employ specialized engineering groups responsible for working both with potential coatings suppliers to select appropriate coatings for their company's needs and with applicators to specify the application parameters. These groups also serve as inspectors to see that the coatings are properly applied. Records are kept of surface preparation, application conditions, coating composition, and coating supplier. The performance of coatings on the various installations is monitored. These records are critical factors in future decisions on coating selection. The customers soon learn which suppliers have proven most reliable for performance and service. Specifications are set up for quality control tests that are met by the selected coatings. In most cases, the customers realize that these quality control tests are not performance predictors for other coatings. As a result, it is difficult for a supplier to introduce new types of coatings and it is especially difficult for a new supplier to break into the business.

Composition of the coatings is a major variable, but surface preparation and application procedures are also critical to performance. The Steel Structures Painting Council is a valuable source of information about the effects of such variables as provided, for example, in the *Steel Structures Painting Manual* [2].

The most commonly used method for surface preparation is sandblasting. (As discussed in Section 26.4, many modifications and alternatives to sandblasting are being studied and used because of potential hazard and contamination problems with sandblasting.) Sandblasting can be very effective in removing all surface rust, oil, and other contaminants. Furthermore, it roughens the surface, increasing the contact area with the coating, which in turn promotes adhesion. It is critical to sandblast only a relatively small area and then apply primer as soon as possible.

One of the authors was asked to explain why there had been a massive failure of coatings on a Taiwan oil tank farm located only a few hundred meters from the sea. After much discussion, it turned out that the tanks had been carefully sandblasted and a high performance coating system had been used. However, there was a delay of some days between sandblasting and coating application. Of course, in the meantime spray from the sea resulted in salt deposits on the surface; the high performance primer was sprayed over the salt-contaminated steel. The salt crystals dissolved in water that permeated through the coating, leading in a few weeks to blistering and massive failure.

The coating system must be carefully selected for the particular installation. All systems include at least two types of coatings: a primer and a top coat. Frequently more than one layer of primer and/or top coat is applied, and in some cases, a combination of primer, an intermediate coat, and top coats is used. Primers generally provide the primary corrosion control, but intermediate or top coats can also have significant effects on corrosion control, especially by reducing oxygen and water permeability of the combined films. Top coats also serve to protect the primer and to provide other properties such as gloss, exterior durability, and abrasion resistance.

Three classes of primers are available: barrier primers, zinc-rich primers, and primers containing passivating pigments. If the surface of the installation can be well cleaned and if the probability of gouges through the coating during its lifetime is small, a barrier primer is most appropriate. If the surface to be coated cannot be well cleaned and if the probability of physical abuse is relatively low, then a zinc-rich primer is the primer of choice. If the surface cannot be well cleaned and if substantial physical abuse must be anticipated, then a primer containing a passivating pigment may well be the primer of choice.

36.1.1. Barrier Coating Systems

As discussed in Section 27.3, an essential requirement for barrier coating systems is that the primer have excellent wet adhesion. Thus, the primer should have a low viscosity continuous phase with slow evaporating solvents to permit as rapid and complete penetration as possible into microscopic cracks and crevices in the metal surface. Based on present knowledge, amine-substituted binders are particularly resistant to water displacement. Phosphate esters such as epoxy phosphates also enhance adhesion in the presence of water (see Section 11.6). The T_g of the fully reacted binder should generally be a little above the ambient temperature at which the curing is to be carried out. If the T_g of the fully reacted cross-linked film is too high, the rate of the cross-linking reaction will become mobility controlled and the reaction may stop prior to complete conversion (see Section 3.2.3). Coatings designed for use on a North Sea drilling rig will not be appropriate for a drilling rig off the cost of Sumatra, or vice versa.

An appropriate class of primers is exemplified by two-package (2K) primers formulated with BPA and/or novolac epoxy resin in one package and a polyfunctional amine in the other (see Section 11.2). Pot life is enhanced by using relatively high molecular weight resins and amine cross-linkers at relatively low solids since this minimizes the concentrations of reactive groups. Especially when the solvents are slow evaporating, the vehicle can penetrate well into surface irregularities. The wet adhesion of the cured coating is excellent. Pigmentation should be high (but somewhat below CPVC) to reduce permeability to oxygen and water and to give a low gloss surface which enhances adhesion of top coat to the primer. Intercoat adhesion is further enhanced by applying top coat before the cure of the primer is complete. Only a thin coat of primer is needed, but it is generally desirable to apply multiple coats of primer to assure that the entire surface of the metal has been coated. Sometimes two different epoxy formulations are used, one as primer and the other for intermediate coats.

The poor exterior durability of BPA and novolac epoxy/amine coatings limits their use as top coats to interior applications where their generally excellent chemical resistance makes them very useful for some applications. An exception is that most epoxy/amine coatings do not resist acetic acid (or similar organic acids) very well. Apparently, in contrast with inorganic acids, acetic acid dissolves in the films, which is promoted by the presence of amines and by inadequate cross-linking. Novolac epoxy resins have higher average functionality and generally provide greater organic acid resistance relative to BPA epoxies.

Generally, top coats have a different composition than the primer. This permits the formulator to design a primer for its main purpose, complete coverage of the surface with a coating that will be very resistant to displacement by water. The remaining requirements of the coating system can then be built into the top coat. Halogenated polymers such as vinyl chloride copolymer resins and chlorinated rubbers are frequently used as top coat vehicles since they have low moisture vapor and oxygen permeabilities (see Section 4.2). Chlorinated resins require stabilization against photodegradation (see Section 25.4). Since they are not cross-linked, they remain solvent sensitive and are, therefore, not appropriate in applications such as petroleum refineries and chemical plants. The use of chlorinated resins is being limited by the high VOC required as a result of their relatively high molecular weight.

Two-package urethane coatings are increasingly used as top coats because of the lower VOC and high solvent resistance of the cured films; urethanes are particularly useful when abrasion resistance is an important requirement. For exterior installations, aliphatic isocyanate systems are used. Alkyds are still used to a significant degree. Coatings based on alkyds are generally lower cost and have intermediate VOC emissions. As a result of their low surface tension, alkyd-based coatings are less likely to show film defects on application. Latex-based top coats are being used more and more frequently and their use can be expected to increase further as limitations on VOC emissions become more stringent.

Corrosion protection generally increases as water vapor and oxygen permeability of the coatings are reduced (see Section 27.3.3). High levels of pigmentation reduce permeability and cost but, of course, also reduce gloss. Sometimes low gloss intermediate coats are applied over the primer, followed by a high gloss top coat. Platelet pigments such as mica, which can orient parallel to the surface of the coating as the solvent in the coating evaporates, are particularly effective in reducing permeability. Pigmentation of the final top coat with leafing aluminum flake pigment is desirable since, during film formation, an almost continuous layer of aluminum forms at the surface of the coating. Such coatings must be carefully formulated to avoid removing the surface treatment from the aluminum flakes that cause leafing.

36.1.2. Systems with Zinc-Rich Primers

As discussed in Section 27.4.3, zinc-rich primers are excellent primers for use when either the surface cannot be completely cleaned or when, as is the case with latex paints, complete penetration into surface irregularities cannot be achieved. The zinc acts as a sacrificial metal, protecting the steel from corrosion. To be effective, the level of zinc pigmentation must be such that PVC > CPVC. The high level of pigmentation provides for electrical contact between the zinc particles and the porosity permits water to enter the film, establishing a conductive circuit with the steel surface. Even after a significant fraction of the zinc is consumed, the primer continues to provide protection since the pores fill with ZnO [also $Zn(OH)_2$ and $ZnCO_3$], whose alkalinity may provide passivation. The primer films are brittle and will not withstand much physical abuse. In some applications, zinc-rich primers do not need top coats; but in most applications, top coats are required to protect the zinc against corrosion, to reduce the probability of mechanical damage, and to provide the desired appearance.

Zinc-rich primers are especially useful for applications like bridge superstructures. It is difficult to clean the steel completely because of the inaccessibility of crevices by joints. There are likely to be some residual rust deposits despite careful sandblasting. The external phase of the zinc-rich primer has a low viscosity and can penetrate through rust quite well, provided it is not oily. There is little likelihood of physical abuse (except in the part of the structure immediately by roadways, where a different paint system can be used).

There are two classes of zinc-rich primers: inorganic and organic zinc-rich primers. In inorganic zinc-rich primers, the binder is a prepolymer derived by reacting tetraethyl orthosilicate with a limited amount of water. The solvent is ethyl alcohol, usually with some other alcohol having a lower evaporation rate to permit better flow. After application, the prepolymer reacts with further water from the air to give polysilicic acid with some of the acid groups converted to zinc salts by the zinc hydroxide on the surface of the pigment. The inorganic primers generally provide better protection than organic primers. For example, it has been reported that in a seacoast environment, a six-year service life is estimated for inorganic primers compared with three years for the organic type [3].

The binder in organic zinc-rich primers is usually an epoxy resin. Such organic primers command a share by their advantages of greater tolerance to incomplete removal of oils from the substrate, easier spray application, and better compatibility with some top coats. For example, they are said to perform well on bridges in Michigan, an inland environment, but one in which salt is used to deice bridge decks [4].

A major challenge in the use of zinc-rich primers is in the proper selection and application of the top coat. The activity of the primer depends upon maintaining its porosity by having PVC $>$ CPVC. If the top coat vehicle penetrates through the pores in the primer, in effect, the PVC of the primer will be reduced to the level where PVC approximately equals CPVC, sharply reducing the activity of the primer.

Penetration of the continuous phase of a top coat into the pores is primarily controlled by the viscosity of the external phase. Penetration is most commonly minimized by applying a *wash coat*, that is, a very thin coating, of top coat first. The solvent evaporates rapidly from a thin coating so that the viscosity of the continuous phase increases rapidly, minimizing penetration. Proper application requires considerable skill since excess wash coat in an area will lead to increased penetration into the pores, but lack of coverage of an area will mean that the surface of the primer is not completely sealed so that the vehicle of the next coat of top coat can penetrate into the pores. It is desirable for the color of the wash coat to contrast with that of the zinc-rich primer to aid the sprayer in seeing that coverage is complete but not excessive.

It has been proposed that top coats for zinc-rich primer should have high conductivity to avoid adverse effects of top coat on corrosion protection [5]. It may be that there would be less adverse effect from the penetration of a conducting top coat than the penetration of a nonconducting top coat, but it would appear preferable to avoid penetration into the pores.

Sealing the pores with a wash coat also minimizes the fairly common problem of blistering with application of thick layers of top coat. After the wash coat is applied, application of further coats requires no more spraying skill than any other spray operation. Since alkaline zinc oxide, hydroxide, and carbonate are present on the zinc, top coats in contact with the primer coat must be stable against saponification. Two-package urethane, vinyl, or chlorinated rubber coatings are widely used. Sometimes, an intermediate epoxy coat is applied to the primer, followed by a urethane top coat.

Since the continuous phase of latex paints contains no binder to penetrate into the pores, it would appear that latex paints should be ideal for at least the first coat over a zinc-rich primer. In general, latex polymers also meet the need for saponification resistance. Latex paint films have high moisture vapor and oxygen permeability. Possibly, high permeability is desirable for the first coat to permit the water to diffuse out of the primer pores relatively rapidly. However, the balance of the top coats should have as low a permeability as possible. Permeability can be reduced by use of platelet pigments.

The use of vinylidene chloride/acrylate copolymer latex (see Section 5.2) as the binder for latex paints appears to be a promising approach [6]. The moisture vapor permeability through such films is substantially lowered by the vinylidene chloride content. Stabilization of vinylidene chloride copolymer films against photodegradation over long periods can be a major problem. Use of latex paints for corrosion protection is discussed further in Section 36.1.3.

36.1.3. Systems with Passivating Pigment Containing Primers

When extensive film damage must be anticipated and when the substrate cannot be completely cleaned (especially when oily rust will remain), or when the coating cannot penetrate into the surface irregularities, passivating pigment primers are the primers of choice (see Section 27.4.2). In contrast to the situation with baking OEM product coatings where passivating pigments are seldom appropriate, there are many cases in field applications where they are preferred, despite the fact that blistering is more likely to occur.

A wide range of primer vehicles has been used. Alkyds have the advantage that they are relatively low cost and will wet oily surfaces, but they are deficient in saponification resistance. Epoxy/amine primers are widely used because of their greater saponification resistance and good wet adhesion. Epoxy ester binders are intermediate in cost and in performance.

Zinc yellow (see Section 27.4.2) was for many years the passivating pigment of choice. But zinc yellow is carcinogenic and care must be taken to avoid inhaling or ingesting spray dust, sanding dust, or welding fumes; in some countries, its use has been prohibited. Zinc yellow is being replaced by other passivating pigments. Possible replacement pigments are discussed in Section 27.4.2. As of 1992, published field data is too limited to compare the effectiveness of the replacement pigments with each other or with zinc yellow.

For application over oily, rusty steel, red lead-in-oil primers have been used for many years and are still rated as giving the best protection by many users. The surface to be painted should be cleaned as well as possible, but the primers can be used over oily rust sections that have not been removed. The drying oil vehicle has a very low surface tension and can wet the oily surface of the rust. Furthermore, the drying oil can remove thin layers of oil from the rust surface by dissolving it. It is important that the viscosity of the external phase is low, the solvents used have slow evaporation rates, and the rate of cross-linking is very slow, so that the vehicle can surround the rust particles and penetrate through the rust aggregates to the substrate surface. While the mechanism of action of red lead in protecting against corrosion is not agreed upon, its effectiveness has been established by many years of field use. There are not regulations prohibiting the use of red lead in such applications, but there is a strong trend to avoid lead pigments. Not only may lead pigments be banned by future regulations, but there is increasing concern about possible contamination problems when the paint must be removed.

In some cases, a thin *shop primer* is applied to steel beams and components at the steel mill to retard rusting before the steel is fabricated. Since protection is only required for a limited time and frequently the structure is sandblasted after construction, alkyd-based primers have been used. Sometimes a poly(vinyl butyral)-based wash primer with zinc tetroxychromate pigment is used (see Section 36.2).

In recent years, the use of latex paint systems for maintenance paints has increased. When water is applied to a freshly sandblasted steel surface, there will be almost instantaneous rusting—called *flash rusting*. Flash rusting occurs with latex paints. In order to avoid it, the formulation should include an amine such as 2-amino-2-methyl-propan-1-ol (AMP). The use of mercaptan-functional compounds as additives to prevent flash rusting has also been recommended [7].

Since the latex particles are large compared to the size of many of the crevices in the surface of steel and since the viscosity of the coalesced latex polymer is extremely high, complete penetration into the crevices cannot be expected. Therefore, it is essential to incorporate a passivating pigment into a latex primer. The passivating pigment must be chosen such that the concentration of polyvalent ions in its composition is low enough, so that package stability of the latex is not adversely affected, but still high enough that it can serve its passivating function. Strontium chromate is less soluble than zinc yellow and has been preferred. Zinc phosphate has been used. Among the newer passivating pigments, zinc/calcium molybdate and calcium borosilicates have been widely recommended.

Acrylic, styrene–acrylic, and vinylidene chloride–acrylic latex polymers are completely resistant to saponification. Special proprietary latexes are sold that provide enhanced adhesion to metal in the presence of water. It has been suggested that amine-functional latexes are especially appropriate for wet adhesion. Use of 2-(dimethylamino)ethyl acrylate or methacrylate as a comonomer is a way to incorporate amine groups on the latex polymer. Another approach to enhancing wet adhesion is the use of an alkyd or a modified drying oil to replace part of the latex polymer. The alkyd is emulsified into the coating. After application, as the water evaporates, the emulsion breaks and some of the alkyd can penetrate into the fissures in the steel surface. Epoxy esters are more hydrolytically stable than alkyds and would be expected to provide better corrosion protection.

Owing to the high moisture vapor and oxygen permeabilities of most latex paint films, it is desirable to use some platey pigment in the formulation. Mica is widely used in both primers and top coats. Final top coats using leafing aluminum are particularly appropriate. To avoid problems of reactivity of aluminum with water, a special grade of leafing aluminum is mixed into a latex base paint just before application.

The need for reducing VOC emissions has been particularly critical in California. Since commercial paint suppliers did not have latex paints available for maintenance coatings for bridges, the California State Office of Transportation of Laboratory undertook the formulation and application of latex paints to several highway bridges to evaluate their potential utility [8]. Latex primers were used in some cases, inorganic zinc-rich primers in other cases; in all cases latex top coats were used. In this preliminary report, it was stated that some systems were still standing up well after over five years' field exposure. Based on the experience obtained, it was recommended that application should only be done when the temperature is above 10°C and when the relative humidity is less than 75%. The use of latex systems in maintenance applications can be expected to expand as requirements for reduced VOC emissions become more stringent.

A relatively recent publication [9] discloses work on a proprietary water-borne coating. The moisture permeability is reported to be a tenth that of coatings with which it was compared and the oxygen permeability was so low that it could not be measured. The authors propose that a passivating layer forms under the film. They report that none of the field exposure panels or structures have shown any sign of failure after exposures of up to five years.

36.2. MARINE COATINGS

The sizable marine coatings market includes coatings for pleasure craft as well as naval and commercial ships. Coatings for pleasure boats are commonly sold on a retail basis to the individual owner. A range of products is available; a larger volume one is *spar varnish* for wood. The original spar varnish (see Section 9.3.2) was a phenolic/tung oil varnish; the tung oil provides high cross-linking functionality, and the phenolic resin imparts hardness, increased moisture resistance, and exterior durability. While some phenolic/tung oil is still used by traditionalists, the bulk of this market has shifted to uralkyds (see Section 12.3.1) which provide greater abrasion and water resistance. Pricing in this field is interesting since, in contrast to most other consumers, boat owners commonly pick the highest priced material offered. This, of course, is welcomed by coating suppliers.

The larger part of the market is for commercial and naval shipping, and some seaside installations that are exposed to marine environments. A wide variety of substrate surfaces are coated, many require special coatings. Reference [10] provides a useful overview of the market and products.

Coatings for exterior surfaces above the waterline are generally analogous to those supplied to the heavy duty maintenance field discussed in the previous section. Wash primer is frequently applied to the newly sandblasted steel to protect the activated surface against flash rusting. Wash primers are low solids poly(vinyl butyral)/phenolic coatings pigmented with zinc tetroxychromate and catalyzed with phosphoric acid. In view of the carcinogenicity of chromates, chromate-free wash primers for this purpose are being developed [11]. It is of interest that the proposed wash primers not only do not contain zinc tetroxychromate, but they also do not contain any passivating pigment. It seems possible that an important reason for the utility of wash primers is the excellent adhesion obtained with the poly(vinyl butyral) and the low viscosity which permits penetration of the vehicle into the pores and crevices in the steel surface. However, other workers have reservations about the effectiveness of such chromate-free primers.

Inorganic zinc-rich or passivating primers are used. A 1982 report provides data on the performance of a wide range of coatings over zinc-rich primers [12]. Based on results of 10 years exposure of test panels in Jacksonville, Florida, inorganic zinc-rich primer with an intermediate epoxy coating and an aliphatic polyurethane top coat gave the best overall performance. Inorganic zinc-rich primer with an epoxy second coat and a vinyl top coat received the next highest overall exterior performance rating. Where the requirements are somewhat less demanding, a two-component epoxy primer with a silicone alkyd top coat may perform adequately.

The exterior durability requirements for ship superstructure painting are severe since, in addition to the direct UV, further UV is reflected off the water surface and the humidity is also generally high. Alkyd coatings have been the standard; however, urethane coatings are being used increasingly because of their greater durability. Latex paints are being increasingly used. Although the initial gloss is lower with latex than with alkyd coatings, the gloss retention and resistance to cracking are superior. Furthermore, the low solvent content of latex paints substantially reduces fire hazard during paint storage on the ship.

Deck paints have the further requirement of being skidproof. Two-package aliphatic urethane coatings satisfy the exterior durability and abrasion resistance requirements. The skid resistance is obtained by mixing coarse sand into the paint just before application.

For many ship bottom areas, which are immersed in water, 2K coal tar/epoxy coatings have been found to be particularly effective. One package contains coal tar pitch, amine-terminated polyamide resin, and pigment; the other contains the epoxy resin. For good performance, the metal surfaces must be carefully sandblasted before application. Multiple coats are applied to give a total film thickness of 400 μm. Coal tar/epoxy films have high chemical resistance and high dielectric strength. This latter property is especially important for ships that use aluminum or magnesium sacrificial metal anodes to provide cathodic protection against corrosion (see Section 27.2.2). The coatings must be handled with care in view of the potential carcinogenicity of components of coal tar. Over properly prepared surfaces, the expected lifetime in sea water immersion is about seven years [10].

Perhaps the most challenging class of marine coatings is *antifouling* paint. Spores and larvae of a wide variety of plants and animals can settle on the underwater hull of a ship; they range from algae to barnacles. The growth of plants and animals on ship bottoms increases the roughness of the hulls, which in turn increases the turbulence of water flow over the surfaces and, hence, speed decreases and fuel consumption increases. Removal of such growths requires putting the ship into dry dock. The economic penalty of such marine growth is very large, providing strong incentive for innovative research programs to address the issues.

The major approach has been to apply coatings containing biocides that leach out of the coating. The biocide must be a very general toxic agent. The leaching rate must be above a critical level required to kill all the organisms settling on the surface, over a long period of time. Critical leaching rates are on the order of 1–20 micrograms per square centimeter per day [13]. Leaching rates decay exponentially in accordance with a first-order release of biocide.

The first biocide used on a large scale was cuprous oxide. Typical binders were solution vinyl resins containing rosin salts. The binder was water sensitive so that loss of cuprous oxide would continue after the surface layer was depleted of cuprous oxide. In order to have sufficient cuprous oxide at the end of the life of the coating, the rate of loss of cuprous oxide in the early stages of use had to be excessive. The maximum antifouling lifetimes of such coatings in actual use were 7–24 months. The economic need was for a coating with a lifetime of 48 months or longer. Organotin toxicants like tributyl tin oxide initially looked promising, but were found in practice to be little, if any, better than cuprous oxide.

A breakthrough came with the development of organotin substituted polymers that release tin compounds on hydrolysis [13,14]. Copolymers of organotin esters of methacrylic acid with conventional acrylates gave excellent antifouling paints for aluminum pleasure craft when used in white paints pigmented with zinc oxide. Unfortunately, it was found that such coatings stripped off the bottoms of commercial seagoing vessels, sometimes in even a single voyage. Copolymers were then developed which provided the necessary adhesion. They are designed so that the copolymer remaining after loss of the tin biocide is sufficiently water-soluble to be washed off the surface of the film. The rate of hydrolysis is controlled by the design of the copolymer and also by the incorporation of some slightly water-soluble pigment. The leaching out of the pigment from near the surface of the film increases the contact area between polymer and water, increasing hydrolysis rate. Part of the pigment used can be cuprous oxide to provide further biocidal activity. Since the rate of loss of biocide (both tin and copper) is controlled by the rate of hydrolysis at the surface, the rate of leaching is approximately linear with time rather than exponential. Furthermore, the slow dissolution of polymer keeps the surface of the film smooth, which favorably influences speed and fuel consumption. The coatings are sometimes called *self-polishing*. Depending on conditions, service lifetimes up to 5 years are obtained.

When ships are in port, leaching of toxicants, of course, continues. In some harbors, it is reported that concentrations of toxicant can build up sufficiently to affect marine life. There is special concern about the potential of heavy metal entrance into the food chain affecting fish that might be caught for human consumption. Regulations controlling biocide release can be expected to get more stringent as time goes on. See reference [14] for further discussion of current antifouling paints, regulations, and approaches to alternative toxicants.

Research in recent years has been aimed at attempts to find means of controlling fouling other than by use of toxicants. A promising approach is the development of coatings to which adhesion of fouling organisms is so limited that ship movement through the water, or underwater brushes or hoses, are sufficient to remove any fouling without dry docking. Studies have shown that silicone elastomers with added methyl phenyl silicone fluids give surfaces that exhibit considerable fouling resistance over a period in excess of 10 years [14]. The poor abrasion resistance and tear strength of silicone elastomers limit their utility as antifouling coatings; however, the results show that the principle is valid.

While other coatings to which fouling organisms cannot stick will be needed for ships, perhaps silicone elastomer coatings will be valuable for offshore rigs and coastal power stations. A variety of other approaches to release coatings is possible, and it seems probable that antifouling coatings of the future will be based on principles other than the use of toxic agents.

36.3. REFINISH AUTOMOBILE PAINTS

The value of coatings for cars and trucks outside of assembly plants is almost as large as the market for OEM automotive coatings (discussed in Section 33.1). The worldwide production of refinish paints is about 500,000 tons per year [15]. Some of these coatings are used for overall finishing of cars and new trucks in special colors, but the largest segment of the market is for repair.

When a car has been in an accident and a fender is straightened or a new door is installed, these parts must be painted so that the color matches the color of the original paint. There are major technical problems in meeting the demanding application and performance requirements and also major marketing and distribution problems. There are hundreds of kinds of cars made each year, each in as many as a dozen different colors. While most cars on the road are 5 years old or less, many are well over 10 years old. If someone has an accident in a 15-year old Jaguar, the owner expects to be able to bring it to a repair shop and have that section of the car painted to match the rest of the car. Furthermore, the *bump shop* expects to be able to call the paint distributor and have the necessary liter of paint delivered by not later than the next day. At the other extreme, the owner of a brand new Cadillac doesn't look where he/she is going as he/she leaves the car dealer's lot with his/her newly purchased car and collides with a truck, he/she wants it fixed immediately.

In some cases, paint manufacturers make and stock small containers of paint to match the colors. These are ready to be shipped to the dealers and repair shops before the first new car comes off the assembly line. However, in most cases, the manufacturers provide formulations of coatings and tinting color pastes to the distributors (or the larger repair shops) to permit them to match any original color. Establishing these formulations requires great color matching skill, especially since many are metallic colors. There is an advantage to being a supplier of both OEM and refinish coatings since the exact pigments used in the new car colors are known ahead of time, which simplifies color matching.

Preparation of the surface is critical. The old paint surface must be cleaned to remove dirt and wax by scrubbing with detergent, rinsing thoroughly, and drying. Scuff sanding may be required to remove chalky pigment and degraded polymer. If the old paint is cracked, it must be sanded down to bare metal. Any breaks through to metal require that the edge of the painted area be feathered out, that is, sanded with a bevel so that there is a smooth change in film thickness. Generally bare metal is exposed or new metal replacement parts are installed during repair. All bare metal must be washed free of grease with solvent. After drying, the bare metal should be treated with phosphoric acid surface treatment solution, rinsed, and wiped dry. This treatment step is critical—excess deposition of phosphate crystals as well as areas without treatment can lead to failure.

Both air dry and force dry (temperatures in the range of 65–80°C) coatings are used. The heat in force dry coatings makes a large difference in the quality of the repair so that superior performance can be achieved. In Europe and Japan, most shops are equipped with force dry ovens. In the United States, use of force dry ovens is increasing, but there is still a significant fraction of repair shops that use air dry coatings.

If, as is commonly the case, bare metal must be covered, a primer–surfacer must be applied. Before applying the primer–surfacer, the ability of the coating to withstand the primer–surfacer solvent should be checked. Current OEM coatings give no problem, but if the car has been refinished in the relatively recent past, the coating might be only partially cross-linked and may lift when solvent is applied to it. The most commonly used primer–surfacers are made with nitrocellulose/alkyd binders. In order to achieve fast dry, a medium oil length, rosin-modified, tall oil alkyd can be used [16]. The rosin modification increases the T_g of the alkyd.

The primers are highly pigmented, for example, having a PVC of 38% [16] and sometimes even higher. The use of nitrocellulose, rosin-modified alkyd, and high levels of pigmentation permit sanding of the primer–surfacer within 30 min of application. The primer is sanded smooth with fine grit paper and the spray dust removed. Owing to the nitrocellulose, the application solids of such primers is low. Stricter controls will force the use of lower VOC primers; epoxy and urethane primers are being increasingly used. It is expected that latex-based primers will be more widely used in the future.

While the primer–surfacer is often applied directly over phosphate-rinsed bare metal, application of an anticorrosive *adhesion primer* before the primer–surfacer has been recommended [15]. One adhesion primer uses a phenolic/poly(vinyl butyral) binder pigmented with a zinc tetroxychromate and talc; phosphoric acid is used to cross-link the phenolic with the poly(vinyl butyral).

Two broad classes of refinish top coats are used: lacquer (thermoplastic) and enamel (thermosetting). Lacquers can be used to refinish either lacquer or enamel OEM painted cars. If the car may have been previously refinished, the resistance of the coating to lifting should be checked with lacquer solvent before coating. The OEM acrylic lacquer coatings show a degree of color flop that can only be matched with a refinish lacquer. Since use of acrylic lacquers in OEM has now stopped, this will be a factor which decreases the use of refinish lacquer. The major application advantage of lacquer is the fast dry; they have short *out-of-dust times* (i.e., the time required for the film to become sufficiently dry that dust particles do not adhere to the surface). There tends to be a great deal of dust in the air in repair shops, and fast dry is a major advantage, reducing contamination of the freshly applied coating. An important disadvantage of acrylic lacquers for refinish is that, when dried at room temperature, their gloss is not high enough. In order to match the gloss of the OEM applied lacquer, refinish lacquers must be polished with rubbing compound, adding to the cost.

The composition of refinish lacquers is similar to that of OEM lacquers (see Section 33.1.3). Thermoplastic acrylic polymers with high methyl methacrylate content are blended with cellulose acetobutyrate (CAB) and plasticizer. Since the films are not baked, efficient monomeric plasticizers such as butyl benzyl phthalate, that would partially volatilize in high-temperature baking ovens, can replace the polyester plasticizer used in OEM lacquers. This substitution helps the gloss problem somewhat. Gloss retention is excellent, equaling that of the original OEM coating. Just as with OEM lacquer, the application solids are low—10–12 NVV. Thus, the VOC emissions are very high. Even in California, use of these low solids lacquers continued since some colors could not be matched with higher solids repair coatings. While the VOC from lacquer was high, the amount of lacquer used in repairing one or two panels in a car gave less emission of VOC than repainting the whole car with higher solids enamels. Now that OEM lacquers are no longer used, replacement of refinish lacquer with lower VOC enamels can be expected.

The other broad class of refinish top coats is enamels. Several types are sold. The oldest and lowest cost are alkyd enamels. The vehicle is a medium oil isophthalic soy or tall oil alkyd (see Section 10.1). These enamels have several advantages besides low cost. Fewer film defects such as crawling or cratering occur during application, probably owing to low surface tension. The gloss is high enough to match the OEM coating without polishing. The VOC emissions are substantially less than with lacquer. On the other hand, the out-of-dust time approaches half an hour. Furthermore, if recoating is necessary, it must be done within 4 hours after application or after 24 hours. At intermediate times, when the film is partially cross-linked, another coat of enamel can lead to lifting. The gloss retention of alkyd coatings is significantly poorer than that of OEM coatings or refinish lacquer.

The properties of alkyd enamels have been improved in several ways. Triisocyanates, such as the isocyanurate derived from isophorone diisocyanate (see Section 12.3.2), can be added just before spraying. They serve as auxiliary cross-linkers, reacting with free hydroxyl groups of the alkyds. This supplemental cross-linking significantly reduces the out-of-dust time and reduces the recoating problem. Another approach is the use of methacrylated alkyds (see Section 10.4). These alkyds give significantly shorter out-of-dust times, but methacrylated alkyds cross-link less rapidly, so that the films remain sensitive to gasoline for a protracted time. Another class of acrylic alkyds is made by reacting an acrylic resin made with glycidyl methacrylate as a comonomer with drying oil fatty acids. The films made with such a vehicle show out-of-dust times approaching acrylic lacquers and still cross-link in time to achieve durability approaching the lacquer and superior to alkyds. Since the relatively high T_g acrylic backbone provides the necessary initial dry, driers are not needed to accelerate the oxidative cross-linking. The absence of driers markedly improves exterior durability.

Two-package urethane enamels have become important, and their usage as refinishing top coats, especially as clears for base coat/clear coat finishes, is increasing rapidly. Thermosetting hydroxy-functional acrylic resins similar to those cross-linked with MF resin in OEM coatings are used together with polyfunctional isocyanate cross-linker. The durability of these coatings is in the same range as OEM coatings, especially when they are force dried at 60–80°C. They give high gloss without polishing. The out-of-dust times are intermediate between lacquers and alkyd enamels. The out-of-dust times are reduced by using more dibutyltin dilaurate catalyst but this, of course, also shortens the pot life. Refinish shops cannot generally afford dual-mixing guns, so the pot life has to be at least long enough to permit finishing the spraying of the car—over an hour. Isocyanurate cross-linkers, such as HDI and IPDI trimers, have been favored (see Section 12.3.2). While paint suppliers emphasize the need to wear masks and to mix and apply the paint in a well-ventilated spray booth, some refinishers do not follow recommended safety precautions. Especially if cross-linkers with some volatile diisocyanate components are supplied, sprayers can develop respiratory problems. For this reason, some shops will not use 2K urethane coatings even from suppliers who have been careful in the selection of cross-linkers.

Some reduction in VOC emissions from repair shops has resulted from changing from conventional air spray guns to high volume, low pressure air guns (HVLP). These guns permit better transfer efficiency, hence lower paint usage and lower VOC emissions (see Section 22.2.1).

Research on ways to reduce VOC emissions from urethane enamels is underway (see Section 28.2.2). Lower average molecular weight polyisocyanates with lower viscosity (see Section 12.3.2) are becoming available. It is important to recognize, however, that VOC is dependent not only on the viscosity of the polyisocyanate, but also on its equivalent weight [17]. If the equivalent weight is low, the formulation will contain a greater fraction of the relatively high viscosity polyol, so that lower

VOC can sometimes be obtained with a higher equivalent weight polyisocyanate even though its viscosity is higher. Research is now directed toward synthesizing low viscosity polyisocyanates that also have relatively high equivalent weights. The use of low viscosity coreactants, commonly referred to as reactive diluents, such as aliphatic diols, low molecular weight ester or urethane diols, hindered diamines, and ketimines, is being investigated [17] (see Section 28.2.2).

Recently it has been reported that water-borne 2K urethane coatings can be formulated [18]. The binders are water-dispersible polyesters with hydrophilically modified polyisocyanates. Because of phase separation and the relatively slow reaction of isocyanates with water, only about 5% of the isocyanate groups react with water during a 4-hr pot life period.

All refinish paint suppliers are actively pursuing nonisocyanate cross-linking systems. It is important to bear in mind that any reactant that can cross-link hydroxyl, carboxyl, or amine groups on a synthetic polymer can also cross-link proteins. Any such cross-linker will be toxic. A new system must be designed so that there is no volatile reactive component and so that the molecular weight is high enough to minimize permeation through the skin and membranes. Major emphasis on safety equipment and training refinishing shop personnel is needed. Use of force dry systems significantly reduces the problems.

36.4. AIRCRAFT COATINGS

The total U. S. market for aircraft and aerospace coatings in 1988 was estimated to be $60 million [19]. By far the majority of the business is in exterior primer and top coats for aircraft—roughly 50:50 for commercial and military aircraft in 1988. Excellent adhesion and corrosion protection are required for a range of substrates, primarily aluminum and aluminum alloys and, increasingly, composite plastics. The coatings must resist swelling by phosphate ester hydraulic fluids, lubricating oils, and fuel. Swelling resistance is usually obtained by relatively high cross-link density. At the same time, coatings must have good flexibility at low temperatures and excellent abrasion resistance so that coatings will not abrade away while the aircraft is flying through dust, rain, or sleet at high speed. Flexibility and abrasion resistance tend to decrease with increasing cross-link density.

Epoxy/amine (see Section 11.2) primers are generally used. For interior spaces such as hold or cargo areas, internal structural members, wheel wells, and the like, the primers are not top coated. The coatings are expected to last for the life of the aircraft and are not stripped and recoated as is the case with exterior coatings. They are generally formulated with polyfunctional novolac epoxies (see Section 11.1.2) and amido-amines or amine adduct curing agents to give relatively high cross-link density and maximum resistance to lubricating and hydraulic fluids. Passivating pigments are used to control corrosion in case of breaks through the film. Strontium chromate is preferred over zinc yellow (see Section 27.4.2), since the solubility in water is somewhat lower and, therefore, the rate of leaching of the pigment out of the film by water is lower. Strontium chromate also has greater heat resistance than zinc yellow.

Primers for exterior surfaces are generally designed with lower cross-link density by using only, or primarily, BPA epoxy resins. Amine-terminated polyamides (see

Section 11.2.2) are usually used as the cross-linking agents. The long-chain fatty acid parts of the molecule provide good wetting to the metal surface and flexibility to the final film. The lower cross-link density also provides greater flexibility and permits easier stripping of the coatings for repainting. Again strontium chromate is used as a pigment. The cross-linking reactions are catalyzed with 2,4,6-tris(dimethylamino)phenol.

Top coats for aircraft are almost always 2K urethane coatings. Most commonly, hydroxy-terminated polyesters are used with isocyanurate trimers from hexamethylene diisocyanate or isophorone diisocyanate; commonly a high ratio of isocyanate to hydroxyl is used so that the coatings are partially moisture cured (see Sections 12.4.2 and 12.4.3). Cross-linking is catalyzed with organotin compounds.

Aircraft coatings have had relatively high solvent contents, but there is pressure to reduce VOC emissions. Water-borne epoxy/amine primers are increasingly used [19] (see Sections 11.2.6 and 29.3). Some reduction of solvent in the top coats has been achieved by using lower molecular weight resins and polyisocyanates, but progress has been limited by the requirements for rapid dry at ambient temperatures which is harder to achieve as molecular weight is reduced.

The exteriors of airplanes are frequently repainted. The biggest problem is not application of the new paint but the removal of the old paint. The paint strippers that have been used for many years contain substantial amounts of methylene chloride. Methylene chloride is toxic and environmentally undesirable. Studies are underway to eliminate the use of strippers by using mechanical paint-removing approaches. The problem is made acute since the removal method must not weaken the metal or composite plastic substrate. For example, sandblasting will remove coatings but also erodes the metal surface. Alternate blasting media like plastic beads, dry ice pellets, and crystalline starch are being investigated (see Section 26.4).

REFERENCES

1. M. S. Reisch, *Chem. Eng. News,* **70** (41), 36 (1992).
2. J. D. Kearne, Ed., *Steel Structures Painting Manual,* Vol. I, *Good Painting Practices,* 2nd ed., 1983, Vol. II, *Systems and Specifications,* 6th ed., 1991, Steel Structures Painting Council, Pittsburgh, PA.
3. J. Paert, *J. Prot. Coat. Linings,* February, 50 (1992).
4. E. Phifer, *J. Prot. Coat. Linings,* February, 48 (1992).
5. A. Koopmans, *XVIIth FATIPEC Congress Book,* Vol. III, 409 (1984).
6. H. R. Friedli and C. M. Keillor, *J. Coat. Technol.*, **59** (748), 65 (1987).
7. G. Reinhard, P. Simon, and U. Remmelt, *Prog. Org. Coat.,* **20**, 383 (1992).
8. R. Warness, *Low-Solvent Primer and Finish Coats for use on Steel Structures,* Technical Report, FHWA/CA/TL-85/02 (1985).
9. F. L. Floyd, R. G. Groseclose, L. A. Sabo, and R. D. Brown, *Polym. Mater. Sci. Eng.*, **58**, 158 (1988).
10. H. R. Bleile and S. Rodgers, *Marine Coatings,* Federation of Societies for Coatings Technology, Blue Bell, PA, 1989.
11. T. Foster, G. N. Bleninsop, P. Blattler, and M. Szandorowshi, *J. Coat. Technol.*, **63** (801), 91 (1991).
12. C. M Sghibartz, *XVIIth FATIPEC Congress Book,* Vol. IV, 145 (1982).

13. A. Milne, "Self-Polishing Coatings in Marine Antifouling Paints," Royal Institute of Chemistry Conference, London, 1980.
14. E. B. Kjaer, *Prog. Org. Coat.*, **20**, 339 (1992).
15. F. R. J. Willemse, *Prog. Org. Coat.* **17**, 41 (1989).
16. B. N. McBane, *Automotive Coatings*, Federation of Societies for Coatings Technology, Blue Bell, PA, 1987.
17. S. A. Jorissen, R. W. Rumer, and D. A. Wicks, *Proc. Water-Borne, Higher-Solids, Powder Coat. Symp.*, New Orleans, LA, 1992, p. 182.
18. P. B. Jacobs and P. C. Yu, *Proc. Water-Borne, Higher-Solids, Powder Coat. Symp.*, New Orleans, 1992, p. 363.
19. A. K. Chattopadhyay and M. R. Zentner, *Aerospace and Aircraft Coatings,* Federation of Societies for Coatings Technology, Blue Bell, PA, 1990.

CHAPTER XXXVII

Perspectives on Coatings Design

Formulation of coatings is a very challenging technical assignment. Sometimes scientists look down their noses at the lowly formulator without realizing that the task of formulating a new coating can be more technically challenging than much so-called pure research. Any given coating must meet a multitude of requirements. There are innumerable possible raw materials, combinations, and proportions. Test methods are generally subject to large ranges of error, and frequently do not give results that predict use performance well. The formulator is faced with variable substrates and application methods. There are commonly severe cost constraints. Frequently, the volume of any one coating is too limited to justify the expenditure of large amounts of time; and, furthermore, the time said to be available to solve the problem is limited. In fact, as one looks at the complexities of the field, one sometimes wonders how a useful coating ever gets formulated.

Historically, the problems of formulation were somewhat eased by following a procedure of making only small modifications of coatings that were known to be satisfactory. Over time, excellent coatings were formulated in this way. This approach of continuous improvement with focus on the customer's needs and with close contact between the formulator and the customer's engineering group continues to be of critical importance. In recent years it has been necessary to make drastic changes in formulations in less time than formerly used in making even small changes. This need results from a variety of factors, but particularly the introduction of VOC emission controls and the increasing number of raw materials that are identified as having potentially serious toxic hazards.

Most of this chapter is written from the perspective of the United States, but experience teaching courses in many countries around the world indicates that the differences in the challenges to coatings design in different countries are small compared with the common denominators.

The challenge of formulation is intensified by the need for increasing productivity and creativity. We all want more of the good things of life, but some are not willing to face the fact that the only way this is possible for the population as a whole is to increase its aggregate productivity. Some politicians and some uninformed segments of the population tend to associate productivity problems with production line workers, especially union workers. Clearly, there is need for increased productivity by labor, but labor unions represent less than 20% of the U. S. working population. Overall productivity depends on the efficiency with which management, sales personnel, accountants, clerical personnel, and, yes, laboratory personnel, including formulators, work. In the inimitable words of Pogo, we "have met the enemy and they is us."

At least as important, and perhaps more important, is the need for increased creativity in formulation. Some of the problems facing the coatings formulator are "impossible." Some of them are in fact impossible, but many so-called impossible problems aren't really impossible but require creative visualizing of ways around the supposedly insuperable hurdles. Enhanced creativity should be a critical component of most aspects of most people's lives, but the need for creativity is especially obvious in the case of technical people, including formulators.

The coatings industry has been known as a relatively low profit margin industry. It is commonly said that "we cannot invest more money in research and development since the profit margins are so low." Actually, the inverse is probably truer: Profit margins are low because a large fraction of the technical effort is spent trying to copy competitors' products and/or applying the same old ideas that every other formulator is trying rather than on innovative research and development.

The need for increased productivity and creativity is indeed important in all fields and in all countries. It is particularly critical in the United States now since we are, as a whole, having difficulty competing in world markets. Part of this problem results from the United States devoting a large fraction of its limited technical manpower to military projects rather than to projects that could aid the competitiveness of U. S. industry. Coatings formulators are in a good position to contribute to increases in productivity and creativity.

There is no magic single route to increasing creativity and productivity. This chapter provides some ideas based on the accumulated experience of three technical people who have worked in different aspects of the field. Others will have other suggestions; the particular approach is not important; the end result is critical.

There are several important aspects involved in working on a formulating problem. While to a degree these are sequential steps, continuous review and reevaluation of these aspects is essential as the work progresses.

DEFINE THE PROBLEM. The first stage in working on a formulating problem is to define the problem. This seems so obvious that one would think that it need only be mentioned in passing. However, experience shows that inadequate definition of the problem is a large factor in unsuccessful technical projects. Defining a formulating problem appropriately is in itself a difficult and relatively time-consuming effort, but effort invested at this stage will often save time overall.

It is particularly critical to define a coatings problem in terms of performance requirements. It is important to note that the statement, "Formulate a coating to meet Specification Number XXXX with a 30% reduction in VOC emissions," does not define the performance requirements in most cases. The specification defines quality control tests that a product must pass. Commonly, these quality control tests are not satisfactory predictors of performance for a new product. Consequently an effective formulator must know or find out what the actual end use requirements are. One should question the statement of a problem that he/she receives. On one occasion, a laboratory was asked to formulate a "harder" coating; this was done; the customer accepted the new formulation; but after some months of use, the customer complained loudly that the new coating had a poor wear lifetime. He had thought that greater hardness would give greater abrasion resistance. The customer was mistaken, but the fault was not really his, the fault was that of the coating formulator who didn't ask: "Why does the customer think he wants a harder coating?" There is great need, if at all possible, for direct contact and interaction between the customer and the formulator who will actually work on a problem.

It is not adequate to define a problem as: "Match competitor A's coating No. YYY at a lower cost." First of all, this definition propagates a self-fulfilling prophecy—that profits in the paint industry will continue to be low. Second, by the time the work has been finished, the competitor may have produced a better product than YYY and a match for YYY is no longer adequate to get the business. Third, the difficulties of analysis, especially for additives, are such that one can seldom have an adequate analysis to duplicate the competitor's formulation. As has already been said, matching the laboratory test results of the competitor's product does not by any means assure that the product performance will be equal. The only satisfactory definition of the technical aspects of a project is a listing of detailed performance requirements.

In any listing of performance needs, there is a range of degrees of importance of the various needs. Some are essential requirements, others are important, still others would be "nice if we could do it," and still others are in a category "as long as we are pipe dreaming why not put this nice goal down too." The laboratory worker should clearly understand how each of the performance needs fit into such a scale.

Applicable regulations must be known; not just current regulations, but best estimates of what they will be over the lifetime of the potential project. No one can accurately predict future regulations, but a choice as to goals from this point of view must be made. After all, assuming that current regulations are going to continue is a prediction of the future. Toxicity problems should be assessed taking into consideration possible future developments. Starting a new study of a corrosion protective primer based on only zinc yellow pigment with a target date for significant sales five years or so ahead is risky since it is known that zinc yellow is a human carcinogen.

Test methods must be agreed on. Few laboratory tests are adequate predictors of use performance. Since coatings are complex compositions and end use requirements are variable, there is a real danger in relying on any one test. The danger is analogous to the blind man feeling the elephant's tail and concluding that the animal is snake-like. Some widely used tests such as salt spray tests for corrosion protection have been repeatedly shown not to correlate with end use results (see Section 27.5). It is absurd to base the decision on a major research project on whether or not the product will or will not pass a salt spray test. The problem of evaluation should be faced in advance. A variety of tests relating to mechanical, spectroscopic, and thermal properties should also be carried out using coatings with known field performance as standards for comparison. Dickie [1] has published a methodology for integrating the results of laboratory performance tests, field history, environmental factors, design parameters, and the fundamentals of degradation to predict service performance.

Cost requirements are an integral part of the definition of the project. The formulator must know what the real permissible upper cost is, not just what some salesperson hopes the cost can be kept down to. Similarly, the timing requirement must be known. Some projects have to be finished by a certain date. Some do not have a specific deadline, although obviously the sooner the better. Unrealistic cost and time goals can lead to wrong decisions in project planning. The technical person has the obligation to let any possible time slippage or cost increase be known to others as soon as he/she is aware of it.

The potential value of the project should be compared with the estimated total cost of the project. Some companies use elaborate discounted cash flow methods of analyzing the potential value of a project. Our experience has been that such analyses have killed more good projects than bad ones. It should be recognized that estimates can be used to "prove" that any project will or will not be economically sound. On the other hand, there have been too frequent occasions when a little thought would have shown that there was no sensible possibility of a return approaching the potential cost of the project.

Unfortunately misunderstandings about the definition of projects are all too common. A way of minimizing such misunderstandings is to have the person who will be doing the work write out his or her understanding of the project. It is far better to take the time required to reach consensus on project definition at the outset than to risk disagreements midway or even near the project's end.

BACKGROUND SEARCH. Too often, laboratory workers jump on their horses and ride off in all directions. First, one should assess the available knowledge. Review the pertinent scientific literature; review (with appropriate concern for bias) supplier's technical data bulletins; review with fellow workers and in the company's files any pertinent background. Discussions with customer's technical and engineering personnel can be very useful. Increasingly companies are accumulating computer data banks comparing actual field performance with composition variables; such data can provide very useful ideas. It is particularly critical to be sure that a problem is not impossible. Some problems are impossible, and there is no point in working on them. No amount of wishful thinking will ever permit developing a flat, jet black paint or a white paint with equal gloss to a high gloss jet black paint (see Section 17.2). No one can match the color of a gloss coating at all angles of illumination and viewing with that of a low-gloss coating (see Section 16.9.1). No one will ever make a kinetically controlled one-package coating with 6 months package stability at 30°C that will cure in 30 min at 80°C (see Section 3.2.2). Stop such projects before they start.

IDENTIFY APPROACHES. Commonly, a technical person starts working on a project based on the most obvious approach to the problem. This is probably the same approach that his or her counterpart in a competitor's laboratory will start with. The greatest opportunity to apply creativity comes early in a project. Devise all the approaches you can to solving the problem; solicit ideas of fellow workers on how to solve the problem. Don't be dissuaded by veterans who say "we tried that 20 years ago." Assess the merits of their comments, but in 20 years many other things have changed—what didn't work then might work now. Set the problem aside for a few days and try again to come up with really different ways by which the problem might be solved. Open up your mind; get out of ruts. Having accumulated a variety of possible approaches to work on, try to assess their merits and pick the one or two most promising. Obviously, sometimes the one picked will be the first approach identified; but commonly it is not.

As the basic understanding of the factors controlling performance of coatings increases, the opportunity for basing experimental approaches on sound scientific understanding increases. It is sad to see the failures to apply known understanding. To take an obvious example, it is well established that in room-temperature cure coatings, cross-linking may become limited by the availability of free volume. If the T_g of the fully reacted system is significantly above room temperature, the cross-linking reaction will slow and probably stop before completion. Why bother to prove this again? Why not start out with raw material combinations that can react fully? Yet it is common for paint formulators to be given a sample of a new raw material, say a cross-linker, and substitute it in a formulation they are using only to find out that it doesn't work. In many cases, just looking at the formulations will tell you not to bother even running the experiment but rather to change the resin being cross-linked to fit with the characteristics of the new cross-linker.

Understanding the principles controlling exterior durability permits a better estimate of the durability of a new coating than any of the laboratory tests available. Using this knowledge permits one to concentrate efforts on compositions that have a reasonable chance of being appropriate. A person who understands recent work on corrosion protection can give a better prediction of corrosion protection by some new formulation by looking at the formulation than can be predicted by running salt fog chamber tests. Understanding the scientific principles gives a head start on formulation of a coating with improved performance.

It is also appropriate to repeat a theme that has recurred through the text—volume relationships rather than weight relationships are almost always the critical ones. A good motto for a coatings formulator is "think volume."

Understanding of the relationship between properties and compositions is approaching the stage where one will be able to design the binder for a formulation from first principles. We know what factors control T_g; we know what factors control cross-link density; we have some good leads as to the factors that control the breadth of the glass transition region; and we have a fairly good understanding of the relationships of these three characteristics with performance of films. The day is not far off when it will be possible to design a resin and cross-linker for an end use and not have to rely as much as today on trial and error. But this will only be true for the formulator who studies the advances in basic understanding and then tries to apply the principles.

EXPERIMENTAL APPROACHES. Except in the simplest projects, there will be two major factors facing the experimental worker: most test methods used in the coatings industry are subject to considerable variation in results and there are many possible variables and different possible responses in many different end results. Almost always there are conflicting requirements, so compromises are needed. Statistical experimental design and statistical analysis of data are particularly applicable to the coatings formulation field. A great contribution to increasing productivity can be made by increasing use of statistics. Proper experimental design permits learning more information, with a higher degree of confidence in the results, with fewer experiments.

People are still being taught in our schools that "Thou shalt **never** change more than one variable at a time." This commandment should have just one word changed so that it reads "Thou shalt **always** change more than one variable at a time." There are two major problems with changing only one variable at a time. First, in a field like coatings formulations, there are so many potential variables and levels of those variables that one could still be changing single variables years later. Second, changing one variable at a time does not permit identification of interactions among variables. For example, the "best" catalyst at one temperature in not necessarily the "best" catalyst at another temperature. The "best" pigment for a coating that must have excellent exterior durability made with one class of binder is not necessarily the "best" pigment to use with a different class of binder.

The strength of experimental design is that it permits changing multiple variables simultaneously in ways that permit separation of the effects of the different variables and of interactions between the variables. Many texts are available to provide background in experimental design. Reference [2], although relatively old, is a useful introduction because it uses examples from the coatings industry to illustrate the advantages and limitations of various types of experimental designs. Reference [3] provides more extensive coverage. References [4–9] are specific examples from recent coatings literature where experimental design techniques have been applied to formulation problems. Courses in experimental design are available; every coatings formulator should take such a course. Any plan for a significant size project should include statistical experimental design.

Data analysis is of critical importance. How many replicates are needed to obtain a test result with a 90 or 95% confidence limit? Chemists, particularly, are used to standard deviations and do not seem to realize that 10 ± 2 means that there is a 1/3 chance that the "real value" is greater than 12 or less than 8. Very commonly people use test results subject to wide variations to decide between two formulations, the differences between which are small compared to the errors in the test results. The worrisome thing is not only that poorer materials may be accepted for further development, but also that good ideas that are discarded because of erroneous test results. It is highly desirable to analyze the data obtained from statistical experimental designs since this approach gives the opportunity of allocating the differences in results between the different variables, the different interactions, and "error." Error is the unexplained remaining difference. If it is large, there can be two possible explanations: the test methods may have error ranges larger than the differences being investigated, or there may be one or more important uncontrolled variables. In either case, before proceeding further, one should do something about the test results (perhaps, increasing the number of replications) and/or about identifying the other variable(s).

WORK. As can be seen from the foregoing discussion, one should think first; one must also work, and work efficiently. One can think through the day before what has to be done the next day so as to mesh as many different tasks together as possible. In general, it is desirable to identify the most difficult goal of the project and concentrate initial efforts on this aspect of the problem. If the most difficult problem cannot be solved, there is no need for the solutions to the relatively easy parts of the task. Time management is one of our most critical needs. It is fashionable to attend seminars on time management and indeed they may help, but primarily each individual needs to think through how he/she can spend his/her time most efficiently. Plans and the problem definition should be frequently reviewed; the situation may have changed and you may end up solving a problem that isn't there any more. In complex problems involving several people from different disciplines, planning procedures should be used that permit monitoring of progress in all aspects of the project together.

REPORTS. Most technical people detest writing reports, but in a discipline like coatings there is a need for continuous accumulation of data. The most valuable parts of reports can be those on the experiments that did not work. Successes lead to production formulations, so there is no great need to have them in a report other than to ask for a pat on the back. The hundreds and thousands of unsuccessful experiments represent a wealth of information that can be used to minimize future work or that may be just the thing to use to meet the requirements of a coating for a different application. It is particularly critical to get into the data base the results of actual field uses and the performance obtained.

The report on a new formulation should spell out the reasons for the inclusion of each of the particular components in the formulation. When others look at the formulation years later, they may not have any idea why a component is there. This situation will complicate reformulation necessitated by a new development.

The coatings field is frustrating because there are so many variables to deal with, but this is also what makes it fun and challenging. The primary factor controlling success is enthusiasm to tackle and solve complex problems.

REFERENCES

1. R. A. Dickie, *J. Coat. Technol.*, **64** (809), 61 (1992).
2. H. Grinsfelder, *Resin Review,* **19**, (4), 20 (1969); **20**, (1), 20; (2), 25; and (3), 16 (1970).
3. H. E. Hill and J. W. Prane, *Applied Techniques in Statistics for Selected Industries, Coatings, Paints, and Pigments,* Wiley, New York, 1984.
4. D. A. Holtzen and W. H. Morrison, Jr., *J. Coat. Technol.*, **57** (729), 37 (1985).
5. K. K. Hesler and J. R. Lofstrom, *J. Coat. Technol.*, **59** (750), 29 (1987).
6. G. A. Cooney, Jr. and R. P. Verseput, *J. Coat. Technol.*, **59** (754), 65 (1987).
7. M. Broder, P. I. Kordomenos, and D. M. Thomson, *J. Coat. Technol.*, **60** (766), 27 (1988).
8. K. J. H. Kruithof and H. J. W. van den Haak, *J. Coat. Technol.*, **62** 790, 47 (1990).
9. V. A. Skormin and R. J. Siciliano, *J. Coat Technol.*, **63** (792), 103 (1991).

Combined Index for Volumes I and II

This index includes citations in Volumes I and II. The appropriate roman numeral precedes each group of page numbers.